B-52 Stratofortress

The Iron Fist of Strategic Air Command

H.J. Campbell

Electrikbooks

Copyright © 2023-2025 by Harold J. Campbell

All rights reserved. No part of this publication may be reproduced or utilized in any form or by any means electronic, mechanical, photocopying, recording, or otherwise, stored in an information and retrieval system, or transmitted in any form without express written permission of the publisher.

The views expressed in this publication are those of the author and do not necessarily reflect the official policy or position of the Department of Defense or the U.S. government. The public release clearance of this publication by the Department of Defense does not imply Department of Defense endorsement or factual accuracy of the material.

On the cover. A B-52H carrying a load of JDAMs over Afghanistan during Operation ENDURING FREEDOM. (U.S. Air Force)

Table of Contents

Introduction vii
Design and Development 1
Models and Variants 33
    Prototypes 33
        XB-52. 33
        YB-52. 37
    Bombers 38
        B-52A. 38
        B-52B. 41
        B-52C. 47
        B-52D. 50
        B-52E. 60
        B-52F 63
        B-52G. 67
        B-52H. 79
    Strategic Reconnaissance 92
        RB-52B. 92
        RB-52C. 95
    Experiments & Tests 95
        NB-52A. 95
        NB-52B. 97
        NB-52C. 99
        NB-52D 100
        NB-52E 100
        NB-52H. 102
        JRB-52B. 102
        JB-52B. 104
        JB-52C. 104
        JB-52E. 104

JB-52F.	105
JB-52G.	105
JB-52H.	105
RB-52B.	106
B-52D.	106
B-52F.	106
B-52G.	107
B-52H.	107
B-52H/D-21.	107
Armament and Weapons	**109**
Defensive Armament	109
Nuclear Munitions	111
Mk-6 Gravity Bomb.	112
Mk-15 Gravity Bomb.	112
Mk-21 Gravity Bomb.	112
Mk-28 Gravity Bomb.	113
Mk-36 Gravity Bomb.	113
Mk-39 Gravity Bomb.	113
Mk-41 Gravity Bomb.	113
Mk-43 Gravity Bomb.	114
Mk-53 Gravity Bomb.	114
Mk-57 Gravity Bomb.	114
B-61 Gravity Bomb.	114
B-83 Gravity Bomb.	115
Nuclear Stand-Off	115
Quail.	116
Hound Dog.	118
SRAM.	121
ALCM.	123
ACM.	126
Conventional Munitions	127
Mk-81 Gravity Bomb.	132

Mk-82 Gravity Bomb. 132
M117 Gravity Bomb. 133
Mk-83 Gravity Bomb. 133
Mk-84 Gravity Bomb. 133
Mk-36 Sea Mine. 134
Mk-52 Sea Mine. 134
Mk-53 Sea Mine. 134
Mk-55 Sea Mine. 134
Mk-56 Sea Mine. 134
Mk-60 Sea Mine. 134
Mk-62 Sea Mine. 134
Mk-63 Sea Mine. 134
Mk-64 Sea Mine. 135
Mk-65 Sea Mine. 135
M-36 Cluster Bomb. 135
CBU-24 Cluster Bomb. 135
CBU-49 Cluster Bomb. 136
CBU-52 Cluster Bomb. 136
CBU-58 Cluster Bomb. 136
CBU-71 Cluster Bomb. 136
CBU-87/103 Cluster Bomb. 136
CBU-89/104 Cluster Bomb. 137
CBU-97/105 Cluster Bomb. 137
CBU-107 Cluster Bomb. 137
GBU-10 LGB. 138
GBU-12 LGB. 138
GBU-28 LGB. 138
GBU-31 JDAM. 138
GBU-38 JDAM. 139
GBU-39 SDB. 140
GBU-54 LJDAM. 140

Conventional Stand-off. 140

Harpoon.	142
CALCM.	145
HAVE NAP.	146
JSOW.	147
JASSM/JASSM-ER.	147
MALD/MALD-J.	147
Organization and Basing	**149**
Firsts and Records	**163**
Record Flights	163
Quick Kick.	163
Power Flite.	164
Long Legs.	165
Long Jump.	165
Persian Rug.	165
Other Flights.	166
Operations	**170**
Cold War Operations	171
Ground Alert Program.	171
Airborne Alert Program.	175
Single Integrated Operational Plan (SIOP).	179
Low-Level Operations.	180
Operational Readiness Inspections (ORI).	181
Maritime Mission Support.	181
Vietnam War	182
Arc Light.	182
Commando Hunt.	189
Freedom Train.	191
Linebacker I.	192
Linebacker II.	194
Gulf War	199
Desert Shield.	199

- Desert Storm. 199
- Iraq Operations 203
  - Desert Strike. 203
  - Desert Thunder. 204
  - Desert Fox. 204
- Balkan War. 205
  - Allied Force. 206
- Afghan War 207
  - Enduring Freedom Phase I. 207
  - Enduring Freedom Phase II. 209
  - Jagged Knife. 212
  - Withdrawal. 212
- Iraq War 213
  - Iraqi Freedom. 213
  - Inherent Resolve. 215
- Competitions and Exercises 216
  - SAC Bomb Comp. 216
  - RAF Bombing Competition. 225
  - High Noon. 226
  - Red Flag. 226
  - Global Shield. 227
  - Busy Prairie. 228
  - Bright Star. 228
  - Busy Brewer. 229
  - Gallant Eagle. 232
  - Other Exercises. 232
- Nuclear Tests 234
- Broken Arrows 237
  - B-52F (57-0036). 237
  - B-52G (58-0187). 238
  - B-52F (57-0166). 238

B-52D (55-0060)	239
B-52G (58-0256)	239
B-52G (58-0188)	240
Displays and Disposition	241
Displays	241
B-52D (55-0100)	241
NB-52B (52-0013)	242
B-52D (56-0687)	243
Disposition	245
Appendix A: Specifications	249
Appendix B: Colors and Markings	259
Appendix C: Unit Assignments	267
Appendix D: Tail Numbers	277
Bibliography	i
Notes and Citations	iii
About the Author	vii

Introduction

The "Iron Fist" in the Strategic Air Command (SAC) shield depicts strength, power, and loyalty, and the science and art of employing far reaching advantages in securing the objectives of war – characteristics which are embodied in the powerful and venerable B-52 Stratofortress. The B-52 quickly became the mainstay of the SAC nuclear deterrence strategy beginning in the mid-1950s and continued in this role, as well as new conventional roles, until SAC's demise in 1992. Amazingly, the B-52H still continues in active service as a critical element in the Air Force Global Strike Command arsenal.

The B-52 is a formidable weapon system capable of carrying the widest array of weapons and armaments of any Air Force aircraft. Initially designed as a standard-wing, six-engine turboprop bomber capable of carrying nuclear weapons, it eventually evolved into the swept-wing eight-engine jet bomber we know today This evolution brought continuous improvements in aircraft performance and capability, culminating in the TF-33 powered B-52H.

In this configuration, the Stratofortress performs multi-mission capabilities including long-range precision strike, close air support, air interdiction, defense suppression, and maritime surveillance. It has a maximum take-off weight of 488,000 pounds and carries up to 70,000 pounds of armament and weapons. The maximum speed is 650 miles per hour, and unrefueled range is 8,800 miles. The remaining 76 aircraft continue to receive regular upgrades and are expected to remain in service until the 2050s – nearly 100 years since they were first introduced.

The B-52A first flew in 1954 and the three aircraft produced were used for test purposes. The B model was the first operational aircraft and entered service in 1955. A total of 744 B-52s were built, with the last B-52H delivered in October 1962. The B-52 served gallantly with SAC for over 35 years in every major conflict throughout the Cold War period. They flew more than 2,000 airborne alert sorties during the Cuban Missile Crisis equipped with nuclear weapons. They flew thousands of sorties carrying conventional iron bombs during the Vietnam War and finished out their service with SAC by launching conventional cruise missiles and dropping thousands of precision weapons and iron bombs during the Gulf War in 1991.

The B-52 Stratofortress was once considered an interim weapon system, which was expected to be replaced by faster and more capable aircraft such as the B-58, B-70, and B-1 within just a few years. But it flies on, defying all odds, and continues to amaze even the most diehard fans with its continuing evolutions. It has truly become the most versatile strategic bomber ever built!

Design and Development

Many aviation enthusiasts are familiar with the story of the Boeing engineers who were holed up for the weekend in a Dayton, OH hotel room and emerged three days later with the design for the most formidable and longest serving jet bomber ever produced. But the real story is far more complicated. The B-52 design process spanned more than five years, through stops and starts, and multiple configuration changes from a standard-wing, six-engine turboprop to the swept-wing eight-engine jet we know today.

During World War II (WWII) the United States Army Air Forces (AAF) bomber inventory consisted of thousands of aircraft including the B-17 Flying Fortress, B-24 Liberator, B-29 Superfortress, and a small number of B-32 Dominators. Several long-range bomber aircraft designs were on the drawing boards and in prototype development. Most of these designs were based on WWII technologies and capabilities. Some incorporated radical new designs and reflected the advancements developed through U.S. Army Air Corps "Giant Bomber" projects.

For example, Northrop submitted a bid in May 1941 for its XB-35 "Flying Wing" design capable of carrying a 2,000-pound bombload over 8,000 miles. Douglas Aircraft was working on its XB-19 long-range bomber, the largest American aircraft ever built at that time, and it made its first flight on 27 June 1941. Consolidated Aircraft won the competition in October 1941 for an intercontinental bomber, designated XB-36, capable of carrying a 10,000-pound bombload over 5,000-mile combat radius, and by August 1944 they had received a definitive contract for 100 B-36 Peacemakers. But none of these aircraft had the speed and combat radius to carry the huge bombloads over intercontinental distances the AAF envisioned.

(Left) The XB-35 introduced a radical new approach that challenged the accepted aviation design concepts. (Right) The XB-36 dwarfed the "heavy" bombers of WWII but failed to meet the speed requirements envisioned by AAF leaders. (U.S. Air Force)

On 15 August 1944, the Engineering Division of the Air Technical Service Command (ATSC) at Wright Field, OH recommended experimental projects to be conducted in

Air Corps "Giant Bomber" Development Projects

The Air Corps began its quest for an ultra-long-range "Giant Bomber" in 1934 under the top-secret program called Project-A. This project required a bomber capable of achieving a 5,000-mile range with a 2,000-pound bombload. Despite this range requirement, the stated mission of the bomber was strictly costal defense of the United States and protection of the Western Hemisphere including Alaska, Hawaii, and Panama. Both Boeing and Martin aircraft companies submitted designs, which were designated XB-15 and XB-16 respectively by the Air Corps. After a review, the Air Corps authorized contracts on 12 May 1934 for prototypes. On 16 May, they revised the mission to include the destruction of distant land and naval targets, as well as the reinforcement of Alaska, Hawaii, and Panama without intermediate servicing facilities. Ultimately, the Air Corps selected the Boeing design and an XB-15 prototype was completed in 1937.[1]

The Boeing XB-15 first flew on 15 October 1937 and was the first of the giant bomber experiments. (U.S. Air Force)

Project D, established in 1935, was a follow-on to Project A. Its goal was to study the maximum range that an ultra-long-range bomber could achieve. There were two major manufactures selected under this program, Douglas Aircraft and Sikorsky. Both companies submitted design proposals and built mock-ups for evaluation. In the spring of 1935, the Air Corps decided to merge Project A and D into a single project called Bomber, Long Range. Under this program, the Boeing XB-15 design was given the designation XBLR1 for eXperimental Bomber, Long-Range, while the Douglas and Sikorsky designs were designated XBLR2 and XBLR3 respectively. In mid-1936, the Air Corps selected the Douglas Aircraft design. It was truly innovative and included an all-metal aircraft, and low mounted wings with thick roots. It featured four 1,600 horsepower XV-3420-1 Allison engines that were later replaced with 2,000 horsepower Wright R-3350-5 engines. By 1938, a production contract was established for the Douglas design now redesignated B-19, and one XB-19 prototype was ordered.

On 11 April 1941, the Air Corps issued a specification for the first intercontinental strategic bomber that took its quest for a Giant Bomber to a whole new level. The specification required a bomber capable of launching from the United States and reaching Europe non-stop. It called for a 12,000-mile range at 25,000 feet altitude, and 275 miles per hour (mph) cruise speed at 45,000 feet. Combat radius was specified at 5,000 miles carrying a 10,000-pound bombload. The expected maximum speed was 450-mph at 45,000 feet. The maximum bombload was 72,000 pounds. These specifications challenged the state-of-the-art and far exceeded anything currently flying or on the drawing board. This was before the advent of aerial refueling, which made the feat even more challenging. By 3 May, Boeing, Consolidated Aircraft, and Douglas Aircraft submitted preliminary designs, but the companies found it difficult to meet the specifications and each of their designs fell short.

Meanwhile, on 27 June 1941, Douglas Aircraft made the first flight of the XB-19. Its four 2,000 horse-power Wright R-3350-5 engines produced a maximum speed of 224 mph at 15,700 feet altitude. It had a 212-foot wingspan, 132-foot length, and 160,000 pounds gross weight. The aircraft carried a normal bombload of 18,700 pounds and had maximum capacity of 37,100 pounds with reduced fuel load. It had a 16-man crew and featured sleeping and galley accommodations for a second crew to allow the aircraft to remain airborne for 24 hours. Although relatively successful, only the one XB-19 was built.

On 19 August 1941, after the preliminary designs submitted failed to fully meet the previous specifications the Air Corps revised the specifications and issued a new solicitation. It now required a 10,000-mile range, 240 to 300-mph cruise speed, and a 4,000-mile combat radius. Given the size and range of modern-day aircraft, it is difficult to understand how far-sighted even these reduced specifications were at the time. It was essentially like going to the moon in the 1960s. Nonetheless, Consolidated Aircraft won the competition in October 1941 and began development of the largest piston engine bomber ever built, the B-36 Peacemaker.

1946 through 1950 to determine requirements for a post-war, high speed, long-range, bomber. Germany had already advanced turbojet engine technology, having made the first flight of an He-178 in August 1939 and the first jet fighter flights of the Me-262 in June 1942. Despite these advancements, American engineers believed the turbojet was only useful for short-range fighter aircraft and light to medium bombers. They considered jet engines less practical for long-range heavy bombers because the high fuel consumption translated into shorter ranges. They saw the turboprop as the logical next step up from reciprocating engines for long-range bombers due to similar fuel consumption and better high-altitude efficiency.

By April 1945, the AAF requested from Boeing and other contractors a design study of a heavy bomber using turboprop engines. However, they all declined to submit a proposal because the characteristics were "so completely out of line with the state of the art."[2] This response from industry spurred the AAF to reassess its bomber requirements. In June 1945, they directed ATSC to formalize the military characteristics for postwar bombers. Bomber development and production was under the leadership of Lieutenant Colonel Pete Warden, Chief of the Bombardment Branch, ATSC, and his civilian deputy J. Arthur "Art" Boykin. Both men had considerable experience with bomber development programs. Warden was head of both the XB-35 and XB-36 programs. Boykin was his deputy for the XB-36 program.

Warden and Boykin took a revolutionary view of bomber development. They saw that WWII had required at least nine aircraft types including the B-17, B-18, B-24, B-25, B-26, B-29, B-32, A20, and A-26, with various range and bombload capabilities, to cover a 2,000-mile combat radius. Warden and Boykin intended to end the development of multiple aircraft to support a single mission requirement. AAF post-war budgets could no longer support it. Their objective was to develop a range of aircraft to enable "all weather bomb delivery to any target up to and including global coverage from operating bases within the continental limits of the United States."[3] This required a combat radius of up to 5,000 miles.

They cautioned that the effort would take several years due to the time needed to develop and make available high horsepower gas turbines to drive propellers and obtain the desired speeds and efficiency. They also cautioned that each experimental (prototype) bomber must use sufficiently advanced technology to prevent early obsolescence while at the same time resist using radical approaches that increased development risks. They believed that each protype must be a potential production article to reduce delay time for fielding any new aircraft selected (this proved correct as the Air Force geared up for the Cold War strategic bombing mission).

Warden and Boykin envisioned a radical three-bomber concept consisting of light, medium, and heavy bombers. These new designations were based on mission requirements rather than weight and were significantly different from the same designations used in WWII. For example, the medium bomber in this concept would carry twice the bombload at

The B-45 Tornado was the first American jet-powered bomber aircraft. It was fielded as a Tactical Air Command (TAC) nuclear capable light bomber and saw service in Korea. (U.S. Air Force)

2.5 times the cruise speed and range as the WWII heavy bomber designation. They envisioned the light bomber primarily as a ground support aircraft. It required high speed and maneuverability, with short range (approximately 460 miles radius). It would replace the B-26 and the B-45, which was the first AAF jet bomber scheduled for production. The Martin B-51 (originally designated XA-45) was their proposed aircraft for the light bomber role.[4] However, the XB-51 never made it past the prototype development stage and was later cancelled in favor of the British-designed B-57 Canberra.

(Left) The XB-51 first flew on 28 October 1949 and was proposed to fill the light bomber role. (Right) However, it was cancelled in favor of the B-57 Canberra which was used extensively by TAC in Vietnam and became the first jet bomber to drop bombs in combat. (U.S. Air Force)

They considered the medium bomber to be the workhorse of the three-bomber concept. It would be an all-weather, high-speed, high-altitude aircraft with 10,000-pound bombload and 2,300-mile radius to replace the B-29 and B-50. The Boeing B-47, six-engine jet bomber would become their chosen aircraft. However, it was initially only considered an interim aircraft until a replacement could be developed. The replacement was the planned Boeing XB-55, which was a larger version of the B-47 with four Allison T40 turboprop engines and two counter-rotating propellers per engine. The aircraft was expected to have a higher gross weight and longer range than the B-47. The aircraft was eventually used to study turbojet bomber configurations including swept-wing designs. But by 1949, the B-47B was being fielded and the aircraft was proving more than capable of performing the medium bomber role so the XB-55 was cancelled.

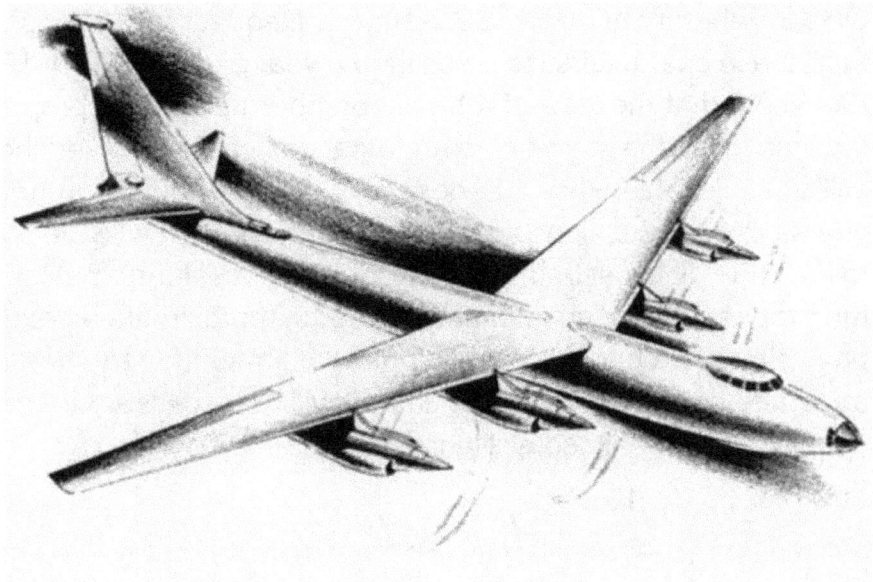

The XB-55 turboprop bomber was intended to replace the B-47 in the medium bomber role. (Boeing)

The heavy bomber was the most important of the three and would deliver a "special bombload to the strategic target system" with a radius of 5,000 miles. The XB-36 was originally selected to fill this role but concerns over its speed and combat radius meant it would always require fighter escorts. The XB-52 ultimately became their preferred replacement to fill this role.

The AAF, Assistant Chief of Air Staff for Operations and Training (AC/AS-4), issued the Military Characteristics for Heavy Bombardment Aircraft on 23 November 1945 that reflected the objectives expressed by Warden and Boykin. It included a 450-mph maximum speed, 300-mph average cruise speed, 35,000 feet operating altitude, 40,000 feet service ceiling, 5,000-mile operating radius with 10,000-pound bombload, 80,000-pound maximum bombload, and accommodations for 12 crew and a six-person relief crew. Due to issues with B-36 structural limitations and the need for faster bombers, AAF added to the characteristic "If, after the above requirements have been met, additional performance may be realized, consideration in the design of this aircraft for utilizing this performance should be given to those features such as high speed, armament, and passive protection which will reduce its vulnerability in penetrating heavily defended zones."[5]

On 13 February 1946, ATSC issued a Request for Proposals (RFP) for an aircraft to fulfill these characteristics. Less than a year earlier, in April 1945, aircraft manufacturers had scoffed at the idea of a heavy bomber using turboprop engines, stating it was not feasible given the state of the art. Most designers believed the new characteristics were still unachievable since the design of suitable turboprop engines could take up to 10 years. As a concession to get industry participation ATSC added the following statement to the RFP, "It is desired that the requirement set forth be considered as a goal and that the proposal be for an interim airplane to approximate all requirements except that emphasis must be placed on meeting the high-speed requirement. Because of the lack of adequate power plants at this time, it will be necessary to make some compromises to design a well-balanced airplane."[6]

The B-47 first flew on 17 December 1947 and was ultimately destined to fill the medium bomber role. (U.S. Air Force)

Boeing, Martin, and Convair all submitted proposals in April 1946. The Boeing Model 462 had a 221-foot wingspan, 360,000-pound gross weight (heavier than the B-36), 35,000-feet operating altitude, 410-mph cruise speed, and 3,570-mile radius. It had six Wright T35 turboprop engines, four remote control turrets (each with twin 20mm cannons), and a tail turret with four 20mm cannons. The Martin Model 236 had a 195-foot wingspan, 275,000-pound gross weight, 35,000-feet operating altitude, 407-mph cruise speed, and 2,147-mile radius. The Convair entry had a 167-foot wingspan, 235,000-pound gross weight, 35,000-feet operating altitude, 364-mph cruise speed, and 3,189-mile radius. All three designs failed to meet the 5,000-mile radius requirement, although the Boeing Model 462, the heaviest of the three, came the closest.

The original XB-52, Boeing Model 462, was a six-engine turboprop with 410 mph maximum speed and 3,570 mile radius. (Boeing)

On 23 May 1946, the AAF accepted the Boeing proposal for the Model 462 saying that it was indeed the closest to meeting all requirements and citing Boeing's extensive experience developing heavy bombers. Boeing was notified on 5 June that the AAF had selected their design and on 14 June the Model 462 was designated as the XB-52. On 28 June 1946, Air Materiel Command (AMC), which had been formed after the redesignation of ATSC on 9 March 1946, awarded Boeing a letter contract (W33-038 ac-15065) for XB-52 Phase I engineering, wind tunnel tests, data, and a mock-up.

Engine and Propeller Development

Engine development quickly became the long pole in the tent in the quest to develop a new long-range bomber. Although Germany had already advanced turbojet engine technology American engineers believed the turbojet was only useful for short-range fighter aircraft. They considered jet engines less practical for intercontinental bombers because the high fuel consumption translated into shorter ranges. AAF and industry leaders remained convinced that the turboprop engine was the key to obtaining the speed and fuel efficiency needed to achieve the required range. However, industry remained unable to meet AAF design goals and schedule.

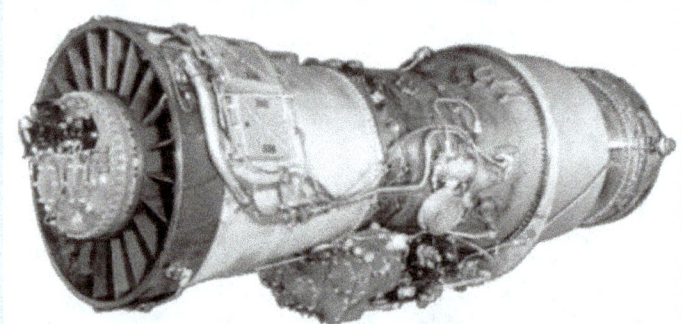

The Pratt & Whitney (P&W) J57 engine was ultimately selected to power the XB-52 after years of development. (U.S. Air Force)

Most aircraft designers believed the AAF military characteristics released on 23 November 1945 were unachievable since the design of suitable turboprop engines could take up to 10 years. Nonetheless, the initial Boeing XB-52 proposal called for six Wright T35 turboprop engines. Engine development continued slowly, and potential propeller manufacturers refused to work closely with other stakeholders to develop a workable solution. Warden and Boykin met at Wright Field on 20 August 1946 with propeller representatives and Boeing to discuss propeller design progress.[7] Boeing had determined a new reduction gear ratio would be needed for the T35 engine to achieve the required propeller speed. Despite Boeing's insistence, the attendees agreed to use a standard reduction gear due to the time required to develop a new one. Both Hamilton-Standard and Curtiss said prototype development would take 15 months and an additional six months to equip one airplane.

Clearly dissatisfied with industry progress, Warden and Boykin met again at Wright Field on 4 February 1947 with the propeller manufacturers and Boeing to discuss propeller design. They all agreed that a four-blade, single rotation, 23-foot diameter propeller was most suitable for the T35-3 engines now planned for the XB-52. Boykin was concerned about development issues (stress and flutter problems) encountered with the B-36 using 19-foot diameter propellers and worried the larger B-52 propellers might have the same problems. Warden suggested using a common hub to allow good hubs and bad blades to be interchanged. The competitors did not want to discuss proprietary data with others in the room so Warden suggested they may switch to turbojet engines, to which they assured they could overcome the difficulties but indicated it would take four years for development (SAC wanted the XB-52 operational by 1954).

On 8 March 1947, P&W began development of the JT3-6 turbojet engine, which along with concurrent development of the XT45 turboprop, evolved into the J57 engine. P&W was originally reluctant to develop a turbojet, believing turboprops were still the future for bomber development. However, they finally got on the bandwagon and realized they needed something the others didn't have - dual spool, axial compression that promised 8,700 pounds of thrust at lower fuel consumption than the J40. The JT3 design gave performance flexibility to allow efficient operation with changes in speed and power settings.[8]

By May 1947, Major General LeMay became convinced that XB-52 development hinged on the T35-3 engine and only about six months remained before a decision would be required to limit funds before cancellation. On 28 July 1947, AMC Engineering Division issued a contract to P&W for concurrent development of the XT45 turboprop and the JT3 turbojet engines.[9] By December 1947, Warden began to push for a jet-engine equipped XB-52 and in May 1948 he asked Boeing to study powering the XB-52 using Westinghouse XJ40 engines. Boeing presented a workable design, but it was never pursued. He then asked Boeing to redesign its prototype using P&W J57 engines. Within days, Boeing presented its new design.

On 26 January 1949, the Air Force proceeded with the development of Boeing's proposed XB-52 turbojet design using eight P&W J57 engines. AMC issued a supplemental agreement to P&W for the development of the XJ57 engine. It included initial design, full scale mock-up, engine specs, components, and parts to convert the XT45 to an XJ57. The new engine would be a "high pressure ratio axial flow turbojet engine employing the two-spool compressor arrangement having a low specific fuel consumption".[10] The engine was in the 9,000-pound thrust class and capable of operation up to 55,000 feet.

Meanwhile, development of competing heavy bomber designs continued. The Northrop XB-35 flying wing made its first flight on 25 June 1946, soon proving the feasibility of the flying wing design, and the Convair XB-36 first flew on 8 August. But the XB-52 design drew almost immediate criticism. AC/AS-3 said in August that the B-52 was unrealistic because of its huge size. They were also concerned that the B-52 did not meet range requirements (because the B-36 did not meet its requirements either).

In response, Warden requested a study from the AMC Aircraft Laboratory to show how speed and range requirements affected aircraft weight and to identify possible weight reductions. He wrote, "Aircraft designs will grow in range from the initial model through the production models. This may be achieved through improved power plants offering better specific fuel consumption and through overloading the airplane... If we are faced with building a given airplane in a given weight class and are too ambitious in the initial range requirement without recognizing the range growth, the speed and altitude performance may well be jeopardized."[11]

The results of the study became available in September 1946. It concluded that "a heavy bomber airplane of the range required by the Air Force cannot be met without prohibitive size unless other currently specified performance criteria are reduced." Consequently, AC/AS-3 recommended a review of the heavy bomber characteristics with "a view of arriving at more realistic performance criteria which are within the capabilities of the industry to meet."

A meeting was held at Wright Field from 17-18 October 1946 to discuss the military characteristics and development of the XB-51, XB-52, and XB-53. AC/AS-3 was clearly dissatisfied with the XB-52 size, but AMC reminded them that the size of the aircraft is driven by one or more characteristics provided by the AAF. Boeing proposed in response a smaller Model 464 with four (verses six) T35 engines, 230,000-pound gross weight, 400-mph cruise speed at 35,000 feet, and 2,500-mile operational radius (only half of the original 5,000-mile requirement). Major General Laurence C. Craigie, Chief of the Engineering Division at AMC, recommended approval but the Air Staff withheld action.

A follow-on meeting was held at the Pentagon on 27 November 1946 with Major General Curtis LeMay, then Deputy Chief of Staff (DCS) Research and Development, Major General Edward Powers, AC/AS-4 (Operations), Brigadier General Alfred Maxwell, AC/AS-3 (Requirements), and Boeing to discuss the XB-52 weight. LeMay outlined the requirement for "a special task force of 5,000-mile airplanes capable of dealing a heavy blow from North American bases..." He also outlined several requirements concessions that convinced Boeing representatives they could construct a bomber weighing not much more than 300,000 pounds that could also achieve the 5,000-mile radius requirement.

Boeing quickly proposed Model 464-16, which was specially designed to carry the atomic bomb over intercontinental ranges. Per LeMay's concessions, they dropped the requirement for an alternate conventional bombload, removed all armament except the tail gun, and reduced crew size, equipment, comfort, and furnishings to the bare minimum. They also considered dropping part of the landing gear after take-off. However, they warned that this new design would require significant design challenges and time. They recommended proceeding with their original Model 464 (four-engine) as a medium bomber replacement to the B-50 in the meantime.

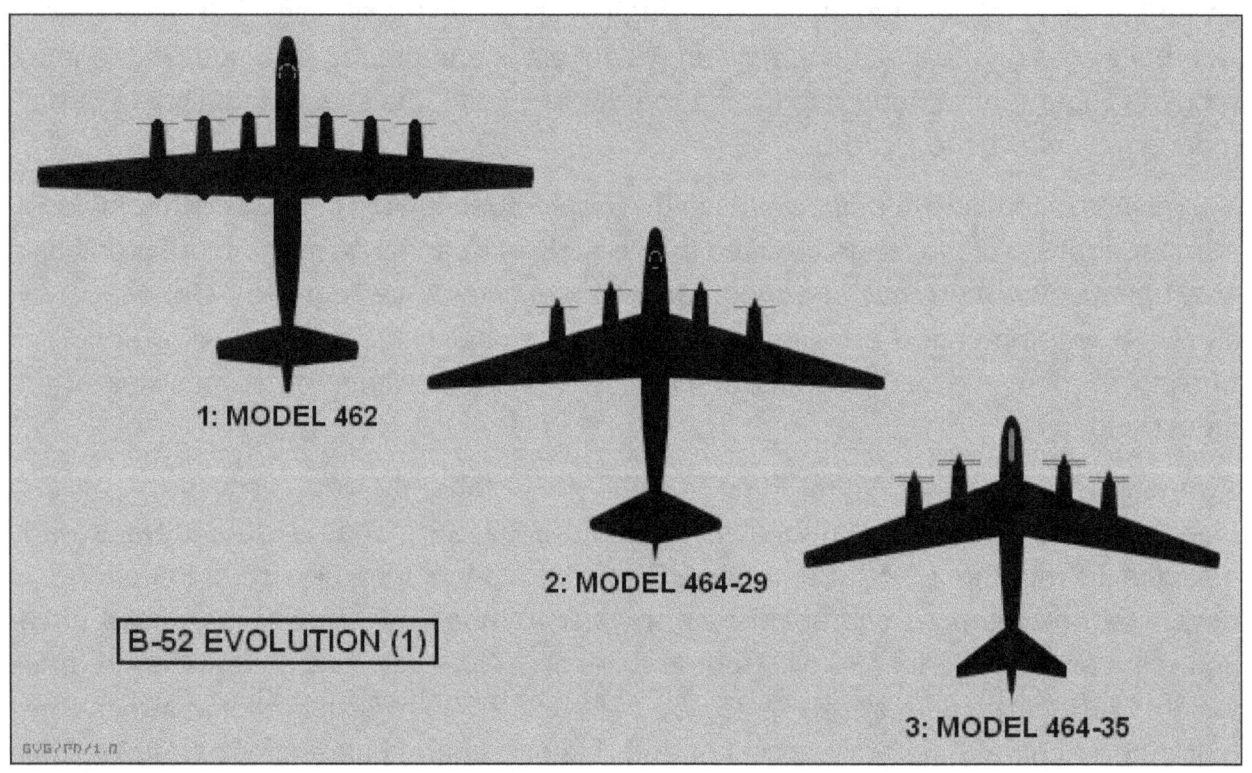

Boeing's XB-52 design evolved significantly from a six-engine turboprop to a four-engine design to reduce size and weight. (Greg V. Goebel)

AMC agreed, but the Air Staff representatives disagreed and instructed Boeing to continue with XB-52 design studies using LeMay's requirements. Boeing was given two months to return with a preliminary design and estimated size of their "minimum design" aircraft. On 7 December Major General Powers requested that AMC change Boeing's contract to reflect the new requirements outlined by LeMay.

On 7 January 1947, a conference with representatives of AMC, Boeing, AC/AS-3 and AS-4 was held in Washington, DC to evaluate the Boeing proposals using the Special

Weapon Bomber requirements outlined by LeMay. Boeing presented the Model 464-16 as the specialized aircraft with four T35 turboprop engines, 480,000 pounds[12] gross weight, 420-mph cruise speed, and 12,000-mile range with 10,000-pound nuclear weapon. Radius was 5,100-miles with no defensive armament. They also presented the Model 464-17, which was essentially the same aircraft with 400,000-pound gross weight and an alternate conventional bombload up to 90,000-pounds, 440-mph (382 knots) maximum speed, and 5,200-mile range (1,950-mile radius). Radius with the same gross weight and 10,000-pounds of bombs was 3,756-miles (3,260 nautical miles), still less than the 5,000-mile requirement.

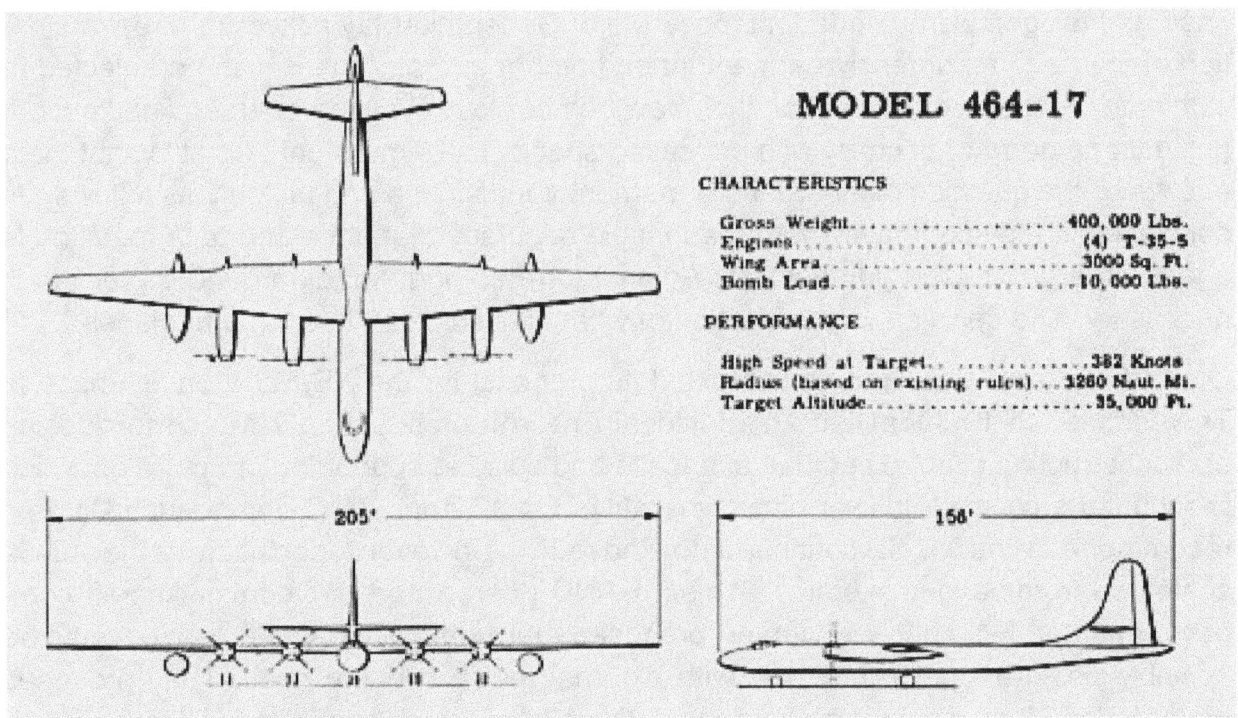

The Boeing Model 464-17 offered a new four-engine design capable of carrying 10,000 pounds of atomic bombs or up to 90,000 pounds of conventional bombs. (Boeing)

Boeing concluded there was little improvement in range, considering larger bombloads, with the minimal approach (Model 464-16). The attendees agreed to continue with the Model 464-17 only and Boeing's contract was subsequently modified. This was a conscious decision that made the XB-52 an alternate conventional bomb carrier (not just an atomic bomb carrier). This was a huge conceptual design change and is one of the decisions that would allow the B-52 to carry a series of new weapons over its service life.[13] But the critics quickly pointed out that the aircraft was too large to evade enemy interceptors, and it still did not meet the 5,000-mile radius requirement. Boeing claimed that the external fuel tanks and landing gear could be dropped to increase range. However, the critics believed this was unworkable and the Aircraft Laboratory indicated that the actual weight savings would be minimal due to the extra weight from the required jettisoning apparatus.

Significant controversy also arose over the need for defensive armament on the XB-52 due to its weight and the growing concern over its obsolescence. Boeing's original design reflected AAF military characteristics, but this was based on the bomber defensive requirements of WWII using turrets all-around the aircraft including the nose and tail. As fighter speeds continued to increase many aviation experts believed that only a nose and tail gun were needed, at most. Major General LeMay asked for a reexamination of the need for defensive armament on bomber aircraft. The British had elected to eliminate defensive armament on their new bomber to allow a more streamlined aircraft (including a narrow tail cone) with increased speed. The Armament Lab at Wright Field answered the question of all-around armament versus nose and tail only as follows, "In conclusion it may be [stated] with reasonable accuracy that nose and tail protection is adequate against fixed gun fighters. Against turreted fighters, all-around protection is necessary. The decision… is a function of what is anticipated in enemy airplanes…"[14]

On 14 February 1947, Boeing submitted its proposal for the XB-52 to enter Phase II. They subsequently submitted an additional Phase II proposal on 7 March for "design and construction of an experimental nacelle and suitable supporting structure on which to mount the complete power plant assembly." On 17 April, AMC Engineering Division recommended the XB-52 continue into Phase II.[15] However, opposition to the XB-52 continued from various skeptics. Project RAND (then a part of Douglas Aircraft company, and hardly a neutral observer) proposed that significant gains in range would be possible using a smaller fuselage with external fuel and bomb pods. They proposed their Model G2 as a possible alternative. It was based on ideas from the Martin Model 263 and combined the flying wing concept with a conventional aircraft fuselage. It included a swept wing design with four T35 engines mounted in a pusher configuration (like the B-36). All fuel tanks and landing gear were also in the wings. They left the entire fuselage for the four-person crew compartment and a bomb bay capable of holding 30,000 pounds of bombs. They estimated the aircraft would have an empty weight of 126,000 pounds and a 2,800-mile range. Unfortunately, the cost was estimated at twice the cost of the Boeing Model 462.

RAND issued its conclusion in a report on 21 April that cast doubt on the XB-52's ability to reach a 5,000-mile radius with 10,000-pound bombload. They claimed that AMC and Boeing had missed the weight estimates by 2 percent and that the aircraft would weigh between 600,000 and 1,000,000 pounds when carrying a 22,000-pound bombload.

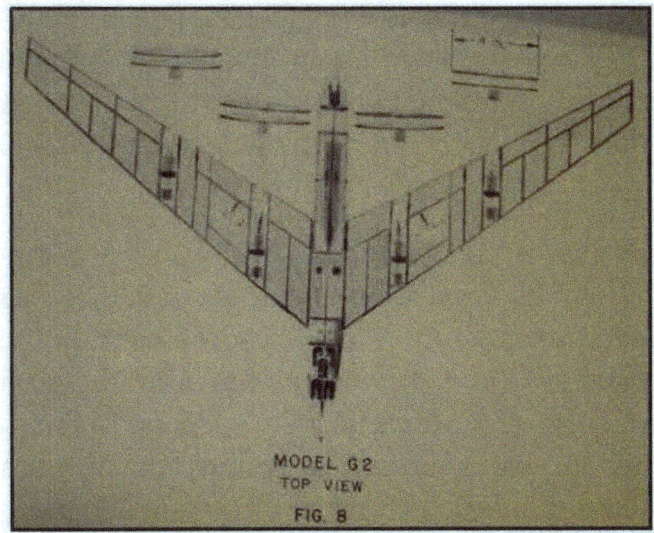 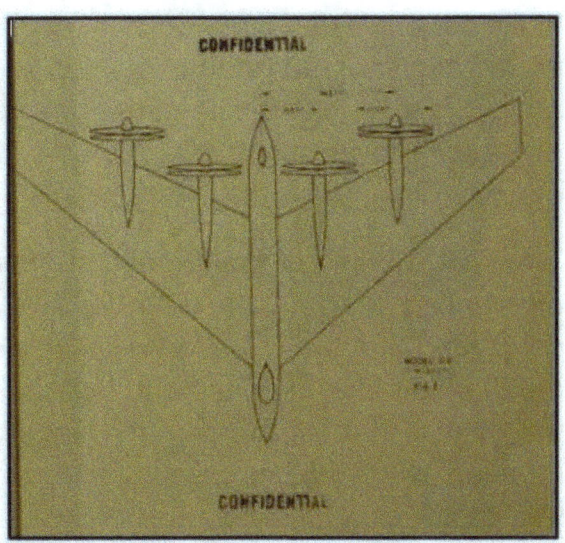

The RAND Model G2 was combination of the traditional bomber fuselage with a wing based on the flying wing designs. All fuel was carried in the wings which provided a larger bomb bay area in the fuselage. (RAND)

Brigadier General Maxwell noted that these estimates differed significantly from Boeing who showed no significant gains from a minimal sized, special purpose XB-52. He also mentioned that Boeing was unwilling to change its fuselage design despite indications that a flying wing or delta wing design would be more efficient. Brigadier General Alden R. Crawford, AC/AS-4, agreed the RAND conclusions were alarming but preferred to wait for comments from AMC (and Boeing) regarding the matter.

On 2 May 1947, the Air Staff informed AMC it would not concur with their 17 April recommendation for XB-52 to enter Phase II until Phase I wind tunnel tests were completed.[16] On 15 May, Major General LeMay sent a letter to General Nathan Twining, Commanding General of AMC, with his views of the XB-52. He believed the B-52 or any other aircraft capable of doing the job would be of such cost and size that neither the industry nor AAF budget could support more than about 100 aircraft. Missions would be small-scale operations because of the small fleet and limitations of the atomic bomb stockpile. He believed that development was hinging on the T35-3 engine and only about six months remained before a decision and commitment would be required to limit funds spent before cancellation. LeMay also recommended that AAF consider proposals from late-starters (Douglas, Northrop, and Consolidated) "to make sure that if the B-52 is the horse we intend to back such action is affirmed after all other possibilities have been considered and eliminated."[17]

AMC issued a 14-page report in June 1947 to rebut the RAND report and Warden traveled to Washington to respond directly. He told the Air Staff attendees, "They are absolutely right, in a sense… [I]f the weight has to be two percent larger, then you have to make the engines bigger, everything else bigger, and that is what winds up [a 600,000-pound airplane]. But in real life, that is not what happens. [If when] you start out, you are going to go 5,000-mile radius and you miss the weight by two percent, you are now only going to go 4,950-mile radius with the same airplane that still weighs 400,000 pounds."[18]

The AMC Bombardment Branch supported the XB-52 design as well-balanced. It recommended an adequate weight control program be implemented and reinforced. It also concluded that the XB-52 was being delayed because of a lack of agreement on the 5,000-mile radius requirement and acceptance of the compromises required to evolve the design to meet the primary range objective. General Craigie agreed and recommended the XB-52 proceed with high priority. He said that temporary deficiencies should not cloud the potential of the design and the possibility of ultimately achieving the design range. Despite this stated lack of agreement on range, the AAF issued a new military characteristic for heavy bombers on 23 June which retained the 5,000-mile radius requirement.

LeMay stirred the XB-52 development pot once again on 14 July when he said the XB-52 was only one method of meeting the strategic bombing mission. He explained that the strategic mission remains firm, but the method of its accomplishment is not fixed. He did not want to slow down XB-52 development but wanted to assure AMC considered all possibilities. He said other means of accomplishing the mission should be explored including a one-way flight to the target, return to friendly territory OCONUS, pre-planned ditching areas, use of lower cost pilotless vehicles, and RAND project recommended solutions such as the Model G2.

Boeing submitted its updated preliminary design on 7 August 1947 based on the revised military characteristic issued on 23 June. The Model 464-25 was a 400,000-pound gross weight aircraft with four main

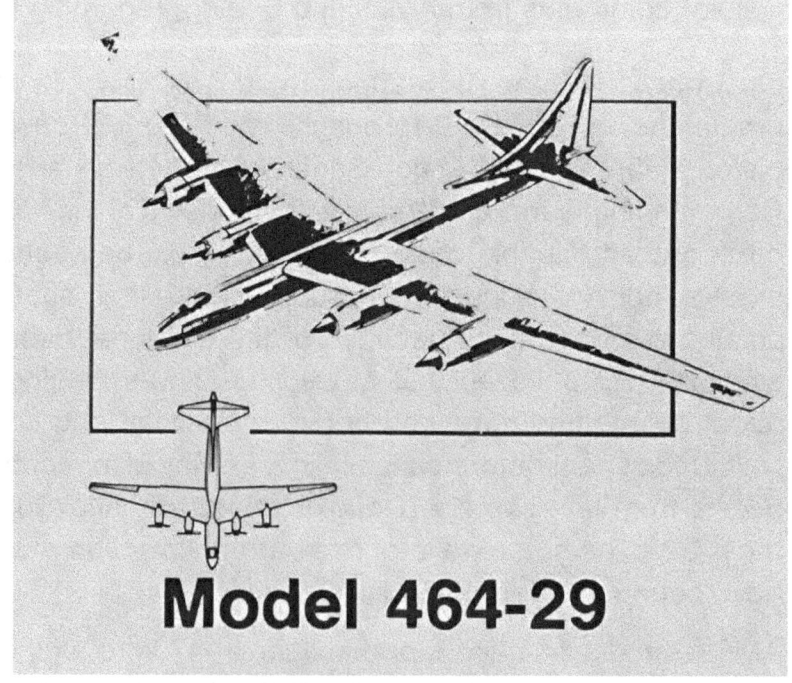

The Boeing Model 464-29 was the first preliminary XB-52 design to meet the 5,000-mile combat radius requirement.(Boeing)

landing gears (and outriggers on the outboard nacelles), elimination of all armament except one .50-caliber tail turret (with radar sighting) and two .50-caliber forward turrets (with optical sighting), and increased bombload to 12,000-pounds. It retained the four T35 turboprop engines now mounted on slightly swept wings. Over the next month Boeing submitted additional preliminary designs including the Model 464-29, a 400,000-pound aircraft with 455-mph speed and meeting the 5,000-mile radius requirement. This was the first submittal that met the radius requirement, but AAF delayed acceptance pending the upcoming meeting of the newly formed Aircraft and Weapons Board.[19]

The board, consisting of experts from across the AAF including two SAC wings, met in September 1947 to consider aircraft requirements including the heavy strategic bomber.[20] The committee recommended adoption of a new military characteristic that reduced range to 8,000-miles (from 10,000), increased cruise speed to 550-mph (from 300), reduced armament to tail only and crew to 5 (from 12), and required aerial refueling capability. The required combat radius was 4,000-miles and the outer 2,000-miles was considered the combat zone. The inner 2,000-miles was considered the non-combat zone, and it was considered a "logistical problem" to get the aircraft to the combat zone. Aerial refueling would be used inside the non-combat zone to extend range.

However, the Air Staff were again expressing doubts about the XB-52 design despite this change in requirements. On 6 November, they directed AMC to withhold further expenditure on the XB-52 while they considered options. Colonel J. S. Holtoner, Chief of the Aircraft Branch, DCS/Material, believed the aircraft in its current state represented little improvement over the B-36C (a Convair proposal with Variable Discharge Turbine engines). He also believed it would likely be a 500,000-pound aircraft and could not meet range requirements. Furthermore, he speculated it would be obsolete before completion. General Earle Partridge, Director of Training and Requirements, DCS/Requirements, argued that the flying wing design may be best suited based on RAND and National Advisory Committee for Aeronautics (NACA) studies. He said Northrop should be allowed to compete in an open competition rather than direct award to Boeing.

On 1 December 1947, several officials met with General Craigie, who had left Wright Field and was now Director of Research and Development, DCS/Materiel, in Washington. Most agreed that the XB-52 should be cancelled. They discussed new characteristics for an airplane with 500-mph cruise speed, 8,000-mile range, and 300,000-pound gross weight. Colonel Warden argued that the range for such an airplane at 500-mph would only be 7,500 miles. He also recommended turbojet engines instead of turboprops, but his suggestion fell on deaf ears.

There was significant discussion of whether the contract should be opened for competition or continued with Boeing. Warden argued for keeping the contract with Boeing for the following reasons, 1) the Air Force would lose one year of development time for competition, 2) they would lose $4 million in development funds, and 3) Boeing had proven itself as the best qualified in the heavy bomber manufacturer field with previous

B-52 and B-36 in-flight near Eglin AFB, FL. Although Boeing was seen as the leader in heavy bomber development, many Air Force leaders understood that other contractors such as Convair were capable of successful bomber development and urged a new competition. (National Archives)

successful bomber designs. Conversely, the group listed the following reasons for competition, 1) unfavorable public opinion if the XB-52 contract was changed (again), 2) if Boeing was really the best they would easily win, 3) Boeing already had most Air Force business, and 4) delay would be negligible since Boeing would continue working. The information was forwarded to the Secretary of the Air Force (SECAF) for a decision.

On 8 December 1947, General Spaatz, Air Force Chief of Staff, approved the September 1947 Aircraft and Weapons Board recommendation for a new heavy bomber military specification. Average cruise speed was reduced to 500-mph at 35,000 feet (per the 1

December meeting in General Craigie's office), range was 8,000 miles as recommended (although no radius was specified it would be about 3,000 miles based on the range requirement), and average bombload was 10,000 pounds. The specification also reduced the armament requirement to tail only and crew to 5 (from 12), and required aerial refueling, as well as all-weather day and night operations.

Lieutenant General Howard A. Craig, DCS/Material, ordered AMC to cancel the XB-52 on 11 December and circulate the newly approved characteristics to aircraft manufacturers. He also ordered AMC to study aerial refueling. The Air Force planned to use the B-50 and B-36 or possibly build half of the XB-52 fleet as tankers. These tankers would use the British hose and drogue method. Warden was sent to Britain to negotiate the sale of equipment for testing. He succeeded, but he was concerned about the slow offload speed, about 200 gallons per minute (gpm), the system provided. It also required the receiver aircraft to slow down to about 200 mph. Meanwhile, Boeing was developing a revolutionary new "flying boom", which offered a 500-gpm offload and allowed quick connection to the bomber. [21]

Boeing president, William N. Allen, strongly protested cancellation of the XB-52. Boeing had originally proposed the Boeing Model 462 airplane with a 3,570-mile radius at 410 mph average cruise speed. He agreed the requirements were now changed, but in any case, the Model 464-29 could be modified to meet 3,000 miles radius at 500 mph cruise speed. He said the current requirements were closer to the Boeing proposal submitted April 1946 than the RFP issued on 13 February 1946, which required a 5,000-mile radius at 300 mph average cruise speed. If Boeing won then, and has continued to perform, he reasoned that it should be allowed to continue to achieve the requirements rather than holding a new competition.

AMC commander, General Joseph McNarney, sent a letter to Air Force headquarters on 30 December requesting reconsideration of the decision to recompete the Heavy Bomber instead of continuing to refine the XB-52 design. He agreed with William Allen's statement that the new military characteristics closely approximated the original Boeing design. He also said the Atomic Energy Commission (AEC) was still developing atomic bombs and their size was not yet determined. He stressed that the size of the bomb bay was a major design parameter for the XB-52. The next day, Air Force Secretary Stuart Symington authorized AMC to continue with the XB-52 instead of an open competition. He had heard pros and cons from various sources, but the arguments of Allen and McNarney seemed to have had great weight.[22] Boeing's ability to use previously obligated XB-52 funds and appreciable time savings due to previously completed design work were also key factors.

AMC issued new XB-52 design specifications to Boeing on 7 January 1948 according to Symington's decision. However, by 15 January the Air Force again discussed cancelling the XB-52 based on a flight test report of the YB-49 flying wing (turbojet version of the XB-35). Some were concerned that the Air Force needed to consider the flying wing concept as a possibility to ensure they did not overlook or miss out on something better. Significant debate then ensued over the benefits and deficiencies of the flying wing concept. SECAF Symington once again stopped the XB-52 development program on 26 January, stating that the 7 January directive was not firm and citing a need to look closer at the flying wing concept.

The Northrop YB-49 first flew on 22 October 1947 and was essentially a jet version of the XB-35 flying wing equipped with eight Allision J35-A-15 engines. (U.S. Air Force)

Of course, William Allen, Boeing CEO, countered that research on the flying wing should be encouraged but its inherent instability would be difficult to overcome and would take significant time. If this could not be solved, then the Air Force would be left without a viable heavy bomber. On the other hand, he concluded, conventional designs such as the XB-52 were well-balanced and should be given priority for development. Brigadier General Donald L. Putt, Acting Assistant DCS/Materiel, supported Boeing stating that development of the flying wing might be the correct answer if there was no time issue. However, if time was critical (and it was) the XB-52 should continue.

Major General Franklin O. Carroll, Director of R&D, Engineering Division, AMC also agreed with Boeing. In addition to stability issues, he believed the flying wing did not have the weapons capacity required (already available in conventional aircraft designs) and the aircraft would have to be modified to carry external stores on wing pylons. He believed this would defeat the purpose of the flying wing design and increase the instability issues. The discussion became moot later in the year when the YB-49 crashed, and the program was cancelled due to instability.[23]

Meanwhile, on 16 January, Boeing proposed Phase I design studies of the XB-52 based on their model 464-35. It had a Gross Weight of 280,000 pounds powered by four T35-W-3 turboprops. The addition of aerial refueling to the military characteristic on 8 December 1947 made it possible to design a lighter aircraft just slightly heavier than a medium bomber with comparable speeds. A mock-up inspection was conducted in Seattle on 30 January. The Boeing model 464-17 was used in the mock-up instead of the model 464-35 to salvage the previous work accomplished and save time. This allowed detailed requirements to be injected into the later model. The mock-up included a (wood) fuselage, empennage, right half of the wing, and nose mock-ups with four different crew arrangements. Three of these were selected for further study with plans for a later conference to select the final arrangement.

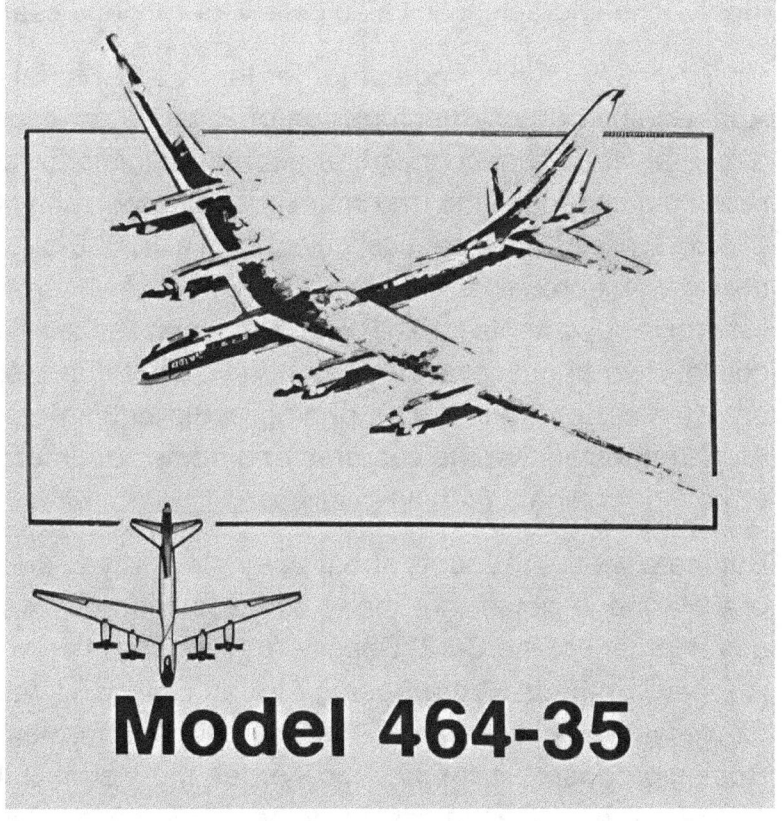

The Boeing Model 464-35 was a significantly lighter aircraft than previous XB-52 designs due to the incorporation of aerial refueling to meet the range requirement.(Boeing)

The crew arrangement became another significant point of discussion among Air Force leaders. On 12 February, the Air Force said that it preferred tandem seating arrangement for strategic bomber cockpits like the XB-47. This allowed for individual crew ejection systems and improved vision for the pilot and copilot. The aerodynamic shape of the fuselage from the streamlined cockpit contributed to increased speed. Also, the bombardier, navigator, and weaponeer would sit side-by-side in the aft section for co-use of equipment. The engineer position was eliminated, and duties were assumed by the copilot. This arrangement was officially chosen by the Air Force on 18 March.[24]

On 14 February 1948, the Air Force made its final decision to continue XB-52 development and not reopen competition. Undersecretary Arthur Barrows said the program should be pushed vigorously. Boeing's XB-52 contract was changed on 1 March to cover preliminary design, wind tunnel tests, data, and a Model 464-35 mock-up. The Air Force issued revised heavy bomber Military Characteristics on 3 March 1948. The desired speed increased to 550 mph, with 500 mph required. Radius increased from about 3,000 to 4,600 miles and the average bombload increased to 15,000 pounds to better approximate the size of atomic bombs being developed. Armament was reduced to only a tail gun turret with 220 degrees of arc. The crew size was increased to six including a "weaponeer" and engineer (eliminated on 18 March as part of the chosen crew arrangement). The characteristics also called for reconnaissance equipment provisions.

On 7 April 1948, a change order for the XB-52 Model 464-35 was approved and on 20 April Boeing proposed completion of Phase II for construction, flight test, and delivery of two aircraft. The much lighter aircraft maintained the four T-35-W-3 engine configuration but increased the maximum speed to 500 mph (435 knots) and achieved a radius of 3,533 miles (3,070 nautical miles) with a 10,000-pound bombload essentially meeting the new Air Force specifications. On 28 April, AMC sent Boeing the proposed equipment list and requirements for an RB-52 reconnaissance version. The crew included the pilot, copilot, two photo-radar-navigators, two photographers, and four radar-countermeasures (RCM) operators. The photographic equipment included seven K-40 cameras. AMC requested that the cameras and bombs be interchangeable so cameras could be removed for night bombing missions.[25]

But there were still doubts about Boeing's ability to deliver an acceptable airplane. General Partridge noted that many concessions and compromises were made to speed development of the B-52. Despite "these concessions and compromises, the Air Force is receiving the B-52 the Boeing Aircraft Company wants us to have rather than the B-52 we want Boeing to build." Partridge argued, further, that if one examined the various Boeing proposals in chronological order, the Model 464-35 was too much like the Model 464 which had been rejected, by the Air Force in 1946. Indeed, he noted the similarity of Models 464 and 464-35 "would lead us to the belief that the Boeing Company is giving us the old B-52 with a new coat of paint. Should this be the case, the intent of the

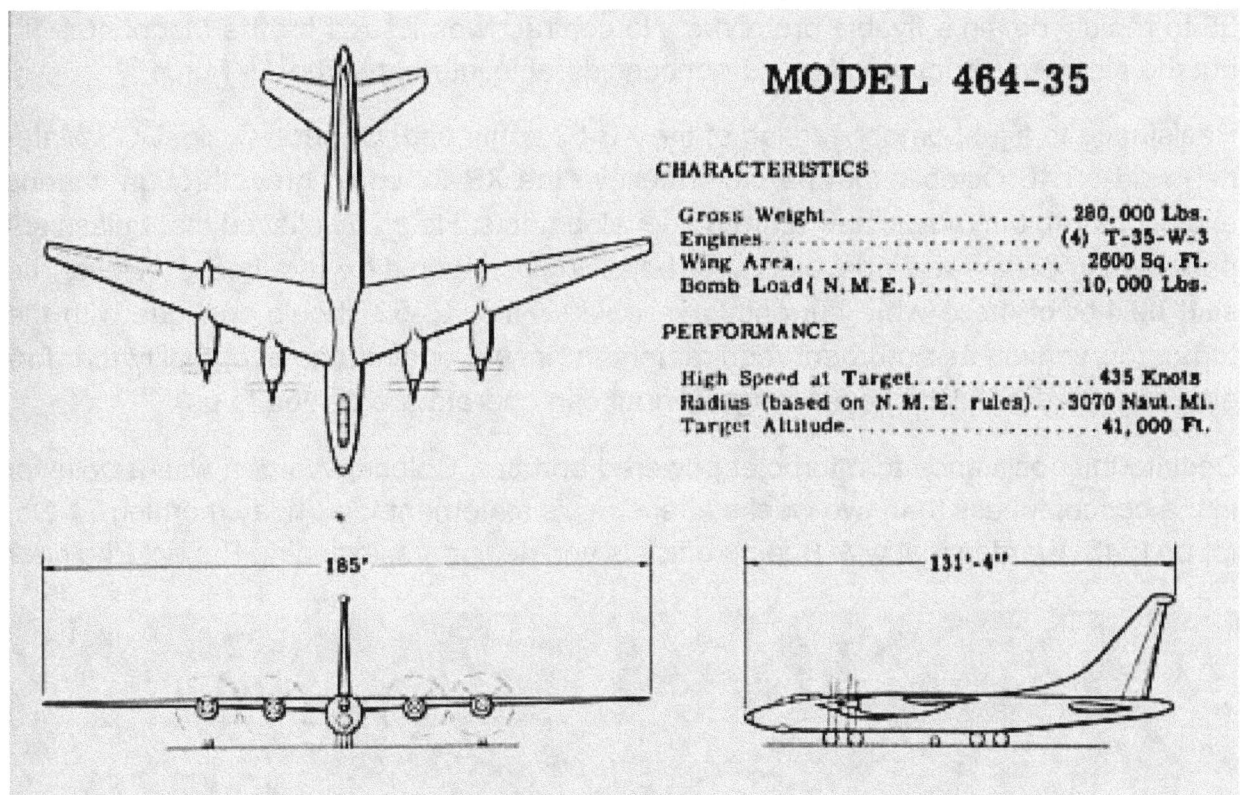

The Boeing Model 464-35 essentially met the new Air Force requirements with a 500-mph maximum speed and 3,533-mile radius. (Boeing)

new characteristics [of March 1948] will have been defeated, and in addition, it would appear that Boeing secured the new contract, without competition, on the basis of unattainable performance figures."[26]

In May 1948, Colonel Warden asked Boeing to consider powering the XB-52 using Westinghouse XJ40 engines. It was now apparent that the development of the turboprop engines and propellers could not keep pace with airframe development. Delays of at least four years were indicated during conferences of propeller and engine manufacturers that same month. AMC could no longer proceed safely with a production program based on availability of a suitable turboprop engine. Although P&W was developing the J57, the J40 had a relatively long development history with the Navy and Westinghouse had the longest jet engine development history (since 1941) in the United States. The J40 had 6,000 pounds thrust and was the most powerful jet available in 1948.[27]

Boeing responded in July with a preliminary study of its Model 464-40. The aircraft was 280,000-pounds gross weight with eight J40 engines. Speed was estimated at 526 (vs 500) mph and altitude was 45,200 (vs. 42,000) feet. However, the range was reduced to 6,750 (vs. 8,000) miles. Boeing made as few changes as possible to the Model 464-

35 to rapidly obtain a flyable prototype. No contract was issued for the Model 464-40, but the study was encouraging and got considerable interest in the Air Force.[28]

Resistance to a jet-bomber version of the XB-52 continued. General Craig, DCS/Materiel, said on 16 October that he didn't believe the XB-52 could grow through various evolutions without radical new airframe developments. He also believed that unless supersonic propellers were developed turbojet engines would be required. However, he said neither of these were currently available so the XB-52 should continue with the turboprop version as an urgent requirement to "insure against the eventuality that foreign bases from which shorter-range aircraft can operate are denied to us".[29]

Despite this resistance to a turbojet powered bomber, Colonel Warden was a believer in the concept. Less than two weeks after Craig's statement, on Friday morning 21 October 1948, Warden met with Boeing officials and designers, including Project Engineer

The XB-52 all jet-version shared several design features with the XB-47 including the swept wings and engine pods. (U.S. Air Force)

Art Carlsen, Aerodynamicist Vaughn Blumenthal, and Chief of Aerodynamics George Schairer at Wright Field to discuss the status of the turboprop XB-52 development. He said the current XB-52 design (Model 464-35) was not a significant enough improvement over the B-36 to justify further development. He also said the engine and propeller manufacturers would not take responsibility for the insurmountable problems presented by the turboprop engines (could not get desired speed, etc.). On the other hand, the turbojet powered XB-47 provided everything the Air Force needed except range which was estimated at 4,500 miles. Warden asked the team to get rid of the turboprops on the XB-52 and replace them with the new P&W XJ57 turbojet engines.

The Boeing team was joined by Bob Withington and Maynard Pennell who were in town to support XB-55 meetings. Boeing's Vice President of Engineering Edward C. Wells flew in from Seattle to support the design effort. They were determined to develop something the customer wanted to buy and told Warden they would have a proposal by the following Monday. So, as the popular story typically goes the team held up in a room at the Van Cleve hotel in Dayton for the weekend and emerged with the design that became the legendary bomber we know today. However, contrary to the typical story, they did not start with a blank page. They built upon the years of effort from the XB-52 turboprop, XB-47 and XB-55 designs, and recent Model 464-40 work from the J40 integration study.

By Monday they presented their work to Warden in a 33-page proposal including inboard profile, three-view drawing, drag polar, and weight estimates. They also presented a balsa model painted silver and mounted on a stand for Warden to use in presentations at the Pentagon. The Boeing Model 464-49 had a 330,000-pound gross weight, 564-mph (490 knots) max speed, and 3,061-mile (2,660 nautical mile) range at 49,400 feet with a 10,000-pound bombload. It also featured Boeing's new flying boom

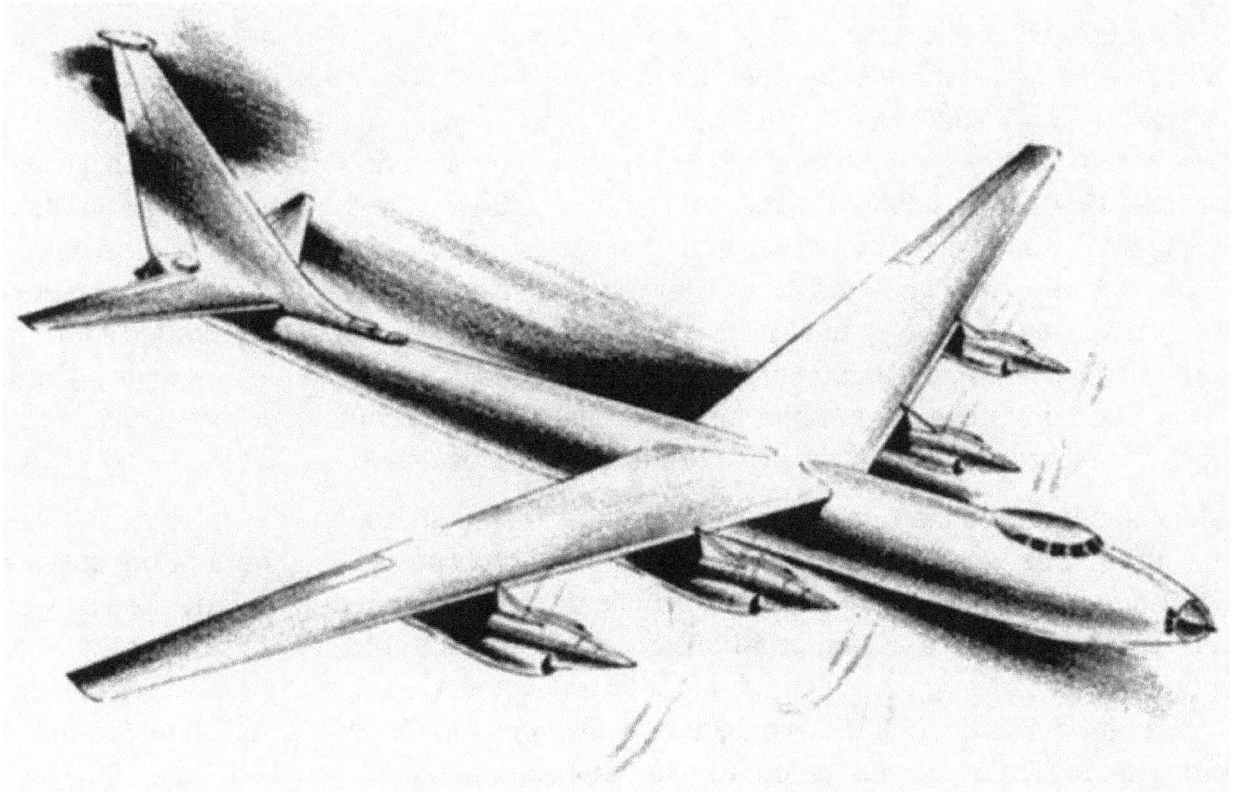

The XB-52 was essentially a doubling of the XB-55 wing area and used the XB-55 variable wing thickness design. (Boeing)

in-flight refueling capability to extend the range. It included the swept wings and engine pods from the B-47. It was essentially a doubling of the XB-55 wing area and engines and used the XB-55 variable wing thickness design. This resulted in a wing that was thicker at the root and tapered at the tip which reduced weight, increased fuel capacity, and increased Mach number. Warden loved the design and reportedly commented, "Now we have an airplane. This is the B-52." By his own authority, he told Boeing to terminate the turboprop and proceed with the jet. He was certain Air Force leaders would approve and promised to get funding. Of course, he was ultimately successful.

As Colonel Warden prepared to present Boeing's new jet bomber proposal to Air Force leaders work on the XB-52 turboprop, and several other bomber projects continued, despite his direction to terminate it. DCS/Materiel directed AMC on 10 November to reduce the weight of the XB-52 by reducing the special atomic bombload from 15,000 to 10,000 pounds, installing a single .50-caliber turret in the tail (with optical and radar gun-laying capabilities), and deleting several electronics and other systems. They also directed adding the capability to carry conventional weapons.

On 17 November, AMC issued a contract modification for two XB-52 aircraft based on the Model 464-35 turboprop design. In January 1949, the Air Force cancelled the XB-55 medium bomber due to lack of funds. It had developed several innovations, and some were used in the XB-52. The follow-on medium bomber would be the B-47B. Warden and his AMC Bomber Branch staff presented its justification for the Boeing Model 464-49, turbojet equipped XB-52, in a meeting with Air Force leaders on 17 December 1948. They said from a tactical standpoint the jet aircraft provides superior flexibility of increased speed for shorter ranges. Cruise control (autopilot) would be easier due to powerplant characteristics. They also anticipated fewer maintenance issues. They made a follow-on presentation in January 1949. This time they argued that the turbojet design took advantage of the latest advances in aerodynamics, propulsion, and military requirements. Even if the turboprop development continued, they argued, turbojet development would still be needed to counter obsolescence issues.

AMC also believed the turbojet presented fewer technical issues. It was now clear that the turboprops were not able to achieve the 500-mph speed requirement because of inadequate propellers. There was still significant development and test required as well as cooperation between engine and propeller manufacturers, which was not strong. Also, the XB-52 design was changed four times without new competition. Development of this new evolutionary turbojet design had saved industry up to $6 million and reduced harm to the Air Force's reputation from several competitions.

There were also time and cost considerations. The Navy was working hard to deliver atomic bombs with aircraft launched from their proposed "Super Carrier" and was fighting against Air Force B-36 deployment. A suitable design for a jet-powered Naval aircraft that would become the A3D Skywarrior was completed in 1949. The swept-wing jet bomber design weighed 60,000 pounds and included a large internal weapons bay with capability to carry 12,000 pounds of conventional or atomic bombs. It had a maximum speed of 600 mph with a ceiling of 41,000 feet. It could be launched from Navy carriers and conduct operations within a range of 1,000 miles or more.[30] The B-36 was quickly becoming obsolete, even as it was being fielded, and failure to replace it with the latest technologies could give the Navy the upper hand.

The Douglas A3D Skywarrior was designed to carry atomic weapons with aircraft launched from the Navy's proposed "Super Carrier". (U.S. Navy)

On 26 January, the Air Force Senior Officers Board decided to proceed with the XB-52 turbojet design and authorized Boeing to proceed without a new competition. Major General Edward Powers, Assistant DCS/Materiel, directed AMC to continue with the XB-52 turbojet conforming generally to the Boeing Model 464-49 in lieu of the turboprop version. AMC subsequently issued a supplemental agreement to P&W for development of the XJ57 turbojet engine. The agreement included initial design, full scale mock-up, engine specifications, and fabrication of components and parts needed to convert the XT45 turboprop to an XJ57. They also cancelled the XT45 and directed P&W to proceed with the XJ57 using as much of the work already completed with the XT45 as possible.

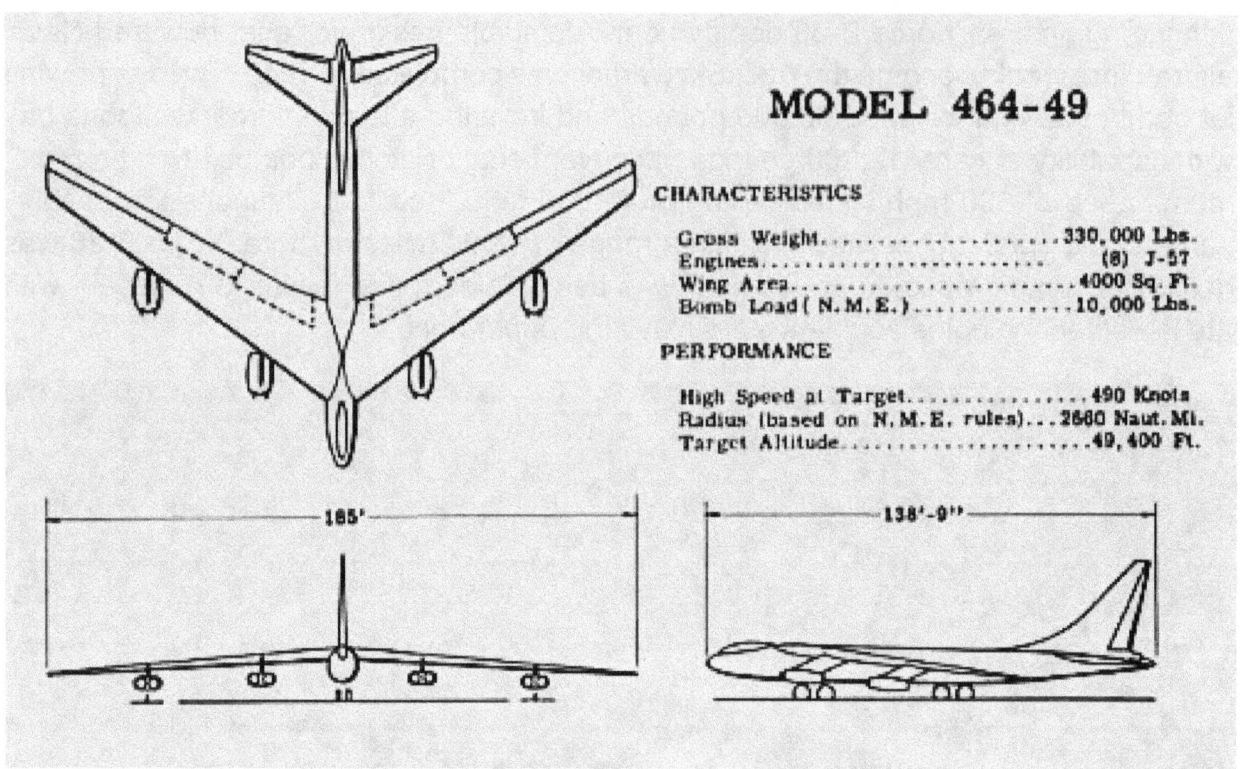

The Boeing Model 464-49 offered the first reasonable solution to the long-range bomber problems using jet engines. (Boeing)

Despite this positive commitment by the board, the critics were still not silenced. General Putt, Director of R&D at AMC, responded on 10 February 1949 to General Craig's (DCS/Materiel) question: "Why are we building the B-52?" Craig had previously resisted the XB-52 turbojet development and had recommended sticking with the turboprop version. Putt responded that the XB-52 was being developed to replace the B-36. The main difference between the two was a matter of time. The B-36 would be obsolete after the next four to five years. "At present," he said, "due to state-of-the-art airframe and power plant capability, the XB-52 turbojet bomber is the best possible replacement. Only by pursuing the B-52 at this time [having a B-36 replacement in development] would the Air Force be able to carry out its strategic mission of delivering the atomic bomb. The B-52 exceeds the B-36B in the following categories: High speed by 204 knots; cruise by 251 knots; altitude by 10,000 feet."[31]

A few days later, on 19 February, AMC Bombardment Branch was once again forced to defend the XB-52 "reasonably conventional approach as compared to throwing away the book and reaching into the blue."[32] Fairchild Corporation had proposed an unconventional bomber using a rail car to launch a large fuel carrying wing. Once the fuel was expended this wing could be jettisoned and the aircraft could proceed. Fairchild had previously presented its proposal in April 1948, directly to General Hoyt Vandenberg,

Air Force Chief of Staff. Vandenberg asked Fairchild if they had presented it to AMC and they said yes, "but those people have got their heads in a bucket of cement." Vandenberg then requested that AMC cancel the XB-52 and adopt the Fairchild proposal.

In reality, Fairchild had bypassed AMC and gone directly to Vandenburg. When AMC was alerted to the proposal in September 1948, Major General Carroll, Director of R&D, informed DCS/Materiel that when subjected to more detailed analysis the design relied on unproven ideas. AMC reiterated its stance about the XB-52 design and said each new development should bite off the largest advancement that can be digested (by available funds, etc.). They said they could only pursue developments with abnormal amounts of risks if they can carry parallel or concurrent developments. The Fairchild program would result in three to six years delay in the available aircraft for SAC and it would lack growth potential (and wings would have to be replaced for each mission causing cost and logistics issues).

AMC recommended to the Senior Officers Board that XB-52 turbojet development continue. The board agreed but also recommended that industry continue to think about advanced designs and, when funds permitted, they be asked for new and possibly unconventional approaches to the intercontinental bomber problem. They suggested a prize should be used as an incentive. AMC received a request from the SECAF in March to "solicit proposals for possible unconventional approaches to the intercontinental bomber problem." AMC did not act due to higher priorities and because this request was never funded.

On 10 March 1949, AMC issued a supplemental agreement to Boeing's contract authorizing Phase II including a mock-up and two experimental XB-52 aircraft based on the Boeing Model 464-54 turbojet design. Four days later Anthony F. Dernbach, Chief of AMC Aerodynamics, Propeller Laboratory, submitted a report of the relative merits of turboprops vs. turbojets for long range heavy bombers. He argued that the turboprop was preferred and would save airframe weight and fuel per mission. He believed it was a serious mistake to curtail turboprop development in favor of turbojets. Turboprops would result in superior performance for speed, range, and altitudes of airplanes being required at the time. He recommended simultaneous development of turboprops and turbojets to reduce risk. Boeing Vice President of Engineering, Ed Wells, disagreed saying the rapid pace of aircraft development leaves no room for the turboprop. It may have had a place three years ago, but three years out the turbojet will fill the needs for most airplanes.[33]

Despite Dernbach's recommendation to continue with both turbojets and turboprops, the Air Force issued a supplemental agreement in June 1949 to P&W for two XJ57-P-

1 engines. The agreement included 50 hours of engine development testing, static loading, and modification of a government furnished B-50 aircraft for a flying test bed. On 1-2 August the XJ57 mock-up inspection was conducted at P&W in Hartford, CT. Project engineer, Frederic G. Hoffman, said "it is apparent P&W is developing an engine for the bomber application. Compression ratio is high, altitude characteristics better than usual, and the specifics are very good."[34]

On 15 August 1949, P&W submitted the specification for an interim engine designated XJ57-P-3 (JT3-10B) for initial XB-52 flights. This was an entirely new design because the previous version (JT3-10A) had less thrust than anticipated, was overweight, and had other faulty design approaches. P&W knew the Air Force was considering the J40 for initial XB-52 tests and were determined to provide a suitable engine. AMC Engineering Division announced on 7 September that P&W had successfully modified the T45 turboprop into a J57 turbojet engine.

Lieutenant General LeMay, now SAC Commanding General, expressed dissatisfaction with the XB-52 (based on the Boeing Model 464-49) in a meeting with Warden at SAC HQ. He stressed that he would not accept range at the expense of speed and recommended the J57 be given precedence over all other engine designs. He believed with proper emphasis it could meet range, speed, and production schedule requirements.[35]

Boeing responded in November 1949 with the new, heavier, Model 464-67 with 390,000-pounds gross weight. They expected the aircraft to achieve a 4,360-mile combat radius for initial production aircraft in 1953. However, they expected to improve the radius to 4,821 miles by 1957. SAC officials, including LeMay, were generally pleased with Boeing's progress and supported the new design.

AMC issued a contract on 13 December to P&W for XJ57-P-3 engines beginning in May 1951 for early XB-52 testing. On 13 December the P&W XJ57-P-3 engine was chosen

The Boeing Model 464-67 included external wing tanks for increased range. (Boeing)

for XB-52 development instead of the J40. Not only was the J57 shaping up to be a more powerful engine, but the Navy Bureau of Aeronautics was reluctant to make changes to the J40 for the B-52.

But the critics were still not silent. A strategic committee meeting, suggested by LeMay, was held at the Pentagon beginning 26 January 1950 to discuss XB-52 progress. Despite LeMay's endorsement of the B-52 as the aircraft best able to meet SAC's requirements, other attendees supported new proposals by the Douglas and Republic Aircraft Companies, Fairchild Corporation's rail-launched design, Convair's YB-60 swept wing jet engine aircraft, the RAND Model G2 turboprop airplane, two new B-47 designs, and several missile aircraft. On 27 January 1950, Colonel Carl F. Damberg, Chief of Aircraft Projects Section, AMC, requested that the Aircraft Laboratory study the feasibility of installing turbine engine supersonic propeller power packages on the B-47 and XB-52. The study would address using propellers on the Allison T40, and possibly the J57, due to time constraints.

RAND had been pushing SAC on the results of its 1946-47 study, and its conclusions had not changed in three years. It suggested a larger number of smaller turboprop bombers instead of a smaller number of high-performance aircraft like the B-52. RAND believed this would allow the Air Force to more quickly create its "air force in being". Warden immediately expressed his distaste.[36] After three days of discussing this concept, he addressed the committee. He said for the concept to work everyone would need to agree on four assumptions: One through three were straight out of the RAND report. Number four was that the aircraft would never carry the H-bomb and will never be a reconnaissance aircraft. SAC and reconnaissance representatives "came alive" and questioned where the RAND report said that. He quickly responded that the aircraft proposed would simply be too small to carry the H-bomb or reconnaissance capsule.

RAND had also suggested that the Air Force could afford 100 wings of its proposed smaller aircraft instead of only 40 of the XB-52. LeMay disagreed with this assumption after some thought, saying he would only get 40 wings regardless because that is how Congress operates. He said he would rather have 40 wings of the XB-52 than 40 of the RAND proposed smaller aircraft. Despite both SAC and AMC rejection of the concept, the Senior Officer's Board supported it and additional studies on the turboprop engines.

In contrast to AMC officers' use of knowledge and analysis to reduce technical uncertainty, it seemed that some Air Staff officers used knowledge and analysis primarily to keep the heavy bomber option undecided until a design could be discovered that would deal with political problems. For instance, RAND reports and studies projecting a high weight for the B-52 and a low probability of meeting its range requirements were used uncritically to argue for a new heavy bomber competition. The faulty premises of these

RAND reports were identified by AMC officers but not by Air Staff officers. The veracity of technical claims made in the RAND reports was not the major issue in the arguments made by Air Staff officers. A larger goal was paramount: to win support for an alternate heavy bomber proposal, such as Northrop's YB-49 Flying Wing.[37]

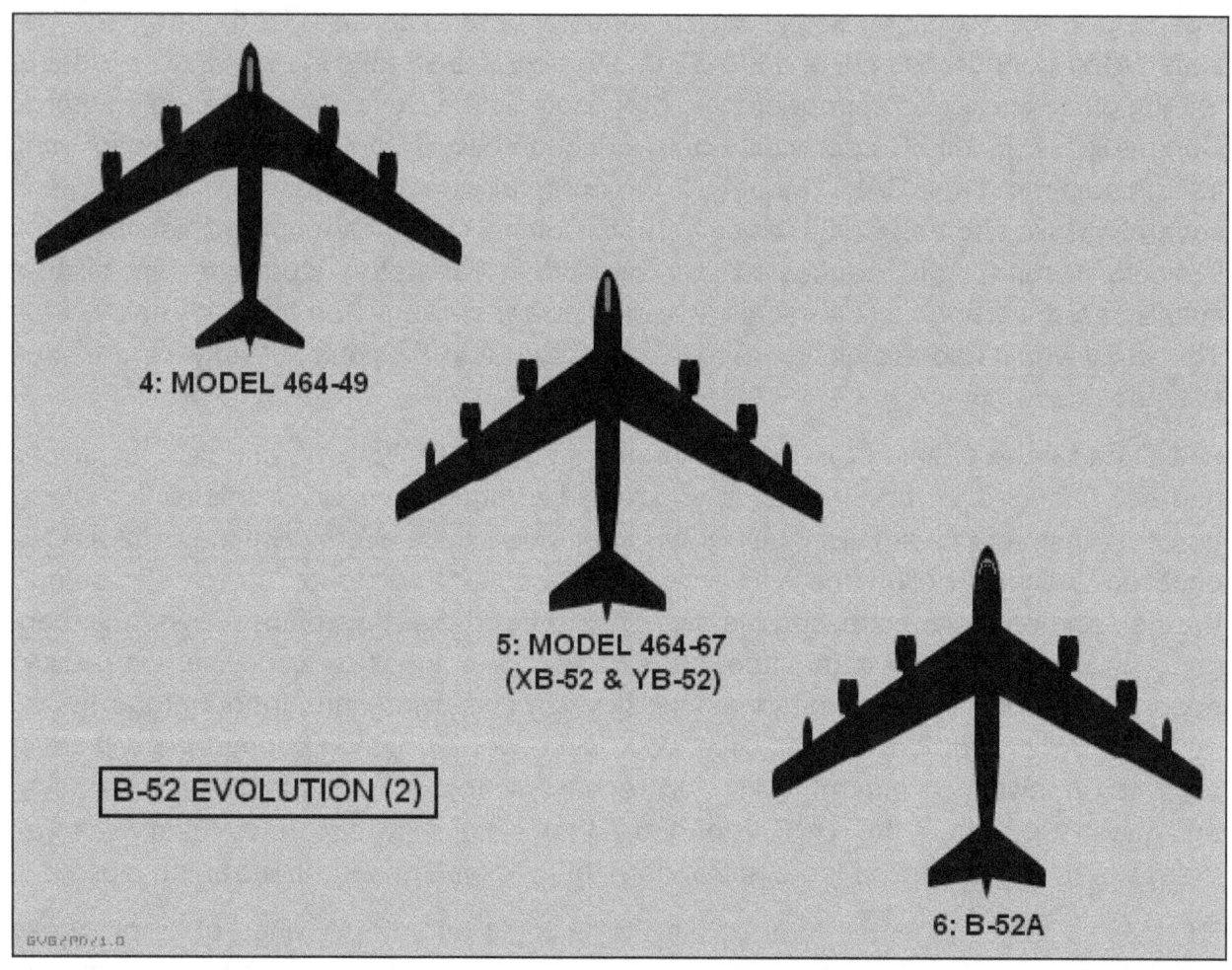

Boeing's XB-52 design evolved significantly, despite many objections along the way, into an eight-engine turbojet aircraft capable of speeds more than 500 mph and worldwide range with the equipped "flying boom" aerial refueling system. (Greg V. Goebel)

However, the formal interaction of these different agencies on heavy bomber development allowed the Air Force to avoid some common disabling behaviors of organizations seeking to innovate. Interaction prevented the Air Staff and AMC from resolving technical uncertainties by simple agreement or contract with a particular manufacturer. As the Air Staff and AMC each marshaled allies to their point of view, their mutual interaction forced them to consider trade-offs and interactions that they would have otherwise avoided. Although, at any one time, the Air Staff and AMC advanced flawed arguments, the process of debate and exchange exposed those flaws and led to a more satisfactory decision.[38]

In February 1950, AMC added another supplemental agreement to the P&W contract for XJ57 development including two more test engines. But by 22 April, the use of turboprop engines with supersonic propellers for the XB-52 was still being argued. Lieutenant Colonel Ernest N. Ljunggren had replaced Warden at AMC as Chief of the Bombardment Branch, and he believed the possible range increase from these engines presented opportunities for possible retrofit in pod-type nacelles.

Although it seemed the controversy would never end, AMC pressed forward with the XB-52 turbojet design. The XJ57 made its first flight on 8 March 1951 mounted to the wing of a B-50 and the first prototype engine was delivered to Boeing three months later. But the Air Force also continued to support alternatives to the XB-52 including the Convair YB-60, which was a B-36F airframe modified with swept wings and eight jet engines mounted on wing pods like the XB-52. This caused significant controversy over the priority for XJ57 engines.

The Convair YB-60 was a B-36F with swept wings and eight J57 engines like the XB-52, but it failed to achieve the speed expected with the XB-52. (U.S. Air Force)

On 14 March, Major General Orval Cook, Director of Procurement and Industrial Planning, informed Boeing that the B-52 would have priority for J57 engines. However, in September, Major General Carl A. Brant, Assistant DCS/Materiel, reversed Cook's decision. He believed the YB-60 had many time and cost advantages over the B-52 and if the idea proved sound the entire B-36 fleet could be converted. Lieutenant Colonel James Murray, B-52 Project Officer, objected believing the YB-60 could use J47 engines since Convair was only testing the airframe modifications (B-36 to B-60). The controversy was resolved on 5 November when Brant decided the XB-52 had priority through December 1951 and then the YB-60 would receive small quantities. Murray was still concerned that a decision was to be made between XB-52 and YB-60 and he feared XB-52 cancellation.[39]

LeMay was still convinced the B-52 was on the right track and asked the Board of Senior Officers to accept the Boeing Model 464-67 as the XB-52 prototype. The board approved the request on 24 March 1950. This action allowed the XB-52 development to continue but production was uncertain. Advocates for the YB-60 and improved B-47Z persisted. But SAC opposed both, believing the new B-60 would be too slow and the B-47Z three-man crew would have inherent limitations for intercontinental operations. A comparative study confirmed the B-52 would be a superior aircraft, but the Air Staff had still made no definite commitment by the fall of 1950.

Apparently becoming impatient with the situation, LeMay again became directly involved. He pointed out that SAC's forward operating bases were becoming vulnerable to attack and modernizing the intercontinental bomber force should be the priority. He also said, "Perhaps even more important is the concurrent requirement for the development of a long-range, high-performance aircraft, such as the RB-52, capable of operating alone over highly defended enemy areas in the performance of the reconnaissance mission." This persuasive argument finally swayed the board, and they agreed the B-52 would be the B-36 successor. They also agreed that since the B-52 was not a radical departure from existing aircraft designs, procurement could start before the XB-52 testing was completed.[40]

Models and Variants

General Vandenberg approved the board's recommendation to proceed with B-52 development on 9 January 1951, and it was subsequently approved by SECAF Thomas K. Finletter two weeks later. The initial letter contract, signed 14 February 1951 and definitized on 7 November 1952, called for 13 B-52A aircraft. Much to his chagrin, perhaps, LeMay's recommendation for the development of an RB-52 was used to justify prioritizing the reconnaissance version over the bomber.

Meanwhile, Boeing had been working on a preliminary RB-52 design and had completed the initial configuration by mid-1950. However, Boeing designers soon realized a better approach would be to incorporate a "multi-purpose" reconnaissance pod that could be easily uploaded in the bomb bay. This pod could also be replaced with a photo or ferret pod depending on mission requirements. AMC agreed and cautioned that the B-52 bombing capabilities could not be jeopardized.

In June 1951, SAC proposed an RB-52 based on Boeing's approach that could be easily converted to a bomber by wing-level maintenance personnel by removing the reconnaissance pod and installing bomb racks in its place. Representatives from Air Staff, Air Research and Development Command (ARDC), AMC, Air Weather Service, Boeing, and SAC met from 29-30 August 1951 to discuss this approach. Ultimately, everyone agreed the B-52A would primarily be a bomber that could be converted for reconnaissance capability using pods carried in the bomb bay. SAC insisted that all pods had to be easily removed to readily facilitate conversion of the aircraft back to the primary bomber role.

However, procurement orders were slow to reflect this change. An amendment to the initial order for 13 B-52A aircraft called for 17 reconnaissance pods and a contract modification in July 1951 increased the number of aircraft to 17, presumably to match the number of reconnaissance pods. Despite the August agreement, Air Staff directed in October 1951 that all aircraft "will be of the RB-52 configuration and there is no requirement for a B-52." Consequently, the follow-on procurement contract designated B and C model aircraft as RB-52. Despite this mix-up, XB-52 development continued as a pure bomber aircraft. Plans for an XRB-52 prototype were discontinued and development of the reconnaissance pods progressed.

Prototypes

XB-52. The XB-52 (49-230) rolled out on 29 March 1951 and made its first flight on 2 October 1952. This flight occurred about six months after the first flight of the YB-52 (49-231) due to required repairs from damage that occurred during full pressure testing of the pneumatic system. Consequently, most Phase I flight testing was completed with the YB-52, and the XB-52 completed only six flights lasting a total of 11.25 hours. However, the XB-52 was used for all Phase II tests conducted by Air Force pilots from 3 November 1952 to 15 March 1953. These tests uncovered several deficiencies including engine surges and shutdowns during normal accelerations at high altitude and low inlet temperatures. The brakes failed to stop the aircraft within the distances guaranteed by Boeing and the aircraft took twice the normal distance to stop. The tires tended to blow out when cross winds shoved the aircraft to one side.[1]

The XB-52 (49-230) made its first flight on 2 October 1952, which was about six months after the first flight of the YB-52 (49-231). (U.S. Air Force)

The XB-52 was the largest jet aircraft built up to that time. It was 152.7 feet long with a wingspan of 185.0 feet, and 48.25 feet to the top of the vertical fin. The leading edges of the wings were swept back 36 degrees and 54 minutes. The empty weight of the aircraft was 155,200 pounds and the maximum takeoff weight was 390,000 pounds. According to the XB-52 Standard Aircraft Characteristics, dated 6 October 1950, the cruise speed was 523 mph, and the maximum speed was 612 mph at 20,000 feet. Combat altitude was 35,000 feet and the service ceiling was 49,900 feet. The combat radius was 3,537 miles. Combat radius, combat ceiling, service ceiling, cruise speed and max speed were calculated with a 10,000-pound bombload at combat weight without refueling (Basic Mission).

The XB-52 was originally powered by eight P&W YJ57-P-3 turbojet engines, with a normal power rating of 8,700 pounds of thrust. However, these were upgraded to XJ57-P-1 engines with a maximum rating of 9,250 pounds of thrust (with variable area nozzle), 9,000 pounds of thrust (with fixed area nozzle), and normal power rating of 8,700 pounds of thrust in March 1950. Assisted take-off (ATO) was provided by four liquid rocket units developing 4,000 to 5,000 pounds of thrust each for a maximum duration of 60 seconds. The aircraft was designed to incorporate two .50-caliber machine guns in a tail turret with 600 rounds of ammunition per gun, although these guns were not installed on 49-230. It had provisions to carry up to 25,000 pounds of conventional or nuclear weapons including a 25,000-pound T-28E2 Samson bomb.

The XB-52 (49-230) with tandem cockpit arrangement flying over Mosses Lake, WA in 1952. (U.S. Air Force)

The B-52 was essentially shaped around its bomb bay. The bomb bay had to be located on the center of gravity to prevent the aircraft from losing control because of the massive shift in the center of gravity when a weapon was released. This meant putting the wing box structure over the bomb bay which resulted in the high wing configuration. The landing gear could not be mounted on the high wing, so the B-52 had a quadricycle landing gear with main gears ahead of and behind the bomb bay. Outrigger landing gears were added on the wing tips to prevent the wings from dragging on the ground when they were heavy with fuel. This landing gear configuration prevented the B-52 from taking off and landing in a crosswind by lowering the upwind wing. Instead, the B-52 incorporated an ingenious crosswind crab landing gear design.[2]

The aircraft incorporated accommodations for a crew of five including the pilot, copilot/flight engineer, bombardier/navigator/weaponeer, and gunner/radio operator using a tandem seating arrangement for the pilots. LeMay felt this arrangement was poor and he believed side-by-side seating provided better crew coordination between the pilot

The XB-52 (49-230) with a P&W J75 engine in a single nacelle on each of the outer pylons circa 1959. (U.S. Air Force)

and copilot. This not only ensured better operational capability but provided significant improvements in safety. In the tandem arrangement the copilot essentially acted as a flight engineer, operating emergency controls and limiting his assistance to the pilot. Furthermore, the smaller instrument panels required smaller instruments that had been unsatisfactory in other aircraft types. The Air Staff agreed with LeMay and initially planned to continue with tandem seating on a few initial production aircraft, which would be retrofitted later bringing all production aircraft to the side-by-side arrangement. However, Boeing was ultimately able to incorporate the side-by-side seating design on all production aircraft from the start allowing SAC to maintain only one configuration and cutting production costs by $17 million.[3]

Although the aircraft had many improvements over previous designs, it failed to meet the required range specifications and SAC continued to push for more. They also wanted a larger bomb bay to carry larger bombloads and the ability to carry large missiles anticipated in the future.[4] Of course, the ability to carry larger loads had a direct negative impact on range.

The XB-52 was returned to the Air Force after completing Phase II testing in March 1953. It was then used in Phase III testing, logging 46 hours in 24 flights. The aircraft was assigned to the Wright Air Development Center (ADC) at Wright-Patterson AFB, OH as a test bed in March 1957. The aircraft flew 893 test hours and was most significantly fitted with a J75 engine on each outboard wing pod making it effectively a six-engine aircraft. The J75 was also initially considered for the B-52G but was eventually rejected due to lack of support in the aviation industry in favor of the J57-P-43WB.

YB-52. The original contract called for two XB-52 experimental prototypes. These two aircraft were initially the same. However, by mid-1949 Boeing suggested installing tactical equipment in the second aircraft to make the aircraft more production representative for testing. The Air Force agreed and redesignated the aircraft as the YB-52 production prototype. As such, the aircraft was fitted with state-of-the-art equipment such as the KA-1 Bombing-Navigation System (BNS), AN/APN-9A LORAN, AN/APX-6 Identification Friend or Foe (IFF) Transponder, AN/ARC-27 Command Radio, MA-2 Autopilot, and space provisions for AN/APQ-27 Electronic Countermeasures (ECM).

The first flight of the XB-52 was delayed primarily due to repair of damage to the wing trailing edge that occurred during pneumatic system testing. It also suffered from lagging engine and pneumatic system deliveries as well as an engineering decision to change the rear wing spar, which was incorporated directly into the YB-52. Consequently, the YB-52 (49-231) was rolled out on 15 March 1952 and flew first from Boeing's Renton Field near Seattle, WA on 15 April 1952, about six months sooner than the XB-52. The 2-hour and 51-minute flight was generally considered a huge success by the test pilots, engineers, and all observers. However, some minor issues were observed including an improperly retracted landing gear and a leaking engine oil valve.

The second flight occurred on 20 April and was again considered successful. The aircraft accelerated to 350 mph despite being restricted to 15,000 feet altitude due to J57 engine qualification issues. The engine issues were soon corrected, and the aircraft continued through its flight test regimen. By October 1952, it had flown 50 hours and reached speeds of Mach 0.84 at altitudes above 50,000 feet. The Air Force accepted the aircraft in place at Boeing on 31 March 1953 and Boeing continued the flight test program logging 738 hours over 345 flights. The aircraft remained at Boeing throughout

The YB-52 (49-231) on its first flight near Seattle, WA on 15 April 1952 fitted with state-of-the-art tactical equipment making it a production representative prototype. (U.S. Air Force)

its short life span, though it was kept in storage throughout most of 1957. The aircraft was donated to the Air Force Museum on 27 January 1958.[5]

Bombers

<u>**B-52A.**</u> The first B-52A rolled out of the Boeing factory in Seattle, WA on 18 March 1954. It was the culmination of years of collaboration and development between the Air Force and Boeing. It was at the time the most advanced bomber aircraft ever built. General Nathan F. Twining, Air Force Chief of Staff, commented during the roll-out ceremony, "To say this is the greatest bomber in the world today is putting it very, very mildly. …And the progress that this airplane has made since the prototype was put on the line is something that has never been equaled. …The long rifle was the great weapon of its day. …Today this B-52 is the long rifle of the air age."[6]

The B-52A featured the new production model J57-P-1W jet engines with 8,250 pounds of thrust each when dry. Thrust could be increased to 11,100 pounds during take-off in warm weather conditions using water injection supplied from a single 360-gallon water tank located in the aft fuselage. However, unreliable pumps and problems with the -1W water injection system meant its use was sometimes limited on the B-52A. The aircraft was equipped with a refueling receptacle compatible with the Boeing-developed aerial refueling boom. In addition to in-flight refueling capability, which gave the aircraft unlimited range, the B-52A carried 35,385 gallons of fuel internally along with a 1,000-gallon droppable auxiliary tank on each wing tip. This allowed the aircraft to achieve a 3,565-mile combat radius unrefueled.

The first B-52A (52-0001) rolled out of the Boeing factory in Seattle, WA on 18 March 1954. (U.S. Air Force)

The aircraft were equipped with essential equipment such as the AN/ARC-27 Ultra High Frequency (UHF) Command Radio, AN/ARC-21 High Frequency (HF) Liaison Radio, AN/APX-6 IFF Transponder, AN/ARN-18 Glideslope Receiver, AN/ARN-12 Marker Beacon, AN/APN-76 Radar Beacon, AN/AIC-10 Interphone, AN/APA-17 Direction Finder, and MA-2 Autopilot. The defensive system included a tail turret with four .50-caliber M-3 machine guns controlled by the installed A-3A Defensive Fire Control System (DFCS) which was capable of both automatic and optical tracking and search. The ECM equipment included an AN/APR-8 Panoramic Receiver, (2) AN/APT-6, (2) AN/APT-9, and (1) AN/APT-16 transmitters along with (2) AN/APR 9 and (3) AN/APR-14 receivers.

The aircraft incorporated the side-by-side seating for the pilot and copilot as requested by General LeMay after the mock-up inspection conducted in May 1951, which he deemed essential for promoting effective crew coordination. The Air Force approved this request and planned to incorporate it into the fourteenth production aircraft, but production delays allowed Boeing to include it on the first aircraft. A forward-facing crew position was also added in the rear of the upper cabin to accommodate the Electronic Warfare Officer (EWO) who was responsible for the operation of the ECM equipment. The navigator and radar-navigator (bombardier) were in the lower compartment, and the gunner occupied a compartment in the tail of the aircraft. This configuration change resulted in a wider fuselage that was also extended by 21 inches compared to the XB-52 and YB-52. In addition, a larger upper radome was added to the nose resulting in a total fuselage length increase of 3.8 feet.

The B-52A fuselage differed significantly from the experimental XB-52 and production representative YB-52 with side-by-side seating, longer nose, and aerial refueling receptacle. (U.S. Air Force)

Although the B-52A was considered the first production B-52 model, it never saw operational service with SAC. The original letter contract signed on 14 February 1951 called for 13 B-52A models with the first aircraft to be delivered in April 1953, but this contract was modified on 9 June 1952 to require only three B-52A models, and the rest were to be completed as RB-52B models. The aircraft was ultimately delayed by 14 months due to various factors including the complexity of the B-52 design and manufacturing techniques, competition with other aircraft programs such as the YB-60 and B-47Z, lack of trained aircraft personnel, Korean War priorities and resulting low priority of the B-52, and incorporation of mandatory changes identified during XB-52 and YB-52 testing.

The first B-52A (52-0001) was accepted in June 1954 and made its first flight on 5 August 1954. The second B-52A (52-0002) was accepted in August 1954, and the third (52-0003) in September 1954. All three aircraft were delivered in place to Boeing and used for Phase IV testing. These tests verified the stability data obtained during Phase II testing and compared performance of the J57-P-29W engine to the J57-P-1W originally installed on the B-52A. The new engine proved superior and would ultimately be used on the B-52B.

The three B-52A aircraft supported critical testing for various B-52 system developments for the next several years. B-52A (52-0001), along with the YB-52, was used for initial testing of the shorter vertical fin eventually used on the B-52G and H models. It was also tested with the J57-P-43W engines that were eventually selected for the B-52F. It was redesignated GB-52A in mid-1959 and used for maintenance training at Chanute AFB, IL. In 1965-66 it was used for a fire fighter training film and training and ultimately destroyed by fire. B-52A (52-0002) was retired in 1960 after flight testing with gross weights up to 415,000 pounds. It was ultimately scrapped at Tinker AFB, OK in April 1961. B-52A (52-0003) was flown from Seattle to the North American Aviation operated Air Force Plant 42 in Palmdale, CA to be modified beginning in January 1958 as the original mothership for X-15 rocket plane tests.

B-52B. This was the first B-52 model to see operational service with SAC. It was outwardly the same as the B-52A. Although originally designed as a bomber, an Air Force directive in October 1951 stated that the B-52 bomber version would be replaced with the RB-52 and the reconnaissance mission would have priority in B-52 development. The original contract definitized in November 1952 called for 17 RB-52B aircraft. A second contract definitized on 15 April 1953 called for an additional 43 RB-52B aircraft. However, this contract was modified in May 1954 and reduced the order by ten aircraft resulting in a total order for 50 RB-52B aircraft.

A total of 27 aircraft (52-0004 to 52-0013; 52-8710 to 52-8716; 53-0366 to 53-0372; 53-0377 to 53-0379) were ordered as RB-52B models and could be fitted with a pressurized reconnaissance pod for two additional crew members. However, a new Air Force directive issued on 7 January 1955 completely reversed the previous direction and once again gave priority to the bombing mission. This resulted in 23 aircraft being delivered as B-52B aircraft that had no capability to upload the reconnaissance pods.

Boeing intended to use a new more powerful J57-P-9W engine using lighter titanium components for all RB-52B and B-52B aircraft to replace the under-powered J57-P-1W engines used on the B-52A. However, cracking of the titanium compressor blades caused delays. A total of 28 aircraft, the first 24 RB-52B (52-0004 to 52-0013; 52-8710 to 52-8716; 53-0366 to 53-0372) and the first four B-52B (53-0373 to 53-0376), were consequently fitted with -1W engines even though water injection reliability issues sometimes limited its use.[7]

P&W returned to using steel parts and quickly produced a new J57-P-29W engine capable of 11,500 pounds take-off thrust with water injection. A new J57-P-29WA engine, able to handle twice the water injected into it compared to the -29W, and capable of 12,100 pounds take-off thrust wet was also introduced. Although about 250-pounds heavier than the -9W titanium engine, the last three RB-52B (53-0377 to 53-0379) and

The first SAC B-52, an RB-52B (52-8711), arrived at Castle AFB, CA on 29 June 1955. The B-52 would ultimately replace the wing's B-47 aircraft. (U.S. Air Force)

the next 14 B-52B (53-0380 to 53-0393) aircraft were fitted with -29W or -29WA engines. Meanwhile, P&W solved their titanium problems and developed a new J57-P-19W engine, which was a more powerful version of the -9W capable of 12,100 pounds take-off thrust with water injection. This engine was fitted on the last five B-52B (53-0394 to 53-0398) aircraft. With the new engines, the B-52B maximum take-off weight increased to 420,000 pounds compared to 405,000 pounds with the -1W engines.

All RB-52B and B-52B aircraft were equipped with essential equipment such as the AN/ARC-34 UHF Command Radio, AN/ARC-21 HF Liaison Radio, AN/APX-25 IFF Transponder, AN/APN-76 Radar Beacon, AN/AIC-10 Interphone, AN/ARN-18 Glideslope Receiver, AN/ARN-12 Marker Beacon, AN/ARN-14 Navigation Receiver, and MA-2 Autopilot. The ECM equipment included an AN/APS-54 Early Warning Receiver, AN/ALE-1 Chaff Dispenser, (2) AN/ALT-7, (2) AN/APT-8, and (1) AN/APT-9 transmitters along with (1) AN/APR 9 and (1) AN/APR-14 receivers.

The Air Force and SAC struggled to find a suitable BNS for the B-52B. They had been working for years to develop an accurate system, with increased automatic operation to reduce human error, for the B-36 and B-47 as well. Two main BNS systems remained under consideration for the B-52B: the K-3A "K-System" which relied on radar and optics, and the MA-2 which combined an optical bombsight, radar target presentation, and an automatic computer together with radar modifications designed for use in high-speed aircraft.[8] Although the MA-2 seemed ideal for both the B-47 and B-52, SAC did not believe it would be fully tested before it was needed for the B-52B.

Some initial RB-52B and B-52B aircraft were fitted with the same J57-P-1W engines used on the B-52A. (U.S. Air Force)

For lack of a better system, early RB-52B and B-52B aircraft were fitted with the same K-3A system used on the B-36. This system was not completely satisfactory and at altitudes above 35,000 feet loss of definition and resolution prevented target identification.[9] A modification to improve power output helped but was more of a band-aid than an improved system. Ultimately, the IBM MA-6A "Brane" BNS, which was an improved version of the K-3A, was installed on the remaining B-52B production aircraft while the Air Force continued to work on improved systems for subsequent B-52 aircraft. The MA-6A system included the APS-23 search radar, Y-3 optical bombsight, and A-1A bombing computer.

Nine of the initial ten RB-52B aircraft (52-0004 to 52-0008; 52-0010 to 52-0013) retained the tail turret with four .50-caliber M3 machine guns used on the B-52A controlled by the same A-3A DFCS, but this system was deficient. One of these initial ten aircraft (52-0009) was equipped for testing with a tail turret containing two M24A-1 20mm cannons

B-52 Stratofortress: The Iron Fist of Strategic Air Command

B-52 K-3A Bombing Navigation System

The following story is from a 1957 edition of *Radio & Television News* report on the K-3A computerized BNS installed on early B-52B aircraft.

Here are the first photos of a system that pinpoints any spot on the globe in any weather. A far cry from the relatively simple "bombsight" of early World War II is the later version of the "K-System" bombing navigation controls in the new Air Force B-52 Stratofortress bombers. These are the first operational views of a master control panel, just released 10 years after the system was initially developed and produced by Sperry Gyroscope for the still-formidable B-47 and B-36 squadrons of Strategic Air Command. First view here reveals, by a selector switch above the navigator's helmet (top right), the transpolar range capability of the bombers.

The K-System automatically measures distance and time to target, computes ballistics of the bomb's curve for existing altitude, temperature, and crosswinds, permits final hairline adjustment via radar or optical sight, triggers bombs away at the proper instant, then helps the navigator to guide the shortest way home.

Originally designed, developed, and produced by Sperry Gyroscope, manufacture of this critical gear was rapidly dispersed through other plants of General Motors, National Cash Register, and IBM. Eastman Kodak, Western Electric, General Mills, Motorola, and Farrand were multiple prime or subsystem sources. Western Electric developed the radar. In all, about one million factory workers and technicians in 36 states, at 3,050 large and small companies, have been directly engaged in this Air Force program to supply needed K-Systems.

More than 70,000 individual parts make up the various computers and other elements of a single K-System for an individual aircraft – about as many parts as a modern automobile with power steering and automatic transmission. Each system contains complex circuitry for hundreds of pretested vacuum tubes, over 50 motors, and about 100 relays, with many sealed amplifiers for quick unit replacement while in the air.

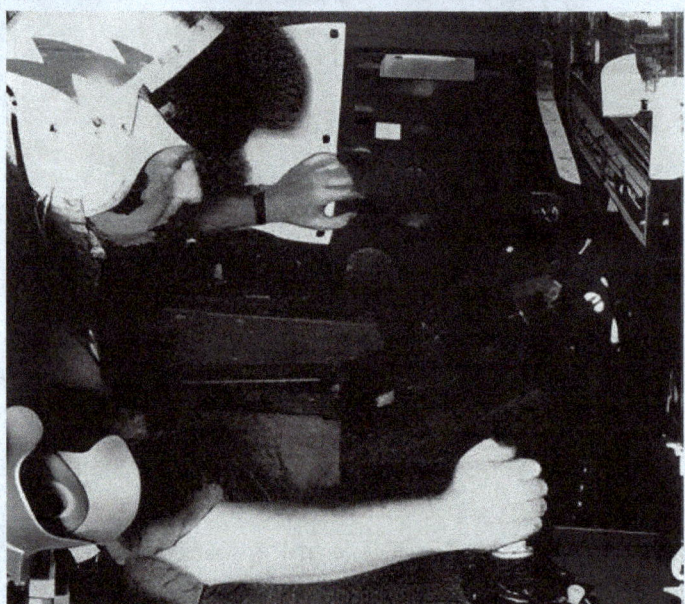

(Top) B-52 bombardier, later called radar-navigator, (left) and navigator (right) operate this two-man station at the master control panel of a later version of the fabulous "K-System" bombing-navigation controls. (Bottom) The Bombardier has complete control of the aircraft for the final moments of the target bomb run.

The original prototype cost nearly a half-million dollars and was reduced to less than half by manufacturing improvements in volume production. Nearly 900 major improvements for system efficiency through all production channels have been made in the last six years. Sperry is now "phasing out" its K-System and turning to more advanced gear. Meantime, the Air Force has announced the later B-52's will replace the K-System with an improved "Brane" bombing system to be produced by IBM.

controlled by a new MD-5 DFCS. This system was also installed in the remaining 17 RB-52B aircraft (52-8710 to 52-8716; 53-0366 to 53-0372; 53-0377 to 53-0379) and the first 16 B-52B aircraft (53-0373 to 53-0376; 53-0380 to 53-0391). Unfortunately, the MD-5 did not meet expectations and was replaced in the last seven B-52B aircraft (53-0392 to 53-0398) with a tail turret containing four .50-caliber M-3 machine guns and an improved version of the A-3A DFCS.

Both Boeing and the Air Force were determined not to repeat the initial deficiencies and delays experienced by the B-47 program. LeMay insisted that several aircraft be dedicated to the test program to ensure issues were identified and corrected early. Consequently, the B-52B underwent extensive testing and the first ten aircraft (52-0004 to 52-0013) were diverted from SAC to support the test program. Two of these aircraft (52-0008 and 52-0013) were redesignated as NB-52B and never saw SAC service. Parts were tested before being installed in production aircraft. Improvements identified in the YB-52 test program were rolled into B-52B production. Additional improvements identified in B-52B flight tests were also rolled into the production line.

However, like most new aircraft, the B-52B had its share of birthing pains. Initial problems included fuel leaks, fuel system icing, water injection pump issues, faulty alternators, and deficient BNS and fire control systems. Uneven cabin heating problems meant the pilots were comfortable while the downstairs crew were forced to wear winter clothing and boots to stay warm. Also, the AN/ARC-21 HF radio first installed on the B-47 was proving unreliable on the B-52. Still, these issues were less severe than those encountered with the B-47 and other new bombers.

Despite an AMC recommendation that aircraft should be perfected before delivery, SAC insisted that the aircraft should be accepted immediately and modified later. LeMay pressed for the best aircraft for his command but after the configuration was perfected as near as possible and approved, he felt too many immediate improvements, refinements and additional requirements were self-defeating. After protesting unnecessary changes SAC requested and received participation in Engineering Change Proposal (ECP) coordination. As a result, some 170 ECPs suggested for the first 20 aircraft were reduced to 60 between February and the end of March 1955.[10]

Seven early B-52B aircraft initially used for testing were modified under Project SUNFLOWER beginning in mid-1956 to bring them up to production configuration before being delivered to SAC for operational service. This required installation of 150 modification kits, and the final aircraft was not completed until December 1957. The B models also underwent many other modifications including such projects as BLUE BAND and QUICKCLIP, all of which were first initiated for the benefit of subsequent models.[11] A special project called HARVEST MOON increased the B-52B combat potential by installing most B-52D modifications on the later B-52B models[12] including the AN/ASQ-48 system which integrated a new AN/ASB-15 BNS and associated systems, BIG FOUR modifications, A/A42G-11 Automatic Flight Control System (AFCS), AN/APN-150 Radar Altimeter, and N-1 Magnetic Compass system.

Despite these successes, an apparently faulty alternator caused the first fatal B-52 crash on 16 February 1956 and 20 B-52B aircraft were immediately grounded. Boeing claimed the alternator problem was solved in May, and the Air Force began accepting aircraft again. However, the problem resurfaced. Fuel system and hydraulic pack issues grounded the aircraft again in July. These issues were fixed quickly but they resulted in

A line-up of B-52B aircraft on the flightline at Castle AFB, CA. The three closest aircraft are a B-52B (53-0396), RB-52B (53-0377), and B-52B (53-0391). (U.S. Air Force)

a delay in crew training, and no crews were available when the 42nd BMW at Loring AFB, ME began receiving their C model aircraft in June 1956.

Like the B-52A, B-52B deliveries were delayed for several reasons including Korean War priorities. Revised schedules called for deliveries to start in April 1954, but this was delayed again. The first B-52B rolled out on 14 July 1954 and the Air Force accepted it in August. Another aircraft was accepted in September and production output continued, but deliveries were delayed until March 1955 while Boeing engineers sought to correct cracking in the landing gear trunnion forgings.[13]

The first ten aircraft were delivered between March and May 1955 and used for testing. The first one, RB-52B (52-0004), was delivered in-place to Boeing on 3 March 1955. The first B-52 assigned directly to SAC, an RB-52B (52-8711), first flew in Seattle on 25 January 1955 and was delivered to the 93rd Bombardment Wing (BMW), Castle AFB, CA on 29 June 1955. It was flown from Seattle by Brigadier General William E. Eubank, Jr., 93rd BMW Commander. The first aircraft produced as a B-52B (53-0373) flew for the first time on 7 July 1955 and was delivered to the 93rd BMW on 9 November 1955. The 93rd BMW ultimately became the B-52 Combat Crew Training Center (CCTC) for SAC and received most of the B-52B aircraft produced.

SAC began phasing out all B-52B aircraft that had reached the end of their service life under an accelerated retirement program beginning in March 1965. The first aircraft retired RB-52B (52-8714) transferred from the 22nd BMW at March AFB, CA to Chanute AFB, IL on 8 March to be used for ground instructional training. The first B-52 assigned to SAC, an RB-52B (52-8711), was donated on 29 September to the Strategic Aerospace Museum at Offutt AFB, NE[14] for permanent display. A small number of B models (52-0011, 52-0012, 53-0377, 53-0378, 53-0383, 53-0387, 53-0388, 53-0392, 53-0395, 53-0396, and 53-0397) were temporarily redistributed to seven bomb wings including the 91st BMW at Glasgow AFB, MT, 92nd BMW at Fairchild AFB, WA, 306th BMW at McCoy AFB, FL, 340th BMW at Bergstrom AFB, TX, 380th BMW at Plattsburg AFB, NY, 494th BMW at Sheppard AFB, TX, and 509th BMW at Pease AFB, NH to act as "bounce birds" for local training beginning in January 1966 before finally being retired. The last two aircraft, RB-52B (53-0378) from McCoy and B-52B (53-0388) from Fairchild, arrived at the Military Aircraft Storage and Disposition Center (MASDC) on 29 June 1966.

B-52C. The first B-52C (53-0399) was delivered in-place to Boeing for testing. It was later designated JB-52C and then NB-52C and used for various test programs throughout its service life. It never saw service with SAC. The second B-52C (53-0400) became the first to fly on 9 March 1956. It was very similar to the B-52B with a few improvements. Outward distinguishing features were the incorporation of larger 3,000-gallon droppable wing tanks, which increased total fuel load to 41,550 gallons, and gloss-white "anti-flash" thermal and radiation reflective paint[15] on the underside. Maximum take-off weight was increased to 450,000 pounds to account for the increased fuel load and paint. The aircraft were fitted with -19W or -29WA engines capable of 12,100 pounds take-off thrust wet which compensated for the increased weight. The water injection system was also significantly improved compared to previous models. Two water tanks, one located

A B-52C inflight with 3,000-pound droppable wing tanks and gloss-white "anti-flash" thermal and radiation reflective paint on the underside. (U.S. Air Force)

in each wing leading edge, with a total capacity of 300 gallons replaced the single tank in the aft fuselage and provided water during take-off for approximately 110 seconds.

All B-52C aircraft were equipped with essential equipment such as the AN/ARC-34 UHF Command Radio, AN/ARC-21 HF Liaison Radio, AN/APX-6 IFF Transponder, AN/APN-76 Radar Beacon, AN/AIC-10 Interphone, AN/ARN-14 Navigation Receiver, AN/ARN-18 Glideslope Receiver, AN/ARA-25 UHF Direction Finder, AN/ARN-12 Marker Beacon, and MA-2 Autopilot. The initial ECM equipment included an AN/APS-54 Early Warning Receiver, AN/ALE-1 Chaff Dispenser, (2) AN/APT-6, (1) AN/APT-9, and (2) AN/ALT-16 transmitters along with (1) AN/ARR-9 receiver.

The AN/ARC-21 HF was subsequently upgraded to the AN/ARC-65. The AN/ARN-12 Marker Beacon was replaced with the new AN/ARN-32 on all aircraft. A new AN/ARN-21 Tactical Air Navigation (TACAN) system was installed on all aircraft. The AN/APX-6 IFF Transponder was subsequently upgraded to the AN/APX-25, and then later replaced with the AN/APX-64 on aircraft incorporating the Automatic Identification and Monitoring System (AIMS) capability upgrade.

While improving, the B-52C had its share of birthing pains. The B-52C retained the four-gun .50-caliber tail turret and improved A-3A DFCS installed on the last seven B-52B aircraft. However, even this improved A-3A system proved unreliable and was replaced on the last aircraft (52-2688) with the MD-9 DFCS. The B-52C had initially retained faulty alternators that caused the crash of a B-52B. While the supplier made some improvements, troubles still occurred due to the use of grease instead of oil in the lubricating system. This problem was corrected by the end of 1956. Also, defective landing gear trunnion fittings were discovered in nearly all B-52C aircraft in 1957.

Initially, the aircraft were fitted with the same MA-6A BNS as used on the last few B-52B aircraft. These aircraft were subsequently upgraded with a new IBM AN/ASB-15 BNS and AN/APS-104 BNS radar. A special project called HARVEST MOON, previously mentioned for later B-52B models, also increased the B-52C combat potential by installing most B-52D modifications including the AN/ASQ-48 system. The AN/ASQ-48 was an integrated system consisting of the AN/ASB-15 BNS, AN/APS-108 BNS radar, AN/APN-108 Doppler Radar, MD-1 Astrocompass, N-1 Magnetic Compass system, and the AN/AJA-1 True Heading Computer. The aircraft were also upgraded with the BIG FOUR modifications, A/A42G-11 AFCS, AN/APN-150 Radar Altimeter, and other modifications that effected the entire B-52 fleet.

B-52 tail gun turret with four .50-caliber M3 machine guns capable of firing 600 rounds per minute per gun. (U.S. Air Force)

The first B-52C (53-0400) to see service with SAC was delivered to 42nd BMW, Loring AFB, ME on 16 June 1956. The wing became fully operational at the end of 1956 and 27 of the 35 B-52C aircraft were initially assigned to it. All B-52C aircraft were produced in 1956 with the last five being delivered in December 1957 to Westover AFB, MA. All B-52C aircraft were phased out of service by 1971. Several B-52C models were redistributed to several bomb wings to act as bounce birds before retirement like the B-52B. The last one, B-52C (53-0402), flew from March AFB, CA to the boneyard on 29 September 1971. Oddly enough, all 35 B-52C aircraft were capable of carrying the multipurpose reconnaissance pod even though the last 23 B-52B aircraft were produced without the capability.

B-52D. The B-52D was the favorite of many BUFF fans[16], with its tall tail, blunt nose, and black-bellied Southeast Asia (SEA) camouflage paint scheme. It was the second most produced model with 170 aircraft, second only to the B-52G with 193 airframes produced. The B-52D was originally delivered with the standard SAC natural metal upper surfaces and gloss-white undersides and was outwardly identical and essentially the same as the B-52C. It retained the same engines and water injection system as the B-52C. It also retained the four-gun, .50-caliber tail turret but replaced the A3A DFCS

A B-52D in flight with its formidable tall tail, blunt nose, and black-bellied SEA camouflage paint scheme. (U.S. Air Force)

with an improved MD-9 DFCS on all except the first eight Wichita aircraft (55-0049 to 55-0056). Unlike its predecessors, the B-52B and C models, the B-52D was built strictly for the bombing mission with no reconnaissance pod capability. As such, it could carry the newest thermonuclear weapons with no bomb bay modifications.

All B-52D aircraft were initially equipped with the same essential equipment as the B-52C consisting of the AN/ARC-34 UHF Command Radio, AN/ARC-21 HF Liaison Radio, AN/APX-6 IFF Transponder, AN/APN-76 Radar Beacon, AN/AIC-10 Interphone, AN/ARN-14 Navigation Receiver, AN/ARN-18 Glideslope Receiver, AN/ARA-25 UHF Direction Finder, AN/ARN-12 Marker Beacon, and MA-2 Autopilot. The initial ECM equipment included (1) AN/APS-54 Early Warning Receiver, (2) AN/ALE-1 Chaff Dispensers, (2) AN/APT-6, (1) AN/APT-9, and (2) AN/ALT-16 jammers along with (1) AN/ARR-9 and (1) AN/APR-14 receiver.

The AN/ARC-34 UHF radio was subsequently upgraded to the AN/ARC-133 and the AN/ARC-21 HF was upgraded to the AN/ARC-65. The AN/ARN-12 Marker Beacon was replaced with the new AN/ARN-32 on later aircraft (56-0580 to 56-0630; 56-0657 to 56-0698). A new AN/ARN-21 TACAN set was installed on all aircraft. The AN/APX-6 IFF Transponder was subsequently upgraded to the AN/APX-25, and then later replaced with the AN/APX-64 on aircraft incorporating the AIMS capability upgrade.

A B-52D flies over a cloud-covered ocean. (National Archives)

Initially, the aircraft were fitted with the same MA-6A BNS as used on the last few B-52B and all B-52C aircraft. However, all aircraft were subsequently refitted with the IBM AN/ASB-15 BNS and then the AN/ASQ-48 system. The AN/ASQ-48 was an integrated system consisting of the AN/ASB-15 BNS, AN/APS-108 BNS radar, AN/APN-108 Doppler Radar, MD-1 Astrocompass, N-1 Magnetic Compass system, and the AN/AJA-1 True Heading Computer. It was highly automatic in operation by interconnection of the various systems. The doppler radar fed ground speed and drift data to the BNS, which in turn supplied data to the other systems.

Latitude and longitude data was sent to the MD-1 Astrocompass which supplied a true heading to the BNS. Alternate true heading could be supplied to the BNS by the AN/AJA-1 True Heading Computer.

The MD-1 Astrocompass system was a very remarkable piece of technology for its day. It could search for and find one tiny, navigator-selected, star among millions of stars in the sky, lock-on, and track it for an entire flight. Using the star as a reference, it calculated, using a series of mechanical computers and electronic components, a true heading accurate to within 1.5 degrees. The AN/AJA-1 True Heading Computer was also very impressive. It received magnetic heading from the N-1 Compass system and converted it, using a magnetic variation "cam", to a very accurate true heading.

The B-52D aircraft enjoyed a long service life, but it was the result of several upgrades and modifications. The B-52D continued to experience many of the issues discovered on previous models. For example, the water injection system, although improved in the B-52C, continued to experience pump failures. The problem was eventually narrowed down to the simple fact that the pumps kept operating even after the water tanks were empty. SKY SPEED field teams quickly installed a sensor to shut off the pumps as part of an overall water injection system improvement program completed by spring of 1959.

Beginning in 1959 and through the early 1960s, increased Soviet missile threats caused SAC to rethink its employment of the B-52 as a high-altitude bomber and instead adopt a new low-level bombing strategy. Aircraft would have to penetrate enemy defenses at 500 feet or lower, at high speed, and in any weather while still retaining its high-speed and high-altitude capability.

This new strategy required modifications to the B-52D and all other models, including later B-52B models, called the BIG FOUR modification package. This package was implemented from 1959-63 and included Hound Dog missiles and Quail decoys previously intended for only the B-52G and H models, and improved ECM equipment. The AN/APS-108 BNS radar was replaced with the AN/APS-81 Advanced Capability Radar (ACR). Other modifications included installation of the A/A42G-11 AFCS and the AN/APN-150 Radar Altimeter used to provide precise altitude for low-level flight.

The AN/APS-81 ACR was a modified version of the AN/APS-81 Seach Radar initially installed in the B-52G. It included a special Terrain Avoidance (TA) mode consisting of an electronic terrain computer, radar scan converter, video distribution unit, and pilot and copilot terrain display indicators. The TA mode provided a radar display of the terrain along the flight path on the pilot and copilot display indicators. By interpreting the display and maneuvering the aircraft accordingly, the pilot was able to fly the aircraft at low absolute altitudes (aircraft to terrain separation distance).

A B-52D (56-0695) with GAM-72 Quail decoy and trailer. (U.S. Air Force)

The original BIG FOUR modification package featured a new AN/ALQ-27 ECM system designed to integrate all functions into one system, but it was quickly cancelled due to lack of funding. Instead, the Air Staff chose a more federated "Quick Reaction Capability" (QRC) using several systems to meet ECM capabilities. The QRC ECM upgrade was completed using a phased approach. Phase I implemented an emergency modification to counter enemy Anti-Aircraft Artillery (AAA) radar and Surface to Air Missile (SAM) threats. The Phase I systems consisted of (2) AN/ALE-1 Chaff Dispensers, (10) AN/ALT-6B, (2) AN/ALT-13, (1) AN/ALT-15H, and (1) AN/ALT-16 jammers along with AN/APR-9, AN/APR-14, and AN/APS-54 receivers.

Phase II was an interim modification and included some PHASE III equipment. Phase III included the most sophisticated equipment which compared to the AN/ALQ-27. Most aircraft were retrofitted to receive all three phases and the B-52H, except for the first 18, were equipped in production. The Phase III systems consisted of (8) AN/ALE-24 Chaff Dispensers, (6) AN/ALE-1 flare dispensers, (5) AN/ALT-6B, (2) AN/ALT-13, (2) AN/ALT-15H, (1) AN/ALT-15L, and (2) AN/ALT-16 jammers along with AN/APR-9, AN/APR-14, AN/ALR-18, and AN/APS-54 receivers.

The A/A42G-11 AFCS was an upgrade of the existing MA-2 Autopilot designed to ease pilot workload during low-level and aerial refueling flight modes. It included a "control wheel steering" capability enabled by installation of roll transducers in the control wheel

B-52D cockpit with AN/APS-81 ACR displays. (U.S. Air Force)

and pitch transducers connected to the control column. The transducers converted the pilots roll and pitch movements into electrical signals that were input into a new Steering Coupler, which processed the signals before sending them to the autopilot amplifier. While most of the system retained the original MA-2 vacuum tube components, the Steering Coupler was a 1960s state-of-the-art solid-state component.

Since these BIG FOUR modifications covered all operational B-52 aircraft, except early B models, they had to be tailored to each model. Several technical problems arose and had to be solved quickly. The Hound Dog missiles had to be integrated on B through F models. Airframe issues varied between models. The B-52B through D models, already in service, cost almost twice as much as other models still in production. Overall costs rose by nearly 30 percent in just over six months from November 1959 to July 1960. Despite these issues, the modifications were completed across the B-52 fleet by the end of September 1963.

Stress was also a key concern in the B-52 service life. The aircraft already faced stress damage from take-off and landings, refueling, wind gusts, maneuver loads, turbulence, and stress corrosion, but introduction of low-level flight was expected to increase stress deterioration by a quotient of eight.[17] Boeing cyclic testing on both the B-52F and G models showed numerous manhours and structural fixes were required to alleviate stress in critical areas on all B-52 models. Consequently, the Air Force introduced the HI-STRESS program designed to address stress problems on B-52B through F models using a phased approach.

Phase I was completed as the aircraft approached 2,000 flying hours and included strengthening the fuselage bulkhead, aileron bay area, boost pump access panels, and wing foot splice plate. Phase II was completed on aircraft approaching 2,500 flying hours and reinforced upper and lower wing panels (supporting the engine pods), upper wing surface fuel probe access doors, and the bottom of the fuselage bulkhead. The first two phases were not allowed to interfere with the BIG FOUR modifications and could not fall behind schedule. They were completed by the end of 1962. Phase III inspected early model aircraft for wing cracks and replaced the vertical fin spars and skins. This allowed removal of flight restrictions from most aircraft. The Air Force also started ECP 1124-2 in September 1963 to reinforce the tail section of all B-52B through F models over several years to withstand turbulence during low-level flight. ECP 1128 was implemented in 1964 to strengthen the upper fuselage and vertical fin.

The B-52D was ordered under four separate contracts. The first signed on 31 August 1954 included 50 aircraft, while the second signed on 26 October 1955 called for 51 B-52D and 26 B-52E models. All were to be built at the Boeing factory in Seattle. The final two contracts signed 29 November 1954 and 31 January 1956, totaled 69 B-52D and 14 B-52E models. These aircraft would be built in Air Force Plant 13 at Wichita, KS which was Government-Owned and Contractor-Operated (GOCO) by Boeing.

The first Wichita built B-52D (55-0049) rolled out on 7 December 1955, the same date the first B-52C came of the Seattle assembly line, and it made its first flight on 14 May 1956. The first flight of a Seattle-built B-52D (55-0068) occurred on 28 September 1956. Both aircraft were immediately dedicated to the test program before delivery to SAC. The first B-52D (55-0049) to enter SAC service arrived at the 93rd BMW, Castle AFB, CA on 26 June 1956. Additional B-52D aircraft began arriving at the 42nd BMW, Loring AFB, ME in December 1956 to replace the wing's B-52C aircraft, which were transferred to the 99th BMW at Westover AFB, MA. B-52D production ended in late 1957 with 101 aircraft accepted from Seattle and 69 from Wichita.

The B-52 quickly proved its versatility and was employed as a conventional bomber when the United States entered combat in SEA. The B-52F was the first to deploy but SAC began implementing the ECP 1224-7 Hi-Density Bombing System, commonly referred to as the "BIG BELLY", modification on its remaining B-52D fleet of about 155 aircraft on 16 December 1965. The modification was initially tested at Air Proving

The first Wichita built B-52D (55-0049) rolled out on 7 December 1955 and first flew on 14 May 1956. It was one of two B-52D aircraft to set world speed records on 26 September 1958. (U.S. Air Force)

Grounds on Eglin AFB, FL in November 1965. Testing proved the aircraft could successfully drop its full load of 84 Mk-82 500-pound bombs on target from an altitude of 25,000 feet flying at 350 knots. Total bombload increased to 60,000 pounds with the addition of wing pylons and B-52D bombers began replacing B-52F aircraft for Operation ARC LIGHT missions beginning in April 1966.

A new Phase IV ECM equipment upgrade was completed from 1964-66 in conjunction with the BIG BELLEY modifications. The upgrade included an AN/ALR-18 Automatic Receiving Set, AN/ALR-20 Panoramic Receiving Set, and AN/APR-25 Radar Homing and Warning System, along with (5) AN/ALT-6B, (2) AN/ALT-13, (1) AN/ALT-16, (2) AN/ALT-15H, and (1) AN/ALT-15L jammers, (6) AN/ALE-20 Flare Dispensers, and (8) AN/ALE-24 and (2) AN/ALE-25 Chaff Dispensers.

A new structural modification was started in December 1966 under ECP 1243 to address stress issues caused by the "BIG BELLY" high density bomb bay, and other stress issues caused by low level flying. The modification added 2,000 hours of service life and applied to all B-52C, D, and F models nearing their maximum flying hours and not identified for phase-out. The program was completed in the second half of 1968 and replaced fatigued structural parts in the most critical wing areas.[18] B-52D models were also selected for special modification to carry extra aerial mines. Modifications were

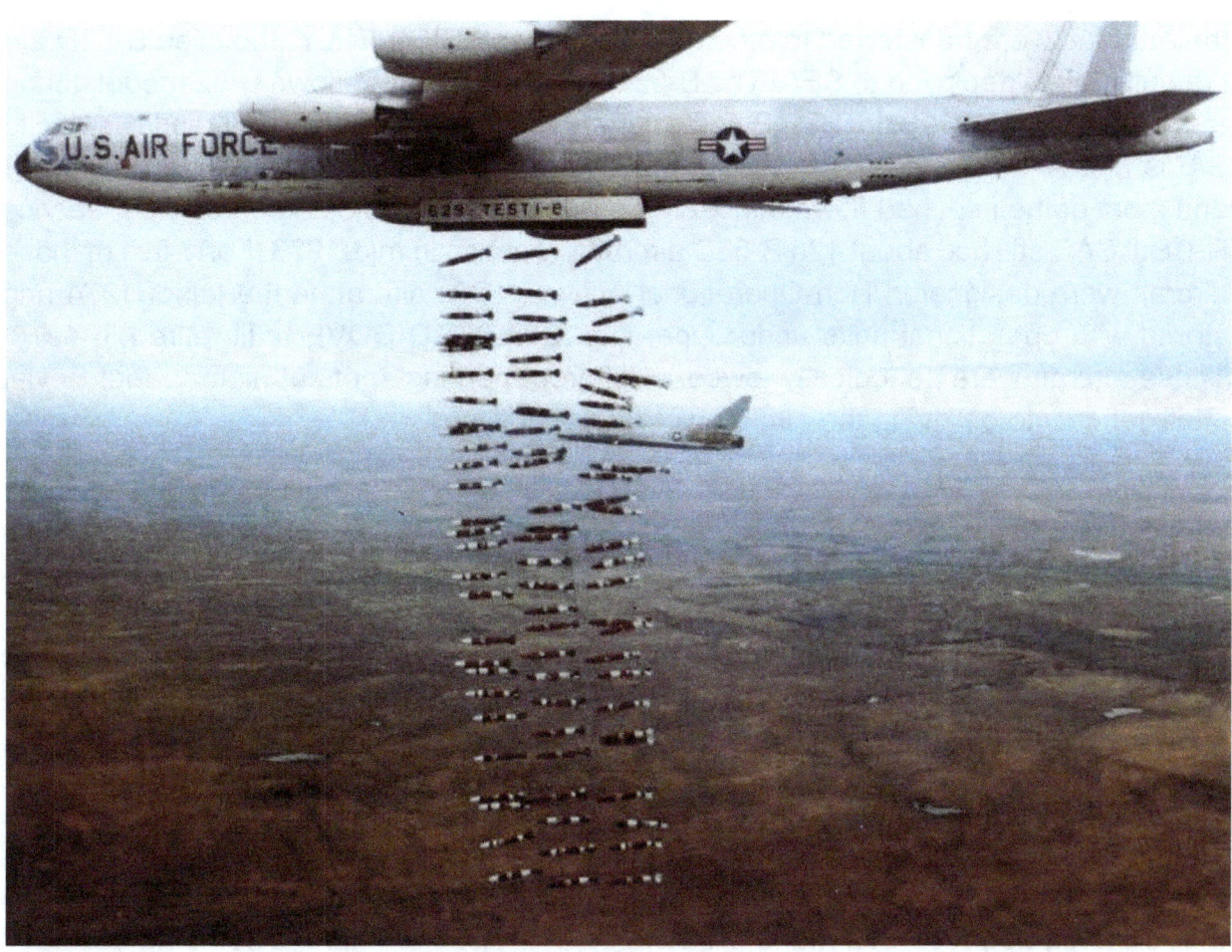

B-52D "BIG BELLY" testing at Air Proving Grounds on Eglin AFB, FL. (U.S. Air Force)

completed in the fall of 1971 and on 8 May 1972 President Nixon ordered the mining of rivers and harbors in North Vietnam.[19]

The B-52D received a new Phase V ECM equipment upgrade from 1967-69 under a program called RIVET RAMBLER. The upgrade was performed to improve combat effectiveness and gave the D models superior ECM capability compared to all other models including the B-52G and H. It included an AN/ALR-18 Automatic Receiving Set, AN/ALR-20 Panoramic Receiving Set, and AN/APR-25 Radar Homing and Warning System, along with (4) AN/ALT-6B or AN/ALT-22, (2) AN/ALT-16, (6) AN/ALT-28, (2) AN/ALT-32H, and (1) AN/ALT-32L jammers, (6) AN/ALE-20 Flare Dispensers, and (8) AN/ALE-24 and (2) AN/ALE-25 Chaff Dispensers. The AN/APR-25 was subsequently replaced by a new AN/ALR-46 Radar Warning Receiver beginning in June 1979.

SAC began deactivating B-52D and E squadrons in 1967 to meet SECDEF McNamara's directive to reduce the strategic bomber force by 1971. However, many of their aircraft were transferred to other units and most "BIG BELLY" modified B-52D aircraft ultimately deployed to SEA. The B-52D became the most-flown B-52 model during the Vietnam War. SAC lost 22 B-52D aircraft in SEA operations including 12 lost to SAMs and AAA and ten lost to accidents. Phase out was originally planned for 1971 and most of the fleet had flown almost twice the originally projected 5,000-hour service life. But SAC still had about 125 B-52D aircraft in service in mid-1973. Forty-five of these aircraft were designated Non-Operational Active (NOA) aircraft in the fall of 1974 and stored with operational units under Operation CRESTED DOVE until retired in 1978. These aircraft were periodically towed to different parking spots at night to fool Soviet intelligence into believing the aircraft were actively flying.

However, SAC decided to retain 80 low-time aircraft and add almost 7,000 hours to their service life under ECP 1581, and a program known as PACER PLANK. This program included replacement of the upper and lower wing skin panels with heavier gauge panels like those used on the B-52G, replacement of spar webs, stiffeners, rib chords, wing center panel, upper longerons, nose pressure bulkhead, and new fuselage side skins mid-way in the bomb bay to the aft wheel well bulkhead. About 30,000 pounds of parts were scrapped and replaced, and the heavier skin added almost 3,400 pounds to the gross weight, but the range and speed were increased due to smoother airflow.

These aircraft also received a new Digital Bombing-Navigation System (DBNS) to replace the aging AN/ASQ-48 system. DBNS was a joint requirement between SAC and the Air Force Logistics Command (AFLC) begun in 1976 to improve bombing accuracy and address ASQ-48 logistics issues. It was a state-of-the-art digital system designed by IBM. DBNS was the heart of the system and received information from original ASQ-48 sensors such as the AN/APN-89 Doppler Radar. It was designed around the Automated Offset Unit (AOU) installed on B-52G and H models, which provided the capability to store up to 200 destinations, targets, and offsets. The DBNS also included new technology, highly accurate Honeywell AN/ASN-131 Inertial Navigation Units (INU) called the Standard Precision Navigator/Gimballed Electrostatically suspended gyro Aircraft Navigation System (SPN/GEANS), typically pronounced "Spin Genes", which fed navigation information to the DBNS. The MD-1 Astrocompass was no longer required and was removed from the aircraft.

The new DBNS was much easier to operate than the AN/ASQ-48 equipment. It used only four control and display panels including a keyboard for data entry and display. Flight testing began with two prototype aircraft assigned to the 7th BMW at Carswell AFB, TX in October 1977 and proved the accuracy of the system for all current and projected B-52D taskings. Production installations began in July 1980 and were completed in May 1982.

The B-52D DBNS was an improved state-of-the-art digital bombing system and was much easier to use than the AN/ASQ-48. (U.S. Air Force)

Some B-52D aircraft were also given the capability to carry and launch the GBU-15 electro-optical guided bomb. The GBU-15 was a 2,000-pound modular glide bomb with an electro-optical television sensor that allowed the bomber to launch the weapon outside the lethal defense range of a ship. The optical sensor presented the radar navigator with a picture of the scene forward of the weapon so he could guide it directly onto the target.[20]

Despite these major upgrades the B-52D was ultimately retired during the early 1980s. It seemed the modifications were just barely completed when aircraft began flying to the boneyard. The last operational mission was flown by the 7th BMW at Carswell AFB, TX on 1 October 1983.[21] The last operational B-52D (55-0674) was flown from the 7th BMW to MASDC on 4 October. A total of 24 aircraft were placed on display at various locations. Seven aircraft remained at Carswell and Andersen AFB, Guam awaiting display facilities and the last B-52D (56-0687) flew from Carswell to Orlando International Airport, FL on 20 February 1984. The last B-52D (55-0094) retired was placed on display at the Kansas Aviation Museum on 11 July 1984.

B-52D mating test of the GBU-15 electro-optical guided bomb at Eglin AFB, FL on 2 August 1977. (National Archives)

B-52E. The B-52E model was essentially a continuation of the B-52D production and was outwardly the same. The B-52E retained the same engines and water injection system used on B-52C and D models. The E model also retained the four-gun, .50-caliber tail turret and MD-9 DFCS. The major internal differences were the relocation of equipment and redesign of the lower compartment for increased crew comfort and better access to instruments for both operations and maintenance ease.

All B-52E aircraft were initially equipped with essential equipment such as the AN/ARC-34 UHF Command Radio, AN/ARC-21 HF Liaison Radio, AN/APX-25 IFF Transponder, AN/APN-69 Radar Beacon, AN/AIC-10 Interphone, AN/ARN-14 Navigation Receiver, AN/ARA-25 UHF Direction Finder, AN/ARN-18 Glideslope Receiver, AN/ARN-32 Marker Beacon, and MA-2 Autopilot. The initial ECM equipment included (1) AN/APS-54 Early Warning Receiver, (1) AN/ALE-1 Chaff Dispenser, (7) AN/ALT-6A and (2) AN/ALT-7 jammers, along with (1) AN/APR-9 receiver. The AN/ARC-21 HF radio was subsequently upgraded to the AN/ARC-65. A new AN/ARN-21 TACAN was installed on all aircraft. The AN/ARN-18 Glideslope Receiver was replaced with the new AN/ARN-31 on later aircraft (57-0014 to 57-0029, 57-0095 to 57-0138).

Initially, the B-52E aircraft were equipped with the AN/ASB-4 BNS in place of the AN/ASB-15 used on the B-52D. However, all aircraft were subsequently modified with the AN/ASQ-38 system. The AN/ASQ-38 was an integrated system like the AN/ASQ-48 consisting of the existing AN/ASB-4 BNS, a new high-speed AN/APS-64 BNS radar, a new AN/APN-89A Doppler Radar in place of the AN/APN-108, MD-1 Astrocompass, AN/AJA-1 True Heading Computer, and N-1 Magnetic Compass system. The AN/ASQ-38 system was highly automatic in operation by interconnection of the various systems like the ASQ-48, although the AN/ASB-4 BNS could still be operated independently if desired.

This new system was not as accurate as anticipated, was difficult to maintain, and parts were in short supply.[22] Since the system was planned to be installed in all subsequent B-52 models as part of the BIG FOUR modifications, improvement efforts received top priority and were incorporated under a special project called JOLLY WELL. This program was completed in 1964, after modification of 480 B-52E, F, G and H models, and resulted in replacement of major parts in the AN/ASQ-38 system. The modification included upgrade of the terrain computer and consisted of changing the design from a calculation of height over terrain, to calculation of angle-of-sight, given a constant input of known height over terrain. All affected vacuum tube circuit assemblies were replaced with solid state devices.

The B-52E was the first model to use the new low-level equipment deemed necessary to elude the ever-expanding Soviet radar and missile network by flying closer to the ground and under radar coverage.[23] This equipment included the BIG FOUR modification consisting of AGM-28 Hound Dog Missiles, ADM-20 Quail decoys, EMC upgrades, and AN/APS-81 ACR with TA capability. The final ECM equipment configuration included (1) AN/APS-54 Early Warning Receiver, (1) AN/ALE-1 or AN/ALE-27 Chaff Dispenser, (1) AN/ALE-20 Flare Ejector, (3) AN/ALT-6B, (4) AN/ALT-13, (2) AN/ALT-15H, (1) AN/ALT-15L, and (1) AN/ALT-16 jammers, along with (1) AN/APR-9, (1) AN/APR-14, and (2) AN/ALR-18 receivers. Other systems included the A/A42G-11 AFCS and AN/APN-150 Radar Altimeter.

The B-52E also received a new AN/AJM-14 Malfunction Detection and Recording (MADREC) system designed to detect and locate malfunctions in the BNS, autopilot, and the Hound Dog missile system. The MADREC system was previously designed to interface with the AN/ASQ-48 and installations were completed by mid-1963. It was then modified to interface with the more complicated AN/ASQ-38 system and installed on B-52E through H models by 1965.

On 20 June 1955, the Air Force Council recommended that the B-52 program be raised from 408 aircraft approved in March 1954 to 576 aircraft. They also recommended that

B-52E (56-0631) in low-level flight. The E model was the first to incorporate the new equipment required for low-level flight. (U.S. Air Force)

the program be accelerated to meet LeMay's objective of equipping 11 bomb wings with 45 Unit Equipped (UE) aircraft and five command support aircraft. The SECAF, Harold Talbott, tentatively approved this plan pending available funding. The president's budget approved in December 1956 funded the plan and reprogrammed funds to acquire 53 B-52E aircraft beginning in mid-1957. Four contracts were subsequently definitized including the B-52D contract signed on 26 October 1955 which added 26 B-52E models, a contract signed on 2 July 1956 for 16 B-52E and 44 B-52F, a contract on 10 August 1955 for 14 more B-52E models, and the final contract on 2 July 1956 for 44 B-52E. The first two contracts called for all aircraft to be produced in Seattle. The last two called for Wichita production and they also addressed further B-52D and F deliveries.

The first Seattle-built B-52E was first flown on 3 October 1957, two weeks ahead of its Wichita counterpart which flew on 17 October. The first B-52E (56-0631) was delivered in place to Boeing on 7 October 1957. The first B-52E (56-0700) delivered to SAC reached the 93rd BMW at Castle AFB, CA on 3 December 1957. Ultimately, 58 of the 100 aircraft were produced in Wichita. Phase out began in mid-1967 when SAC deactivated several B-52D and E squadrons. Although most of the B-52D aircraft were assigned to other units and used in SEA, all B-52E aircraft were permanently retired by March 1970. The last operational B-52E (56-0638) arrived at MASDC from the 22nd BMW at March AFB, CA on 13 March 1970.

B-52F. The B-52F was an improved E model and was outwardly like its predecessor. It also retained the four-gun, .50-caliber tail turret and MD-9 DFCS used on the B-52D and E models. A new "hard drive" alternator, called a Constant Speed Drive, was installed on the left engine of each engine pod to replace the unsafe and unreliable air drive turbines installed in the fuselage on previous models.

The B-52F featured a "blister" on the left engine of each pod covering the new Constant Speed Drive that replaced the unreliable bleed-air turbines used on previous models. (U.S. Air Force)

The water injection system was significantly improved on the B-52F compared to previous models. The two water tanks, one located in each wing leading edge, were replaced by four water tanks, two in each wing located next to its corresponding engine pod, which significantly increased the system's water capacity to 1,200 gallons. This required a slight modification to the wing structure to accommodate the new tanks.

The B-52F featured significantly improved Pratt & Whitney J57-P-43W, -43WA, or -43WB engines instead of the J57-P-19W or -29WA used on previous models. The new engines produced 13,750 pounds of thrust each with water injection, an increase of 1,650 pounds over the previous versions. However, these more powerful engines introduced significant stress on the wing secondary structures and trailing edges. The engines were installed on NB-52C (53-0399), and ground runs at full power with water injection were conducted on a continual basis to determine and cure sonic fatigue problems. Consequently, more than 1,000 secondary structure changes were incorporated into the B-52F.[24]

All B-52F aircraft were initially equipped with same essential equipment as the B-52E consisting of the AN/ARC-34 UHF Command Radio, AN/ARC-21 HF Liaison Radio, AN/APX-25 IFF Transponder, AN/APN-69 Radar Beacon, AN/AIC-10 Interphone, AN/ARN-14 Navigation Receiver, AN/ARA-25 UHF Direction Finder, AN/ARN-18 Glideslope Receiver, AN/ARN-12 Marker Beacon, and MA-2 Autopilot. The initial ECM equipment included (1) AN/APS-54 Early Warning Receiver, (1) AN/ALE-1 Chaff Dispenser, (7) AN/ALT-6A and (2) AN/ALT-7 jammers, along with (1) AN/APR-9 receiver.

The AN/ARC-21 HF radio was subsequently upgraded to the AN/ARC-65. A new AN/ARN-21 TACAN was installed on all aircraft. The AN/ARN-18 Glideslope Receiver

A B-52F takes-off with AGM-28 Hound Dog missiles mounted on pylons under each wing. (U.S. Air Force)

was replaced with the new AN/ARN-31 on all aircraft. The AN/ARN-12 Marker Beacon was replaced with the new AN/ARN-32 on all aircraft. Initially, the B-52F aircraft were equipped with the AN/ASB-4 BNS like the B-52E. However, all aircraft were subsequently modified with the AN/ASQ-38 system.

The F model benefitted from the HI STRESS structural modifications and other fleet wide improvements including low-level capability enhancements. This equipment included the BIG FOUR modifications along with the A/A42G-11 AFCS and AN/APN-150 Radar Altimeter. The B-52F also received the new AN/AJM-14 MADREC designed to interface with the AN/ASQ-38.

Like the B-52E, the B-52F received B-52D ECM upgrades including Phase I through III, as well as other fleet wide improvements. The final ECM equipment was the same as the B-52E configuration and included (1) AN/APS-54 Early Warning Receiver, (1) AN/ALE-1 or AN/ALE-27 Chaff Dispenser, (1) AN/ALE-20 Flare Ejector, (3) AN/ALT-6B, (4) AN/ALT-13, (2) AN/ALT-15H, (1) AN/ALT-15L, and (1) AN/ALT-16 jammers, along with (1) AN/APR-9, (1) AN/APR-14, and (2) AN/ALR-18 receivers.

The B-52F continued to experience the fuel leaks that had plagued the earlier models. The clamps on the interconnecting lines between fuel tanks were prone to failure within weeks of operation causing fuel to gush from the lines. Boeing attempted several fixes including various types of clamps, but none fully solved the problem. Project BLUE

BAND implemented in September 1957 installed new aluminum clamps on all B-52 aircraft, but they quickly showed signs of stress and corrosion. Project HARD SHELL, completed in January 1958, installed new stainless-steel clamps, but several latch pins failed, and the clamps ruptured. New latch pins were then installed on all clamps but were not considered a final fix. Finally, in mid-1958 project QUICKCLIP was implemented which installed a strap around the clamps. Even though some broken clamp latch pins were still reported the straps prevented fuel leaks. QUICKCLIP was subsequently implemented on all B-52 aircraft.

Fuel system icing was also a continuing problem and still present on B-52F models. It had been suspected as a probable cause in several jet aircraft crashes but could not be confirmed. However, the crash of a B-52D (56-0610) from the 28th BMW at Ellsworth AFB, SD on 11 February 1958 confirmed the icing of fuel filters and screens. Several fixes were implemented including immediately installing screens and filters that were less susceptible to icing. New fuel draining procedures were developed and SAC directed use of only the driest fuel available. The most promising fix was the immediate installation of fuel heaters which was implemented in late 1959. Meanwhile, a new fuel booster valve was under development, and a research program was underway to develop new fuel additives. Despite initial problems with the additives damaging the fuel cell inner coating, which was eventually resolved, they proved successful and were implemented across the fleet. Fuel heaters were ultimately disconnected on the B-52H models.

Beginning in June 1964, 28 B-52F aircraft were modified under project SOUTH BAY to carry twenty-four 500- or 750-pound bombs externally on their wing pylons, along with 27 internally in the bomb bay, for a total of 51 bombs. This modification included an I-Beam adapter that could be mounted to the existing AGM-28 wing pylons. Two Multiple Ejector Racks (MER), capable of carrying six weapons each, were then mounted to each I-Beam adapter to control the release of the conventional weapons. One year later, another 46 aircraft were modified in just one month under project SUN BATH. These modifications nearly doubled the B-52F conventional bombload to 38,250 pounds.

SAC deployed two B-52F wings from the 2nd BMW at Barksdale AFB, LA and the 320th BMW at Mather AFB, CA to Andersen AFB, Guam in February 1965 to support Vietnam War operations. The 7th BMW at Carswell AFB, TX relieved the 2nd BMW in May 1965. Consequently, 27 B-52F aircraft from the 7th and 320th BMW were the first SAC bombers to enter combat in SEA on 18 June 1965 for the first ARC LIGHT mission. Unfortunately, two B-52F aircraft (57-0047 and 57-0179) were lost when they collided mid-air enroute to the target.

B-52F (57-0162) "Casper the Friendly Ghost" with external bomb capability drops bombs over SEA. (U.S. Air Force)

All B-52F aircraft originally retained their standard SAC colors for ARC LIGHT operations with natural metal on top and white-gloss anti-flash paint on the underside. However, beginning in late 1965 several aircraft received gloss-black camouflage paint on their undersides while retaining their natural metal tops. Later, after returning from combat and being returned to their strategic deterrence role, some B-52F aircraft received standard camouflage with white-gloss undersides and camo tops like G and H models.

B-52F procurement was ordered through two B-52E contracts, one calling for 44 aircraft to be produced in Seattle and the other calling for 45 aircraft produced in Wichita. The first Seattle-built B-52F (57-0030) flew on 6 May 1958 and the first Wichita-built B-52F (57-0139) flew on 14 May 1958. A production slippage at Boeing caused by curtailment of authorized overtime resulted in a delay in aircraft deliveries. Consequently, the first two B-52F (57-0139 and 57-0140) aircraft assigned to SAC reached the 93rd BMW at Castle AFB, CA on 14 June 1958. The last Seattle-built B-52F (57-0072) delivered to SAC left Seattle for Barksdale AFB, LA on 24 February 1959. B-52F production ended in Seattle when the last Seattle-built aircraft (57-0073) was delivered in place to Boeing on 20 March 1959 and all further B-52 production was then transferred to Wichita.

B-52F aircraft that had reached the end of their service life were permanently retired from 1967 through 1971. A total of 24 aircraft were retained for training with the 93rd BMW at Castle AFB, CA until 1974. These aircraft were then designated NOA in 1974 and stored with various operational units under Operation CRESTED DOVE. They were finally retired in late 1978, and the last aircraft (57-0171) arrived at MASDC from the 2nd BMW at Barksdale AFB, LA on 7 December 1978.

B-52G. The G model was a huge improvement and drastic departure from previous B-52 designs. Integral fuel tanks in the wings, often referred to as "wet wings", increased the internal fuel load by 6,500 gallons. The shorter tail and new wing design, which relied heavily on aluminum structures, reduced weight by 15,400 pounds. Gross weight

The B-52G featured drastic design improvements including integral fuel tanks, shorter tail, and aluminum structures. (National Archives)

increased from 450,000 to 488,000 pounds and the increased fuel loads extended the unrefueled combat radius by nearly 450 miles. The B-52G used only J57-P-43WB engines to accommodate the mounting of new 70/90-kVA generators. The four water injection tanks used on the F model, two located in each wing leading edge, were replaced by one 1,200-gallon tank in the forward body aft of the defensive stations.

Externally, the aircraft was noticeably different than previous models. The vertical fin was significantly reduced in height by nearly eight feet, the tail cone was modified to remove the gunner's station, ailerons were eliminated and the aircraft relied on spoilers only for lateral control,[25] the dual nose radomes were replaced with a larger one piece unit and the nose was more streamlined, the crew entry hatch now faced forward instead of aft, and the 3,000-gallon external droppable wing tanks were replaced with 700-gallon fixed tanks. These wing tanks not only provided additional fuel, but they acted as fuel-filled bob weights to help prevent flutter.[26]

The four .50-caliber machine tail gun turret was retained. Internally, the gunner was moved to the forward cabin which was arranged in the so-called "battle-station" concept with the defensive crew (gunner and EWO) facing aft on the upper deck, the offensive crew (navigator and radar-navigator) facing forward on the lower deck, and the pilot and copilot in the cockpit.[27] The gunner received the new AN/ASG-15 DFCS to control the tail gun turret from the forward cabin using automatic radar search and tracking, or manual operation using closed-circuit television. The new cabin arrangement also came with several improvements including increased cabin height, achieved by lowering the crew deck by two inches, which gave the entire crew more headroom and drastically improved the pilot's aerial refueling visibility.[28] All ejection seats were redesigned for greater comfort during 20-hour missions, and the air conditioning system was improved to provide even airflow between crew stations. New hot cups for making soup and coffee were also added.

These major changes to the B-52 design were the result of a deliberate Air Force decision to develop a more potent version of the original B-52.[29] In January 1955, the Air Force became concerned about the future of the manned strategic bomber force when problems with the B-58 Hustler threatened its production. The ARDC envisioned a new aircraft, sometimes referred to as a "super B-52", with redesigned wing, J75 engines,[30] and other detailed changes. LeMay supported improvements in production as much as possible but was concerned that production might be delayed by these changes.

Despite its apparent urgency, funding shortfalls delayed the beginning of design efforts until June 1956. An initial engineering development inspection was held at Boeing from 16-18 June 1956, and Boeing submitted a formal model improvement program proposal on 15 August 1956. The proposal promised a 30 percent increase range, 25 percent decrease in maintenance man-hours, a 70 percent increase in ECM capability, and a 15,000-pound weight decrease. Air Staff approved it on 29 August 1956 on a minimum sustaining basis until more was known about the status of the B-58 program.[31]

In December 1956, the President set the B-52 program at 11 wings and the Air Force had requested a production rate of 20 aircraft per month. Procurement contracts were modified earlier in the year to increase the quantities of B-52E and F models. However, by early 1957, SECDEF Charles E. Wilson decided that B-52 production would be held to 15 aircraft per month. Progress was being made on the B-58 and the production decision seemed definite. Revised intelligence estimates indicated that Soviet Bison and Bear bomber production was slowing. Although SAC pointed out that slower B-52 production rates would slow its conversion to 11 wings by almost one year, Wilson believed phasing the production out over a longer period would result in cheaper prices. It might also allow time for procurement of more of the improved B-52G models.

B-52G procurement was ultimately ordered in three different contracts. The first, issued as a letter contract on 29 August 1957 and definitized on 15 May 1958, called for 53

B-52G (57-6471) in flight featuring the original SAC colors with bare metal top and white underside. (U.S. Air Force)

aircraft. The second, a letter contract signed on 14 June 1957, called for 101 aircraft and was also definitized on 15 May 1958. The final contract, issued as a letter contract on 5 September 1958 and definitized 28 April 1959, called for 39 more aircraft. The last contract, and subsequent B-52H contracts, were above the 11-wing requirement and were prompted by initial dissatisfaction with and high price of the B-58.[32]

All B-52G aircraft were built in Wichita. The first aircraft (57-6468) rolled off the assembly line on 23 July 1958, and the first flight occurred on 31 August 1958. The aircraft was delivered to ARDC at Wright-Patterson AFB, OH on 1 November 1958. The next seven aircraft (57-6469 to 57-6475) were delivered to Boeing from November 1958 through January 1959 for testing. The first B-52G (57-6478) delivered to SAC entered service with the 5th BMW at Travis AFB, CA on 13 February 1959, one day after the last B-36 retired, making SAC an all-jet bomber force. In May 1959, the 42nd BMW at Loring AFB, ME also began receiving B-52G aircraft and by the end of June SAC had received 41 B-52G aircraft.[33]

Initially, the B-52G aircraft were equipped with the AN/ASB-9 BNS in place of the AN/ASB-4 used on the B-52E and F models. However, all aircraft were subsequently modified with an improved AN/ASQ-38 system which integrated the AN/ASB-16 BNS, AN/APS-81 ACR, AN/APN-89A Doppler Radar, MD-1 Astrocompass, and the AN/AJA-1 True Heading Computer. It was highly automatic in operation, although the BNS could still be operated independently if desired.

B-52G Navigator station with the fully integrated and fully automatic AN/ASQ-38 system. (National Archives)

All B-52G aircraft were initially equipped with essential equipment including the AN/ARC-34 UHF Command Radio, AN/ARC-21 HF Liaison Radio, AN/APX-25 IFF Transponder, AN/APN-69 Radar Beacon, AN/AIC-10 Interphone, AN/ARN-14 Navigation Receiver, AN/ARA-25 UHF Direction Finder, AN/ARN-31 Glideslope Receiver, AN/ARN-32 Marker Beacon, and MA-2 Autopilot. The initial ECM equipment included (1) AN/APS-54 Early Warning Receiver, (1) AN/ALE-1 Chaff Dispenser, (1) AN/ALE-14 Flare Dispenser, (14) AN/ALT-6B jammers, along with (1) AN/APR-9 and (1) AN/APR-14 receivers.

The AN/ARC-34 UHF radio was subsequently upgraded to the AN/ARC-164. The AN/ARC-21 HF radio was subsequently upgraded to the AN/ARC-58, and later the AN/ARC-190. A new AN/ARN-21 TACAN was installed on all aircraft and was subsequently replaced by the AN/ARN-118. The AN/ARN-31 Glideslope Receiver was subsequently replaced with the new AN/ARN-67 on all aircraft. The AN/APX-25 IFF Transponder was subsequently replaced with the AN/APX-64 on aircraft incorporating the AIMS capability upgrade.

B-52G with two Hound Dog missiles installed on specially designed wing pylons. (U.S. Air Force)

The B-52G was the first model designed as a missile platform as well as a gravity bomber. The AGM-28 Hound Dog missiles were produced for use with this aircraft, and it was also the first to incorporate the capability during production. The aircraft could also carry four ADM-20 Quail decoys and had provisions to carry four AGM-48 Skybolt Air Launched Ballistic Missiles (ALBM).[34] It also received fleet-wide improvements including low-level capability enhancements and the new AN/AJM-14 MADREC.

Like the B-52E and F models, the B-52G benefitted from B-52D ECM upgrades including Phase I through III as part of the BIG FOUR modifications. By 1964, the ECM equipment configuration consisted of Phase III components including (1) AN/APS-54 Early Warning Receiver, (2) AN/ALE-1 or AN/ALE-24 Chaff Dispenser, (1) AN/ALE-20 Flare

Ejector, (3) AN/ALT-6B, (4) AN/ALT-13, (2) AN/ALT-15H, (1) AN/ALT-15L, and (1) AN/ALT-16 jammers, along with (2) AN/APR-9, (1) AN/APR-14, and (2) AN/ALR-18 receivers.

Design changes in the B-52G airframe made it more susceptible to structural failure. This was highlighted by the crash of a B-52G (58-0187) with nuclear weapons on board in January 1961, when it experienced structural failure and a massive fuel leak of the right wing during airborne alert. The wing failed when the flaps were engaged during landing. Boeing analysis subsequently determined that the operating stress placed on the B-52G wing structure was approximately 60 percent higher than the preceding B-52 models.[35] The most significant issue was inadequate durability and damage tolerance characteristics of the aluminum used in the wing structures to reduce weight, increase fuel load, and extend range.

Consequently, SAC imposed flight restrictions from February 1962 through September 1964 until B-52G and H model aircraft were retrofitted under ECP 1050 with modified wings. The modified wings included a wing box with thicker aluminum, stronger steel taper lock fasteners, added brackets and clamps to the wing skins, added wing panel stiffeners, and a new protective coating was applied to the fuel tanks. A new trailing edge structure was also introduced to offset sonic fatigue like the F models. ECP 1185 was initiated in May 1966 to replace the fuselage side skins, crown skin fasteners, and upper longerons on all B-52G and H aircraft. These structural modifications were implemented during the aircraft's regular Inspect and Repair As Necessary (IRAN) schedule and were expected to extend the B-52G and H service life through the 1980s.

In addition, modifications began in mid-1969 to install a new Stability Augmentation Sub-System (SASS) to improve aircraft yaw and pitch stability, improve low-level flight characteristics, reduce structural loads, and improve controllability in turbulence.[36] The modification was installed on all B-52G and H model aircraft over the next two years. It was approved as ECP 1195 in October 1967 but had been under study since 1965. It consisted of two main units, the Yaw Electronic Control Unit (YECU) and the Pitch Electronic Control Unit (PECU). These units were designed with solid state components and triple redundant channels according to the 1960s state of the art. They were subsequently redesigned with digital components in the late 1980s.

Beginning on 15 October 1971, the Air Force began equipping B-52G and H model aircraft with the new AGM-69A Short Range Attack Missile (SRAM) to replace the AGM-28 Hound Dog missiles when the first B-52G was inducted for modification at the Oklahoma City Air Materiel Area (OC-AMA). The first SRAM-equipped B-52G was delivered to SAC at the 42nd BMW, Loring AFB, ME in March 1972 and the wing became operational in August. Ultimately, 19 B-52 (and two FB-111A) wings were equipped with

B-52G (57-6518) loaded with AGM-69A SRAMs in the early 1970s. The aircraft could carry a total 20 missiles with eight in the bomb bay and six on each wing pylon. (U.S. Air Force)

SRAMs. Each B-52G and H model could carry 20 missiles, 12 externally on the wing pylons and eight in the bomb bay.[37]

SAC initially resisted deploying B-52G aircraft to SEA to support the Vietnam War but finally committed them in mid-72. These aircraft flew Operation LINEBACKER I and LINEBACKER II missions from Andersen AFB, Guam. They could only carry 27 bombs internally and could not carry conventional bombs on the wing pylons like the B-52D and F models. But they were still effective and added to the overall combat strength in theater. Unfortunately, SAC lost seven B-52G aircraft during this operation with six of the losses blamed on inadequate ECM equipment. Enemy SAMs hit all six aircraft over North Vietnam. Four went down around Hanoi and two eventually crashed in Thailand. The last one crashed into the Pacific Ocean after take-off from Andersen.

SAC continually struggled to keep its ECM system capabilities ahead of improvements in enemy defenses. While all aircraft received ECM Phase I-III systems through the BIG FOUR modifications, B-52D "BIG BELLY" aircraft also received Phases IV and V under Project RIVET RAMBLER to counter SA-2 SAM radars. These modifications were completed on the D models by 1971. However, they were slower in coming on B-52G and

H models and most aircraft were not upgraded until 1973 leaving the G models at a distinct disadvantage when they were finally committed to combat operations in 1972.

Linebacker II B-52G losses validated SAC's existing requirements for advanced ECM systems with advanced transmitters and jammers, and an improved radar warning system capable of detecting SAMs, AAA, and airborne interceptors in the B-52G and H fleets. ECP 2525 was approved in June 1971 and was intended to provide more efficient airborne early warning countermeasures. ECP 2519, also called RIVET ACE, was approved in December 1971 to upgrade the radar warning receivers. Together, they comprised Phase VI+ and became ongoing modifications that began in mid-1973 and were not completed on the fleets until 1984 due to various technical and funding delays. Additional ECM upgrades continued to counter new threats throughout the life of the aircraft.

The Phase VI+ equipment included (8) AN/ALE-24 Chaff Dispensers, (12) AN/ALE-20 Flare Ejectors, (2) AN/ALT-16A, (10) AN/ALT-28, (2) AN/ALT-32H, and (1) AN/ALT-32L jammers, along with AN/ALR-20A and AN/ALR-46 receivers, AN/ALQ-153 Tail Warning System, and AN/ALQ-155 Power Management System (PMS), and (4) AN/ALQ-117 PAVE MINT active countermeasures system, and (1) AN/ALQ-122 Smart Noise Onboard Equipment (SNOE). The heart of the system was the AN/ALR-20 receiver which displayed information on a small screen at the EWO station.

The AN/ALQ-155 PMS was developed in the mid-1970s along with the B-52 Transmitter Improvement program. Both programs were closely related. SAC established requirements in 1974 for a computer-controlled system to manage AN/ALT-28 jammer assets. Although there were several modes of operation ranging from automatic to manual, these systems would remain under the positive control of the EWO. An interface processor brought together inputs from the AN/ALT-28 jammers, AN/ALQ-117 countermeasures systems, AN/ALR-46 and AN/ALR-20A receivers to assist the EWO in applying jammers and maximize utilization of jamming resources. Testing began in November 1976 and was completed in March 1977. Aircraft were subsequently modified during Programmed Depot Maintenance (PDM).

Installations began in mid-1973 and before long, various ECM antennas extended from the fuselage. The rear fuselage aft of the tail was extended 40 inches to accommodate the new AN/ALQ-117 ECM system. A pair of AN/ALQ-117 antennas were also added to the nose and a new AN/ALQ-153 tail warning antenna was installed in the tail resulting in a noticeable bump on each side. Additional ECM antenna arrays were mounted in the wingtips with jammers in the nose and rear fuselage. A Phase VI+ ECM upgrade began in 1988 and replaced the AN/ALQ-117 with a new AN/ALQ-172 countermeasures system.

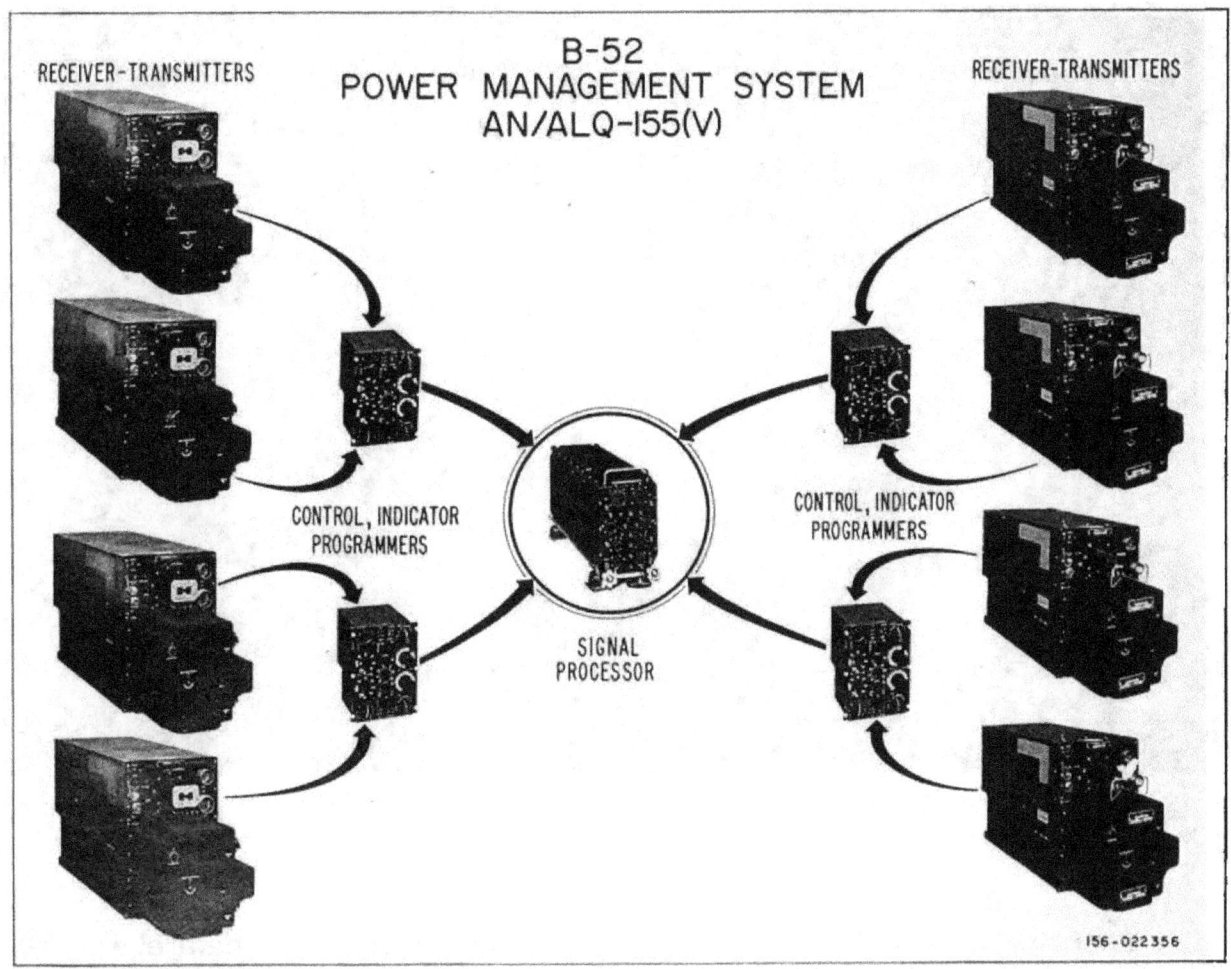

Pictorial display of the AN/ALQ-155 PMS components. Two additional AN/ALT-28 transmitters later installed in the tail and the AN/ALR-46 radar warning receiver, not shown, also tied into the signal processor. (U.S. Air Force)

Also in mid-1973, the Air Force began modifying B-52G and H aircraft with the AN/ASQ-151 Electro-optical Viewing System (EVS). The modification installed Hughes AN/AAQ-6 Forward-Looking Infrared (FLIR) and Westinghouse AN/AVQ-22 Low-Light Television sensors in turrets under the aircraft nose radome and new monitors in the cabin. These new EVS sensors and monitors significantly increased the aircraft's effectiveness as a low-level penetrating bomber and allowed it to fly at altitudes of 200 feet or less to counter new Soviet radar and SA-3 SAMs. This innovative capability allowed accurate images of the terrain in front of the aircraft both day and night without radar. However, the TA radar profiles were also overlayed on the new EVS monitors to replace the relatively crude radar displays previously used. Perhaps most important, the EVS allowed an outside view even when thermal curtains[38] were installed for nuclear bomb runs.

Beginning in the early 1980s, SAC began equipping its B-52G and H fleets with the new AN/ASQ-176 Offensive Avionics System (OAS) capable of launching the Boeing AGM-

B-52G (57-2587) on the ramp at Castle AFB, CA in 1988 featuring EVS "chin turrets" and AN/ALQ-117 ECM antennas on the nose. (National Archives)

86 Air Launched Cruise Missile (ALCM). Development proceeded in parallel with new wing pylon systems able to carry either six ALCMs or AGM-69A SRAMs on each wing. The OAS was designed to replace the AN/ASQ-38 system on both the B-52G and H models. The upgrade was driven by both the need for increased bombing accuracy and the decreasing reliability and supportability of the AN/ASQ-38. CEP for the bomber's missile payload was expected to decrease by 30-35 percent and bomb error probabilities by 20 percent.

OAS was expected to increase mission reliability by 25 percent and Mean-Time Between Failure (MTBF) by a factor of three. It was also expected to save about $65 million annually in operating costs while reducing the weight by 1,800 pounds on the B-52G and 1,900 pounds on the H model.[39] The first flight of the B-52G OAS prototype occurred on 3 September 1980. The first SAC unit equipped with B-52G OAS and ALCM capable aircraft was the 416th BMW at Griffiss AFB, NY. It became operational in December 1982. Ultimately, 98 B-52G and all H models were equipped with OAS and ALCM capability by 1986.

B-52 lower deck with OAS installed whose integrated systems replaced several obsolete systems. (National Archives)

The OAS was a significant improvement as evidenced by B-52 OAS equipped aircraft regaining their edge against the FB-111A in SAC Bombing and Navigation Competitions (Bomb Comp). The OAS and its integrated systems replaced several obsolete systems including the AN/ASQ-38 system, MD-1 Astrocompass, N-1 Compass, and AN/AJA-1 True Heading Computer. It featured a MIL-STD-1553A digital data bus system with dual AN/ASN-131 SPN/GEANS INUs like those used in the B-52D DBNS, AN/AYQ-10 Ballistic Computer Set with two avionics processors and four data transfer units, an AN/AWQ-3 Weapons Control and Delivery System with three missile interface units, the AN/ASQ-175 Control Display Set, a Modified AN/APS-81 ACR (MACR) redesignated OY-73/ASQ-176, a new AN/APN-224 Radar Altimeter, AN/APN-218 Doppler Radar, and a Lear Siegler AN/ASN-134 Attitude Heading Reference System (AHRS). The entire system and its cabling were nuclear hardened to protect against Electro-Magnetic Pulse (EMP). However, the system was designed primarily for launching the ALCM and dropping nuclear gravity weapons. The only compatible conventional weapons were gravity and cluster bombs.

When the B-52D was phased out in the early 1980s, SAC designated the B-52G as its new conventional weapons platform. Although it did not receive the "BIG BELLY" modification like the D model, 46 aircraft were modified with the Integrated Conventional Stores Management System (ICSMS). These aircraft were also fitted with wing pylons to carry up to 24 conventional bombs externally, giving them a total load of 51 bombs. The 42nd BMW at Loring AFB, ME and the 43rd Strategic Wing (SW) at Andersen AFB, Guam were given responsibility for conventional operations, and 30 aircraft were equipped with AGM-84D Harpoon missiles by the mid-1980s. Eventually, the 43rd SW (by then redesignated the 43rd BMW) was deactivated and the 2nd BMW at Barksdale AFB, LA was given its conventional role. Eight aircraft were also modified to carry the AGM-142 HAVE NAP or POPEYE air-to-surface missiles.[40]

The B-52G and H models received several new capability and reliability upgrades during the late 1970s through the early 1990s. These included installation of the AN/ASC-19 Air Force Satellite Communications System (AFSATCOM), replacement of one AN/ARC-164 UHF Command Radio with a new AN/ARC-171 UHF Line-of-Sight (LOS) radio, replacement of the OY-73/ASQ-176 MACR with a new AN/APQ-166 Strategic Radar, and upgrade of the A/A42G-11 AFCS with a new AN/ASW-49 Digital Autopilot system.

As with the MACR, the AN/APQ-166 Strategic Radar included a special TA mode with a radar processor, display generator, video distribution unit, and the pilot's and copilot's

B-52G with the new AN/APQ-166 Strategic Radar integrated into the EVS monitors prepares for take-off. (National Archives)

terrain display indicators (EVS monitors) and their associated electronic components. The TA system provided a radar profile display of the terrain along the flightpath of the aircraft. By interpreting the display and maneuvering the aircraft accordingly, the pilot could fly the aircraft at low absolute altitudes (aircraft to terrain separation distance).

The new Digital Autopilot replaced the obsolete and unreliable AFCS vacuum tube system with a state-of-the-art digital system but retained the same modes of operation including the Low-Level and Aerial Refueling CWS modes. B-52G aircraft assigned to the 42nd BMW and 93rd BMW at Castle AFB, CA were the first to receive the new AN/ARN-151 Global Positioning System (GPS) receiver to drastically increase bombing accuracy. GPS was later integrated into the OAS on all B-52H models.

Perhaps the most significant conventional capability added to the G model in the late 1980s was the ability to carry and launch the Conventional Air Launched Cruise Missile (CALCM). The CALCM was a conventional version of the nuclear armed ALCM with an improved GPS guidance system. It proved highly accurate and was used against critical targets in the opening hours of Operation DESERT STORM in 1991.

SAC began phasing out its G model fleet in 1989 to comply with Base Realignment and Closure (BRAC) committee recommendations. BRAC required closure of several SAC bases and aircraft were realigned to other bases or retired as necessary to accommodate the decrease in infrastructure. The first B-52G (57-6500) retired flew from the 93rd BMW at Castle AFB, CA to the boneyard on 11 May 1989 and 34 aircraft joined it by the end of 1990. SAC got a bit of a reprieve in 1990 when at least 64 B-52G aircraft were committed to the build-up of forces under Operation DESERT SHEILD in preparation for Operation DESERT STORM. However, retirements continued after the war with another 43 aircraft retired before the end of 1991.

On 1 October 1991, SAC discontinued use of tail gunners on the B-52. Initially, the guns remained installed but were deactivated. However, they were eventually removed, and panels were installed over the previous turret locations.

All B-52 aircraft were transferred to the newly formed Air Combat Command (ACC) when SAC was deactivated on 1 June 1992 and ACC continued retiring the G models as previously planned. The last B-52G (58-0240) retired flew from Castle to Davis-Monthan on 3 May 1994. The last G model was chopped up and scrapped at the Aerospace Maintenance and Regeneration Group (AMARG)[41] in December 2013.

B-52H. The B-52H was very similar to the G model. However, the major difference was the replacement of the J57-P-43WB turbojet engines with newly developed TF33-P-3 turbofan engines. The TF33 was a modified J57 that had already been adopted for use on commercial transport aircraft.[42] These new engines were quieter and provided a vast improvement in performance, now producing 17,000 pounds take-off thrust without the need for a water injection system. This equated to significant power compared to the G model and resulted in a 50 percent increase in take-off thrust, which reduced the take-off roll by 500 feet. It also increased cruise power by 20 percent.[43] The TF-33 engines were about 12 percent more fuel efficient than the J57 engines, which increased the unrefueled combat radius by over 700 miles.

B-52H (61-0040) with oversized engine inlets and SIOP-camo paint scheme circa 1975. (U.S. Air Force)

This increase in power also caused some significant new handling problems. The sudden increase in aircraft acceleration during take-off could cause the fuel to slosh in the tanks, causing an unexpected change in the center of gravity and a pitch up beyond the pilot's ability to control with available elevator authority. Boeing installed a new mechanical thrust gate on the throttle quadrant to prevent the pilot from applying too much power. The outboard spoiler segments were also set to extend up to about 10 degrees to allow small movements during aerial refueling without inducing a pitch-up.[44]

These new engines changed the outward appearance of the aircraft with the addition of oversized engine inlets. Another outward appearance change was the replacement of the four .50-caliber machine guns in the tail gun turret with a single 20mm M-61 Vulcan cannon. This cannon was controlled by the new AN/ASG-21 DFCS and had an adjustable rate of fire of between 3,000 and 6,000 rounds per minute.[45] All B-52H aircraft were equipped in production to carry two AGM-28 Hound Dog missiles and four ADM-20 Quail decoys. They and all G models were also equipped with provisions to carry four GAM-87 Skybolt ballistic missiles on the wing pylons. However, the Skybolt program was subsequently cancelled.

B-52H procurement began in 1959 and aircraft were ordered under two separate contracts. The first, a letter contract signed on 2 February 1959 and definitized on 6 May 1960, called for the first 62 aircraft. The final 40 aircraft were ordered under a letter contract signed on 28 July 1960 and definitized in late 1962. Definitization of this final contract was intentionally delayed while the Air Force finalized its B-52 force structure

B-52H and Vulcan Bomber in-flight near Edwards AFB, CA on 10 July 1961. (National Archives)

requirements to ensure no more than 40 aircraft would be needed. This contract marked the end of B-52 procurement. However, the Air Force established an agreement with Boeing on 17 October 1962 to store all B-52H tooling until July 1963.

There was no B-52H prototype, but the TF-33 engines were tested on a JB-52G (57-6471), which is sometimes referred to as the YB-52H prototype. It first flew on 10 July 1960. This successful flight and the following flight test program proved the TF33 engines would provide significant range and power improvements over B-52G performance. However, these tests uncovered some significant problems such as throttle creep, hang or slow start, flameout, and uneven throttle alignment. The engine also consumed too much oil, turbine blades failed, and inlet cases cracked. Even though many of these problems were corrected by mid-1962, the Air Force implemented the HOT FAN modification program to increase reliability and ensure at least 600 hours of operation before engine failure. Curtailed temporarily during the Cuban Missile Crisis in

October 1962, the OC-AMA completed the modifications on all 894 TF33 engines by the end of 1964.[46]

Meanwhile, the first B-52H (60-0001) came off the production line on 30 September 1960 and first flew on 6 March 1961. It was delivered to the 379th BMW, Wurtsmith AFB, MI on 9 May 1961. Five B-52H (60-0002 to 60-0006) aircraft were delivered in place to Boeing from 8 March to 3 April 1961 for further testing before eventually being delivered to SAC.

B-52H (60-0008) in flight. (U.S. Air Force)

All B-52H aircraft were initially equipped with essential equipment such as the AN/ARC-34 UHF Command Radio, AN/APX-25 IFF Transponder, AN/APN-69 Radar Beacon, AN/AIC-18 Interphone, AN/ARN-14 Navigation Receiver, AN/ARA-25 UHF Direction Finder, AN/ARN-32 Marker Beacon, and MA-2 Autopilot like the B-52G. However, the AN/ARC-21 HF Liaison Radio was replaced with the AN/ARC-58 and the Glideslope Receiver was either an AN/ARN-31 (60-0001 to 60-0062) or AN/ARN-67 (61-0001 to 61-0040). The initial ECM equipment was the same as the G model and included (1)

B-52H aircraft on alert at Grand Forks AFB, ND on 1 April 1981. (National Archives)

AN/APS-54 Early Warning Receiver, (1) AN/ALE-1 Chaff Dispenser, (1) AN/ALE-14 Flare Dispenser, (14) AN/ALT-6B jammers, along with (1) AN/APR-9 and (1) AN/APR-14 receivers.

The AN/ARC-34 UHF radio was subsequently upgraded to the AN/ARC-164. The AN/ARC-58 HF radio was subsequently upgraded to the AN/ARC-190. A new AN/ARN-21 TACAN was installed on all aircraft and was subsequently replaced by the AN/ARN-118. The AN/APX-25 IFF Transponder was subsequently replaced with the AN/APX-64 on aircraft incorporating the AIMS capability upgrade.

Initially, the B-52H aircraft were equipped with the AN/ASB-9 BNS. However, all aircraft were subsequently modified with an improved AN/ASQ-38 system like the B-52G models which integrated the AN/ASB-9A BNS designed specifically for low-level flight, AN/APS-81 ACR, AN/APN-89A Doppler Radar, and MD-1 Astrocompass. A new AN/AJN-8 True Heading Computer Group replaced the AN/AJA-1 used on previous models. A new J-4 Compass system also replaced the N-1 Compass used on previous models. Like previous models, the system was highly automatic in operation, although the BNS could still be operated independently if desired.

The first 18 B-52H aircraft were delivered without the capability for all-weather, low-level flying including the A/A42G-11 AFCS, and AN/APN-150 Radar Altimeter. However, they received post-production modifications between April and September 1962 to add the capability. The remaining B-52H aircraft were the first to incorporate this capability in production. The B-52H was also the first to incorporate the new AN/APS-81 ACR in production for TA, an anti-jamming unit, and improved low level mapping capability.[47]

B-52G aircraft undergoing environmental and climate testing at Eglin AFB, FL in 1964. (U.S. Air Force)

All structural modifications identified for the B-52G were quickly applied to the B-52H. In addition, SAC soon discovered cracks where the wings and fuselage joined. Boeing attributed this problem to taper lock fasteners and stress corrosion. They developed a modification package which was implemented immediately under Project STRAIGHT PIN. This project was performed at rework centers in Mosses Lake, WA; Wichita, KS; and San Antonio Air Materiel Area, TX and was completed by the end of 1962. Work included replacing fasteners and cleaning up cracked fitting holes.

In August 1962 stress corrosion caused main landing gear cylinder failures on B-52D and B-52F aircraft. These were the latest in a series of failures going back to 1959 that included B-52G and B-52H aircraft. Air Force engineers recommended using an alloy less susceptible to corrosion in the landing gear cylinders. An anti-corrosion coating was applied to the landing gear components to prevent further failures while a study was completed, and the cylinders were eventually upgraded.

Each B-52H was equipped during production with two cartridge starters to allow faster alert starts and support take-offs from dispersal airfields. These cartridges were installed on all other B-52 models between January 1963 and March 1964 based on successful B-52H operations. This important mission enhancement proved to be expensive and time-consuming, requiring electrical system modifications, new duct covers, and nickel cadmium battery installations. But SAC was still dissatisfied with the engine start times and the Air Force eventually approved Project QUICK START to equip all B-52G and B-52H aircraft with an improved cartridge start system. The improved system installed a cartridge on each engine rather than only two per aircraft and greatly reduced alert response times making the fleet less vulnerable to surprise attacks.[48]

B-52H in-flight during BULL RIDER '88 on 1 May 1988. (National Archives)

The tail gunner was removed from the aircraft on 1 October 1991 like the G model. The B-52H was employed as the primary nuclear bomber within SAC. After SAC was deactivated in 1992 and the G model was retired soon after, the B-52H began to pick up a dual-role responsibility for both nuclear and conventional missions. The Air Force was keen to equip the aircraft with an array of new precision guided munitions in development. But the OAS software was very tightly integrated requiring regression testing of the complete system to qualify any new weapons. Also, the current rotary launcher was created specifically for the SRAM and was too small to accommodate ALCMs and other larger weapons in the bomb bay.

A new OAS Block II software upgrade and Common Strategic Rotary Launcher (CSRL) were implemented to solve the problem. The Block II software repackaged the original navigation software as the Flight Management System (FMS) which would remain essentially stable and would not require retesting when weapons were added. The new weapons would interface with the system through a Stores Management Overlay (SMO) and only the SMO would require testing and certification, greatly decreasing the time required to add new capabilities to the B-52H.

The CSRL accommodated a host of weapons in the bomb bay including eight ALCMs giving the B-52H the capability to carry 20 total. A new AGM-129 Advanced Cruise Missile (ACM) was fitted to the B-52H using the new SMO and CSRL integration allowing carriage of up to 20 missiles. The B-52H was also equipped to carry the AGM-84D Harpoon missile, taking up this role after the G model was retired. While the Harpoon could only be carried on 30 conventional capable B-52G aircraft it could be carried on all B-52H with OAS Block II incorporated. All B-52H aircraft were modified to carry the AGM-142 HAVE NAP. GPS was also integrated into the OAS at this time to allow greater precision for conventional weapons.

By the mid-1990s, the B-52H was in dire need of communications upgrades. The AN/ARR-85V Miniature Receive Terminal (MRT) was installed in the mid-1990s to provide Very Low Frequency/Low Frequency (VLF/LF) receive-only capability of Emergency Action Messages (EAM). This replaced SAC's long-range high frequency communication system to direct B-52H aircraft launched for a nuclear strike to proceed past its positive control point and deliver its weapons on target.

The B-52H conventional mission also required real-time communications and datalink systems to allow in-flight retargeting with data received from a Combined Air Operations Center (CAOC) or Forward Air Controller (FAC). The Rockwell Collins AN/ARC-210 was installed in the 1990s to replace the existing AN/ARC-164 Command Radio and AN/ARC-171 UHF radio to provide VHF/UHF and SATCOM capability. The ARC-210 included secure communications capabilities including HAVE QUICK I and II Anti-Jamming, Single-Channel Ground-Air Radio System (SINCGARS), and SATCOM Demand Assigned Multiple Access (DAMA) to allow secure beyond line-of-sight data transmissions. This real-time data was interfaced with hardened laptops running Falcon View moving map software or Combat Track II displays to provide the crews with secure networking capability and situational awareness displays.

All B-52H aircraft were also fitted with new EVS systems when the G models were modified in the early 1970s. This system proved invaluable over the years but was becoming unreliable and unsupportable by the 1990s when the B-52H FLIR was replaced with a new Loral AN/AAQ-23 and the LLTV was replaced with a BEA Systems AN/AVQ-37. By the late 1990s the OAS was also quickly becoming obsolete, unreliable, and unsupportable. The Air Force launched a new Avionics Midlife Improvement (AMI) program in 2000, which became combat ready on the first B-52H in 2006. It replaced the Avionics Computer Unit with a new state-of-the-art digital computer, and the Data Transfer Units were upgraded from magnetic tape cartridges to new solid state memory units. The 1970s technology AN/ASN-131 SPN/GEANS INUs were replaced with new SNU-84 ring laser gyro units.

The B-52H received ECM upgrades consistent with previous models. It was equipped with Phase III systems by 1963 and its ECM equipment included (1) AN/APS-54 Early Warning Receiver, (1) AN/ALE-1 Chaff Dispenser, (1) AN/ALE-14 Flare Dispenser, (4)

B-52H cockpit with original EVS displays. (U.S. Air Force)

AN/ALT-6B, (6) AN/ALT-13, (1) AN/ALT-15L, (2) AN/ALT-15H, and (2) AN/ALT-16 jammers, along with (2) AN/ALR-18 and (1) AN/ALR-19 receivers. The last 18 aircraft were also fitted with the AN/ADR-8 Chaff Rocket Launcher during production. Twenty of these chaff rockets were contained in AN/ALE-25 Chaff Dispenser pods mounted on pylons installed between the engine pods on each wing. This was subsequently installed on all B-52G and H models as a post-production modification and remained in service until 1970.

B-52H overhead view on 7 September 1988. (National Archives)

The B-52H also received all ECM upgrades through Phase VI+ like the B-52G. However, by 1997, the B-52H ECM systems were major Mission Capability (MICAP) drivers and the leading cause of low Mission Capable (MC) rates. In March 1997, ACC and the Oklahoma City Air Logistics Center (OC-ALC)[49] launched the ECM Support Improvement Program (SIP) to address supply and reliability issues. By April they had eliminated the MICAPs and subsequently established an AN/ALQ-172 upgrade program. This program added a third AN/ALQ-172 system to the ECM suite to handle advanced threats and correct coverage problems. The Air Force continues to evaluate threats and upgrade ECM capabilities as required. For example, a new AN/ALQ-172 upgrade began in 2021 to replace the old analog components with state-of-the-art digital units.

B-52H aircraft were also upgraded with a new communications capability called Combat Network Communication Technology (CONECT). Development began in 2007, and Low-Rate Initial Production (LRIP) approval was given in 2012 after a series of successful flight tests. The modification replaced the monochrome EVS and OAS displays with new flat-panel color displays, added a new client/server architecture, high-bandwidth data network, Link-16, advanced wideband terminal, and new digital interphone system. The CONECT system allows the aircraft to receive updated target coordinates through the datalink and transfer them to the weapon in flight. The target coordinates are displayed on full color moving maps along with the aircraft location and threat information. This greatly reduces the kill-cycle for attacking time-sensitive targets and for supporting ground troops.[50]

Boeing proposed an upgrade of the eight B-52H TF-33 engines in 1996 with four Rolls Royce RB-211 turbofan engines. Boeing expected the more efficient and reliable engines to save about $6 billion in operating costs, provide substantially more power and range, as well as quieter and cleaner operation. They proposed a lease financing plan and privatized "power by the hour" maintenance support that would allow the engines to be paid off from normal operating cost savings. After further study, Boeing revised its estimated savings to $4.7 billion. However, the GAO estimated Boeing's proposal would cost $1.3 billion rather than saving money. The Air Force had other priorities and rejected the plan.[51]

B-52H aircraft are equipped for carriage of the Northrop Grumman AN/AAQ-28 Litening II Targeting Pod. The pod is mounted on the right wing between the two engine pylons. This pod was developed from the Israeli Rafael "Popeye" Litening I. It contains a FLIR sensor to display an infrared image of the target as well as a television camera for real-time viewing of the target. It has a laser designator for laser-guided munitions, a laser spot tracker/rangefinder, and an automatic target tracker. The pod gives B-52H crews the ability to acquire and positively identify targets without being dependent on an external observer to designate targets.[52]

The pod was tested on two B-52H (61-0008 and 61-0021) aircraft at the Utah Test and Training Range in January 2003 and was quickly deployed for combat operations in March 2003. An extended range upgrade with improved FLIR resolution and longer range was soon introduced. Additional upgrades included the AN/AAQ-28A Litening AT/ISR which could generate target coordinates for GPS-guided weapons and incorporated a datalink to transmit imagery to a ground station. The AT became operational in 2006.

B-52H aircraft can also carry the Lockheed Martin AN/AAQ-33 Sniper Advanced Targeting Pod (ATP). Based on the success of the AN/AAQ-28 Litening II testing, the Air Force decided to provide all B-52H aircraft with a precision strike capability. They elected to integrate the ATP beginning in 2008. The integration allows the aircraft to carry either the Litening II or the Sniper ATP. It links the pod control, display and target geo-location with the bomber's OAS via new multi-function color displays and a digital integrated hand controller. The system is operated from the Navigator's station and gives the capability to acquire real-time Intelligence, Surveillance, and Reconnaissance (ISR) data with full motion video. It allows the aircraft to operate as an "overwatch" asset and deliver precision-guided weapons in support of ground forces. ISR data can also be transmitted to forward deployed forces.[53]

The Air Force continues to pursue various B-52H upgrade options including radar modernization and engine replacement to extend service life through the 2050s. (U.S. Air Force)

The last B-52H (61-0040) was delivered to SAC at the 4137th SW, Minot AFB, ND on 26 October 1962. A total of 102 B-52H aircraft were produced, and all were built in Wichita. A total of 76 aircraft remained in-service as of 2025, with 56 assigned to Air Force Global Strike Command, two assigned to Air Force Materiel Command, and 18 assigned to the Air Force Reserve Command. B-52H aircraft paint schemes followed the same path as the B-52G, and all are currently painted with the Gunship Gray paint scheme like most other Air Force aircraft.

The Air Force continues to pursue various upgrade options to keep the fleet in service through the 2050s. One option explored in the mid-2000s was a stand-off jammer EB-52H version with jammer pods replacing the outer wing tanks. However, this option proved too expensive and was dropped. Another option, explored in the mid-2010s, allowed carriage of the AN/ASQ-236 Dragon's Eye radar pod on the wing pylons. The pod contained an active array radar that allowed observations of sea traffic and could potentially enhance search and targeting functions for the AGM-84D Harpoon. This option was also eventually dropped.

The Air Force began a new Radar Modernization Program (RMP) in 2016 to replace the now aging AN/APQ-166 Strategic Radar. The new system will include all current functions plus an active array radar capable of ground mapping, sea search and targeting, and countermeasures capability. It will include two new large screen high-definition displays, hand controllers, and display processors to integrate the radar with other aircraft systems. The Air Force selected an upgraded version of the Raytheon AN/APG-79 Active Electronically Scanned Array (AESA) radar used on the Navy's F/A-18E/F Super Hornet in 2019. It is expected to achieve Initial Operational Capability (IOC) in 2027. The radar upgrade will allow removal of the EVS equipment including the chin-pods mounted on the lower nose of the aircraft. The radar will perform some EVS functions of the FLIR/EO system and the Litening or Sniper targeting pods integrated with the new displays will take on the remaining functions.

A new effort to replace the engines, called the Commercial Engine Replacement Program (CERP), was also launched in 2018. A competition was held between P&W, General Electric (GE), and Rolls Royce and the Rolls Royce F130-200 turbofan was selected. These commercial engines, based on the Rolls Royce BR725 engines, produce 17,000 pounds of thrust each and were already in Air Force service with the C-37 and the E-11 Battlefield Airborne Communications Node (BACN) aircraft. All eight TF33 engines will be replaced with the new F130 engines meaning the B-52H will remain an eight-engine aircraft. This approach significantly reduces the design task and allows reuse of the existing engine pylons, throttle controls, and potentially other engine accessories and leaves the cockpit layout essentially intact. The new engines are expected to increase fuel efficiency by 30 percent and unrefueled range by 2,300 miles. A Honeywell 36-150 Auxiliary Power Unit (APU) may also be installed to eliminate the need for external power carts and starter cartridges.

Plans are in place to qualify the aircraft for even more weapons including the new AGM-154C Joint Standoff Weapon (JSOW), AGM-158C Long Range Anti-Ship Missile (LRASM) intended to replace the AGM-84D Harpoon, the nuclear AGM-181 Long-Range Stand-Off (LRSO) missile intended to replace the AGM-86B ALCM, and GBU-53 Stormbreaker Small Diameter Bomb (SDB) II. Plans to equip the B-52H with the hypersonic Lockheed Martin AGM-183A Air-Launched Rapid Response Weapon (ARRW) were cancelled by 2023 and the new Hypersonic Attack Cruise Missile (HACM) was in development. The development of a new wing pylon and stores adapter was started in 2018 to allow each pylon to carry a single weapon weighing up to 20,000 pounds, which may include the GBU-43 Massive Ordnance Air Blast (MOAB) bomb.

Other planned upgrades include AN/ARR-85V MRT VLF/LF receiver modernization, digital cockpit with color displays for enhanced targeting and situational awareness, a hybrid analog-digital engine control system, and an Advanced Extremely High Frequency (AEHF) SATCOM to allow the aircraft to exchange data with ground stations from ground, air, and space platforms. The B-52H with new engines, radars, and other approved modifications, redesignated as B-52J after the conversion, is expected to achieve IOC by 2033. Fleetwide retrofits are expected to be completed by 2038. Together, these modifications are expected to extend the service life to at least 2050.

Strategic Reconnaissance

RB-52B. On 19 March 1951, the Air Force issued a letter contract for the Phase I study of an RB-52 with a mock-up to be inspected in September 1951. There was growing interest in a reconnaissance version of the aircraft. Initially, plans called for converting one of the production B-52A models with flight test anticipated for 1954. The XRB-52 configuration was decided after much haggling between the Air Force and SAC. It would be a dedicated reconnaissance version with no bombing capability.

However, Boeing designers soon realized a better approach would be a "multi-purpose" reconnaissance pod that could be easily uploaded in the bomb bay. Representatives from the Air Staff, ARDC, AMC, Air Weather Service, Boeing, and SAC met from 29-30 August 1951 to discuss the options. Everyone agreed the B-52 would be a bomber with a reconnaissance capability using pods carried in the bomb bay. SAC insisted that all pods had to be easily removed to readily facilitate conversion of the aircraft back to the primary bomber role.

The group discussed various pod configurations that could support night and special photo reconnaissance, weather reconnaissance, and Electronic Intelligence (ELINT) missions, and a multi-purpose pod. SAC preferred a ferret pod with special electronic reconnaissance equipment that they believed would be able to gather in one flight all the data obtained by three RB-52 flights using multi-purpose pods. Ultimately, the group agreed to phase in two pods with a multi-purpose pod being available in mid-1953 and a ferret pod later the same year. A major redesign would occur in 1955 when new reconnaissance equipment was available.

SAC renewed its request for ferret pods after a successful mock-up inspection of the multi-purpose pod on 12 December 1951. However, the Air Staff saw an opportunity to save money by using the multi-purpose pods and ultimately cancelled the ferret pod in December 1952. The special and night photo pods were also cancelled when it was determined that the multi-purpose pods would meet SAC requirements. SAC made a second request in 1954 but was again unsuccessful. However, by 1955 the multi-pur-

> ### RB-52 Reconnaissance Pod
>
> As XB-52 development continued, Air Force and SAC leadership struggled with the reconnaissance requirements for the planned RB-52 aircraft. Lieutenant General LeMay agreed that a reconnaissance version of the B-52 was important, and he was so persuasive in his support that Air Staff directed in October 1951 that all aircraft should be delivered in the RB-52 version. But the requirements for this version were the subject of continuing disagreement. The Air Staff believed that photography should be the main mission, and any other equipment should be excluded. SAC believed the aircraft should have a full complement of reconnaissance (or ferret) equipment for operation at night or in bad weather. SAC wanted only a minimum of cameras for local photographic coverage when light conditions permitted.[54]
>
>
>
> *The multi-purpose reconnaissance pod could be loaded in the RB-52B and C model bomb bay within four hours. It contained a crew of two and performed long-range reconnaissance missions. (U.S. Air Force)*
>
> Meanwhile, Boeing had been working on a preliminary RB-52 design and had completed the initial configuration by mid-1950. However, Boeing designers soon realized a better approach would be to incorporate a "multi-purpose" reconnaissance pod that could be easily uploaded in the bomb bay. This pod could also be replaced with a photo or ferret pod depending on mission requirements. AMC agreed and cautioned that the B-52 bombing capabilities could not be jeopardized. In June 1951, SAC proposed an RB-52 that could be easily converted to a bomber by wing-level maintenance personnel by removing the reconnaissance pod and installing bomb racks in its place. All key decision makers agreed with this approach during an August 1951 meeting, and that the aircraft should be considered a bomber first with ability to convert for reconnaissance missions. However, procurement orders were slow to reflect this change. An amendment to the initial order for 13 B-52A aircraft called for 17 reconnaissance pods and a contract modification in July 1951 increased the number of aircraft to 17, presumably to match the number of reconnaissance pods. The follow-on procurement contract designated all 50 B model and 35 C model aircraft as RB-52.[55]

pose pods had evolved as planned into "General Purpose Pods" carrying the latest reconnaissance equipment.

These pods contained an array of radar receivers including two AN/ALA-6 Direction Finders, one AN/APR-14 Low Frequency ECM Receiver and two AN/APR-9 High Frequency ECM Receivers, three AN/APR-8 Panoramic Receivers, two AN/ALA-5 Pulse

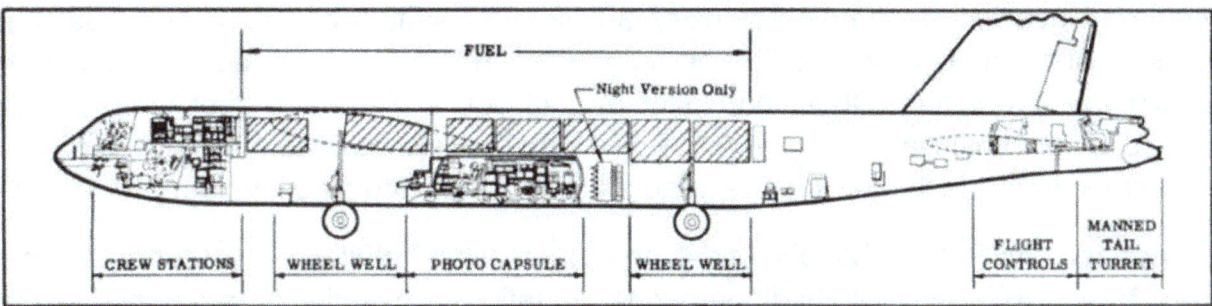

The "multi-purpose" reconnaissance pod could be easily uploaded in the RB-52B bomb bay. (U.S. Air Force)

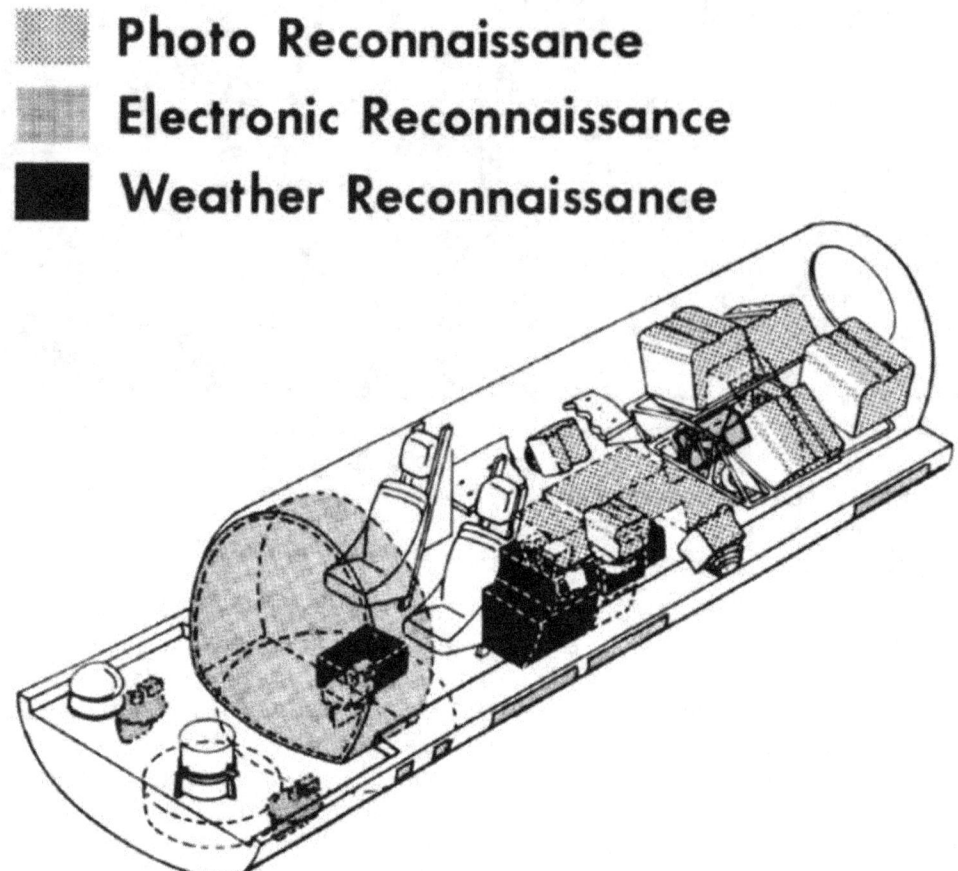

All pods were pressurized, allowing aerial reconnaissance above 50,000 feet, and had an array of recon equipment with accommodation for two additional crew members. (U.S. Air Force)

Analyzers to process the data collected, two AN/ANQ-1 Wire Recorders, and an AN/AIC-10 Interphone system. Each pod also contained an array of photographic equipment including four K-38 Cameras, one T-11 or K-36 Camera, and three T-11 Cartographic Cameras for mapping.

The pods were initially tested on an RB-52B (52-0010) and proved successful. The pods could be installed in the bomb by using a winch in only four hours allowing easy conversion of a standard bomber for the reconnaissance role. The pods and all installed equipment added only 300 pounds to the aircraft weight. They were pressurized, allowing aerial reconnaissance above 50,000 feet, and had accommodation for two additional crew members. Each crew member had a downward facing ejection seat.

A rack containing 24 M-120 photoflash bombs was also installed in the rear of the bomb bay. However, the RB-52B had operational limitations preventing it from operating over heavily defended areas and was never used in operational missions. The Air Force instead relied on the RB-36 and RB-47 during 1954-58 until dedicated U-2 reconnaissance planes became operational. Most RB-52B aircraft were redesignated B-52B upon assignment to SAC units.

RB-52C. All B-52C aircraft were equipped with the capability of uploading the multi-purpose reconnaissance pods used on the RB-52B. In its reconnaissance configuration, the B-52C was superior to the RB-52B due to the increased thrust provided by the J57-P-29W engines and increased range provided by the 3,000-gallon wing tanks. The aircraft were originally contracted as RB-52C but were ultimately redesignated and delivered as B-52C when the Air Force designated bombing as the primary B-52 mission.

Experiments & Tests

NB-52A. The first B-52A (52-0001) was delivered in place to Boeing and used for Phase IV testing. It was redesignated JB-52A on 30 November 1955 and NB-52A on 8 October 1957. It was used for initial testing of the shorter vertical fin eventually used on the B-52G and H models. It was also used for testing the J57-P-43W engines used on the B-52F. B-52A (52-0002) was delivered in place to Boeing and used for Phase IV testing. It was redesignated JB-52A on 30 November 1955 and NB-52A on 8 October 1957. It retired in 1960 after flight testing with gross weights up to 415,000 pounds.

The third B-52A (52-0003) was originally delivered in place to Boeing on 30 September 1954 for Phase IV testing and various service tests. Phase IV testing on this aircraft began on 25 January 1955 and it completed 288 Phase IV testing hours in 60 flights. The aircraft was redesignated as JB-52A on 30 November 1955. The aircraft was modified with B-52B wing tanks and other modifications to support test programs. It was temporarily assigned to the Wright Development Center at Wright-Patterson AFB, OH but then quickly returned to Boeing for more electronics and structural testing.

The aircraft was selected as the mothership for the upcoming X-15 hypersonic research program in 1957. It was redesignated as NB-52A on 8 October 1957 and assigned to the National Aeronautics and Space Administration (NASA) at Edwards. It earned the nickname "The High and Mighty One" during its service as the mothership for the X-15 rocket plane which had over 57,000 pounds of thrust and reached 4,520 mph. The X-15 was conceived as a means of obtaining technical data on hypersonic aeronautics. It was a joint venture between the National Advisory Committee for Aeronautics (NACA)[56], the Air Force, and the Navy. The manned space program became the immediate beneficiary, and the X-15 became the most successful research aircraft.[57]

The X-15 was carried under the wing of the NB-52A, which required significant modifications to accommodate it. These modifications were performed by North American Aviation at Air Force Plant 42 in Palmdale, CA beginning in January 1958 before the

(Left) The X-15 was attached to the NB-52A wing pylon using three remotely-actuated standard Air Force bomb shackles. (Right) Notch in NB-52A wing for X-15 vertical fin. (NASA)

aircraft was assigned to NASA. A 6 by 8-foot section was cut out of the wing to accommodate the X-15 tail and a pylon was installed between the inboard engines and the fuselage. Liquid oxygen (LOX) tanks were installed in the bomb bay. Wires and LOX lines were routed through the pylon to power the X-15 and top off the LOX before release. Television cameras allowed the NB-52A crew to monitor the X-15 and a launch control system ensured the X-15 was launched at the precise time required. The NB-52A with X-15-1 (56-6670) attached first flew on 10 March 1959.[58]

The aircraft launched the three X-15 aircraft on several historic flights including X-15-1 for the first free flight on 8 June 1959. On 12 May 1960, she carried X-15-1 aloft for the first Mach 3 flight. On 17 July 1962, she launched X-15-3 (56-6672) on its first spaceflight and on 26 July she launched Neil Armstrong in X-15-1 on his first free flight. On 22 August 1963 the NB-52A launched X-15-3 when it reached the highest altitude of the X-15 program, 354,200 feet.[59]

On 3 November 1965, the NB-52A launched a newly modified version of the X-15A designated X-15-A2 (56-6671). The X-15-A2

NB-52A (52-0003) prepares for a Mach 2.31 flight of X-15-A2 (56-6671). (NASA)

was previously damaged during a landing accident. It was then subsequently modified with a liquid hydrogen fuel tank and a scramjet engine. The aircraft was dropped from the NB-52A, and the scramjet ran for 84.1 seconds achieving a flight of Mach 2.31 and altitude of 70,600 feet.

The NB-52A participated in 59 of the 199 X-15 flights until the program ended in 1968. The aircraft was subsequently retired on 15 October 1969 and flown to MASDC for storage. The aircraft was acquired by the Air Force Museum but remained in storage until 1981 when it was transferred to the Pima Air & Space Museum in Tucson, AZ. The aircraft remained on display there for many years until it became too deteriorated. It was removed from display in 2016 and entered into restoration. This took several years, and the aircraft was only recently returned to display as of 2025.

NB-52B. One RB-52B (52-0008) was delivered in place to Boeing on 3 May 1955 and used for various tests. It made its first flight on 11 June 1955 and was redesignated JRB-52B on 30 November 1955. It was permanently configured to support NASA testing and was redesignated NB-52B on 12 December 1958. It was reassigned to AFFTC at Edwards AFB, CA on 13 December 1958. It was named "The Challenger" and earned the nickname "Balls 8". It served with AFFTC and NASA for nearly 50 years. It underwent similar modifications to those incorporated on the NB-52A. The aircraft supported X-15 testing and carried an X-15 for the first time on 23 January 1960. It flew 140 of the 199 X-15 flight tests. After being launched from the NB-52B, the X-15 flew to over 250,000 feet and reached speeds exceeding Mach 6.

NASA tried at least four times to launch two X-15 flights in one day. They were never able to pull it off but one of the aircraft was able to conduct research each time. In the 4 November 1960 attempt the NB-52A launched its X-15 (56-6670) which completed its sixteenth flight with a speed of Mach 1.95 and altitude of 48,900 feet.

One of at least four unsuccessful attempts was conducted on 1 November 1960 to launch two X-15 fights in one day using the NB-52A (52-0003) and the NB-52B (52-0008). (NASA)

On 18 November 1966, the X-15A-2 (56-6671) made a flight of Mach 6.33 at 98,900 feet. It was launched for its 53rd and last flight on 3 October

1967 from the NB-52B at 45,000 feet over Mud Lake, NV. Its scramjet engine burned for 140.7 seconds, and the aircraft achieved Mach 6.72 at 102,100 feet setting a speed record that still stands.

NB-52B (52-0008) also participated in the HL-10 lifting body research aircraft program sponsored by the Air Force and NASA from 1966 to 1975. The HL-10 was launched from the NB-52B and tested blended body aerodynamics that were eventually used for both the Stealth Fighter and Space Shuttle programs. The program also included testing of Martin's needle-nosed X-24 aircraft launched from the NB-52B.

NB-52B (52-0008) and TF-104G Starfighter chase plane in flight. (NASA)

The Air Force loaned the aircraft to the NASA Dryden Flight Research Center (DFRC) at Edwards beginning in 1975 for the remainder of its service. The aircraft participated in several critical NASA test programs. It tested solid rocket boosters for the space shuttle, carried the NASA Pegasus launch vehicle to put small satellite payloads into orbit, and played an active role in the remotely piloted research vehicle program.[60] Its last mission was the launch of the third X-43A Hyper-X scramjet-powered hypersonic aircraft on 16 November 2004. The X-43A set an absolute speed record of Mach 9.6 for aircraft with air-breathing propulsion. It was also used to test the IBM navigation systems used in the AN/ASQ-48 and AN/ASQ-38 BNS. It completed its 1,000th flight on 16 September 1999 and was replaced by a NB-52H (61-0025) on 17 December 2004.

NB-52B (52-0008) carries the X-15A-2 on its last flight in which it achieved a speed record of Mach 6.72 that still stands. (U.S. Air Force)

Another RB-52B (52-0013) was assigned to the Air Force Special Weapons Center (AFSWC), Kirtland AFB, NM on 4 May 1955 and used for nuclear testing. It was redesignated JRB-52B on 12 December 1955. It participated in Operation REDWING from 16 April to 16 June 1956. It became one of the few B-52 aircraft to drop a nuclear weapon when it dropped the Hydrogen bomb (H-bomb) during Operation REDWING on 21 May 1956. It was redesignated JB-52B on 16 May 1958 and NB-52B on 4 January 1962. It also dropped several nuclear weapons during Operation DOMINC in 1962.

NB-52C. One B-52C (53-0399) was redesignated NB-52C. This aircraft was redesignated JB-52C on 11 April 1956 and used for various tests at Boeing including J57-P-43 engine testing. The engine nacelle was also modified with a bulge to cover the newly installed alternator that would become a distinguishing characteristic on all B-52F models. It was redesignated NB-52C on 7 May 1961 and assigned to Wright-Patterson AFB, OH and used for Airborne Systems Development (ASD) testing to evaluate and integrate new technologies and systems. It was then redesignated JB-52C on 31 July 1962 and tapped to support testing of the X-20 reusable space plane, but the program was cancelled in 1963. The aircraft was assigned to AFFTC at Edwards AFB, CA on 27 August 1968 where it supported various tests including the B-1A crew escape capsule until it retired on 28 July 1975.

NB-52D. One B-52D (56-0620) was delivered to the AFSWC at Kirtland AFB, NM on 15 November 1957. It used for nuclear testing and dropped nuclear weapons during testing in the Pacific. It was redesignated JB-52D on 25 April 1958 and NB-52D on 4 January 1962. It never saw operational service with SAC and was the only B-52D never painted with camouflage colors.

NB-52D (56-0620) was used for nuclear weapons testing at the AFSWC, Kirtland AFB, NM and was the only B-52D never painted with camouflage colors. (U.S. Air Force)

NB-52E. The second B-52E (56-0632) was delivered in place to Boeing on 21 February 1958 and assigned to major testing programs. It was redesignated NB-52E and used for prototyping improvements of landing gear, engines, and other major subsystems with the results being used for subsequent B-52 fleet improvements. It was used to test a more aerodynamic engine cowling than the one used on the B-52F to cover the J57-P-43WA engines used on the B-52G. It was permanently modified for highly specialized development projects including small swept winglets, called canards, attached to the nose. A long probe extended from the nose, and the wings displayed nearly twice the normal controlling surfaces. The normal control surface linkages were replaced with electronic and electrical "fly-by-wire" systems.[61]

NB-52E (56-0632) was used for extensive testing of aircraft sub-systems upgrades and new technologies including electronic and electrical "fly-by-wire" systems. (U.S. Air Force)

This aircraft also contained special electronic measuring systems used to develop an electronic flutter and buffeting suppression system, called the SASS. This system was designed to reduce crew fatigue and stress during low level flights and was later used on B-52G and H models. The NB-52E also participated in the Load Alleviation and Mode Stabilization (LAMS) project. This project used sensors to sense gusts and activate the control surface to reduce fatigue damage. In mid-1973, it flew 11.5-mph (10 knots) faster than the speed at which flutter would normally disintegrate the aircraft due to the canards and other improvements.[62] Despite its illustrious career, relatively low cost, and major contributions to B-52 fleet improvements, the NB-52E was retired 26 June 1974.

Another B-52E (56-0636) was designated NB-52E on 13 December 1967 and used for P&W JT-9D turbofan engine testing at Bradley Field, CT. One JT-9D was mounted on

NB-52E (56-0636) resting in the boneyard circa 1989. It was one of the last E models with SAC paint scheme. The inboard engine pod was removed, and the aircraft was used to test an installed JT-9D engine for the Boeing 747. (Mike Freer)

the right inboard engine pylon along with the remaining six J57 engines. The JT-9D engine was intended to power the Boeing 747 transport aircraft. This aircraft retired on 30 July 1981 and made the last B-52E flight from Bradley Field to MASDC.

NB-52H. A B-52H (61-0025) was tapped to become the new NASA test bird and given the designation NB-52H. The aircraft was assigned to the NASA DFRC at Edwards AFB, CA on 30 July 2001. This aircraft was extensively modified to act as the mothership for the X-37 Space Maneuver Vehicle and was given a modern NASA gloss white and blue paint scheme. Modifications included an improved and strengthened wing pylon which sat further forward on the wing than the NB-52B and allowed full operation of the flaps. Despite all these efforts, NASA decided instead to contract out the X-37 mothership duties and the NB-52H was retired in 2006 without ever flying NASA research missions. The aircraft was considered damaged beyond economical repair due to the modifications. The Air Force authorized a one-time flight on 9 May 2008 to become a ground maintenance trainer at Sheppard AFB, TX. The vertical fin was removed to comply with SALT requirements.

JRB-52B. One RB-52B (52-0004) was delivered in place to Boeing on 3 March 1955 and used for various tests. It was redesignated JRB-52B on 30 November 1955. It was later assigned to the AFSWC, Kirtland AFB, NM and used for monitoring detonations during Operation REDWING nuclear testing from April to July 1956. RB-52B (52-0007)

NB-52H (61-0025) was tapped as the newest NASA test aircraft and sported an improved wing pylon and a more modern paint scheme than previous NASA aircraft. (NASA)

was initially delivered in place to Boeing for various tests and redesignated JRB-52B on 30 November 1955. It departed from MASDC after retirement to the Missile Development Center (MDC), Holloman AFB, NM on 5 May 1966 and was tested to destruction on 23 June 1966 during the BIG MAMA explosive test.

RB-52B (52-0009) was redesignated JRB-52B on 30 November 1955 and equipped for testing with a tail turret containing two M24A-1 20mm cannons controlled by a new MD-5 DFCS. RB-52B (52-0010) was delivered in place to Boeing on 3 May 1955 to support various tests including initial testing of the multi-purpose reconnaissance pod. It was redesignated JRB-52B on 30 November 1955 and then B-52B on 18 May 1956 after testing. It also departed after retirement, along with 52-0007, from MASDC to the MDC on 5 May 1966 and was tested to destruction on 23 June 1966.

RB-52B (52-0011) was delivered to Eglin AFB, FL on 23 May 1955 to support ARDC tests It was subsequently assigned to Wright-Patterson AFB, OH on 24 August 1955 and redesignated JRB-52B on 1 December. It was redesignated B-52B on 14 November 1957 after testing. RB-52B (52-0012) was delivered to Eglin AFB, FL on 10 May 1955 to support ARDC tests. It was redesignated JRB-52B on 1 January 1956 and B-52B on 6 May 1957 after testing.

JB-52B. RB-52B (53-0379) was delivered to the 93rd BMW at Castle AFB, CA on 3 November 1955 and redesignated B-52B. It supported nuclear tests at the AFSWC, Kirtland AFB, NM from 30 January to 10 August 1962. It was redesignated JB-52B on 2 March 1962 and B-52B after testing. B-52B (53-0383) also supported nuclear tests at AFSWC from 28 December 1955 to 31 December 1957. It was redesignated JB-52B on 4 January 1956. It was assigned temporary duty (TDY) at Eniwetok to support Operation REDWING, Shot Cherokee from 16 April to 21 May 1956 and other atomic tests. It was redesignated B-52B on 27 July 1958.

JB-52C. B-52C (54-2669) was delivered in place to Boeing on 9 August 1956 and used for testing. It was transferred to ARDC at Wright-Patterson AFB, OH on 14 January 1957 for testing and redesignated JB-52C on 15 January. It was then transferred to ARDC at Eglin AFB, FL on 12 June 1957 for additional testing. It was assigned to operational service at Westover AFB, MA on 11 December 1957 and redesignated B-52C. It was subsequently assigned to the AFSWC at Kirtland from 30 July to 16 December 1963 for weapons compatibility testing. It was later used as the FLACHBACK test vehicle in January 1965. B-52C (54-2676) was also initially delivered to Boeing on 26 June 1956 for testing and redesignated JB-52C. It ultimately crashed 15 miles north of Tulsa, OK on 29 March 1957 during a flight test from Wichita, KS.

JB-52E. The first B-52E (56-0631) was delivered in place to Boeing on 7 October 1957 and used for various tests. It was transferred to ARDC at Wright-Patterson, OH on 19 August 1958 and redesignated JB-52E on 25 August. It transferred to Kirtland AFB, NM on 1 July 1959 for additional testing. It was redesignated B-52E on 9 March 1961.

Another B-52E (57-0119) was redesignated JB-52E on 5 October 1966 and transferred to AFFTC at Edwards AFB, CA for special test and development. This aircraft was used to test the GE TF39-GE-1C turbofan which was planned to be used in the new C-5A Galaxy. The test engine was located on the right inboard pylon and replaced the twin Pratt & Whitney J57 engines normally in that location. The single TF39 had about twice the thrust of the twin J57s it replaced.

JB-52E (57-0119) fitted with a single TF39-GE-1C engine. (U.S. Air Force)

JB-52F. The first B-52F (57-0030) was delivered in place to Boeing on 12 May 1958 and used for various tests. It was redesignated JB-52F and transferred to North American Aviation on 27 February 1959 for AGM-28 Hound Dog missile testing. It was finally committed to operational service with SAC on 20 August 1962 and redesignated B-52F. Another B-52F (57-0038) left operational service on 4 December 1963 and was redesignated JB-52F for testing at Eglin AFB, FL.

JB-52G. B-52G (57-6468) was delivered ARDC at Wright-Patterson AFB, OH on 1 November 1958 for testing. It was redesignated JB-52G on 13 November and then transferred to Boeing for various tests on 28 February 1959. It was redesignated B-52G on 22 September 1961.

B-52G (57-6470) was delivered in place to Boeing on 17 December 1958 and then to Air Force Systems Command (AFSC) at Eglin AFB, FL on 5 June 1962 for testing. It was redesignated JB-52G on 19 April 1963 and then back to B-52G on 1 November. B-52G (57-6472) was delivered to Boeing (Seattle) on 15 November 1958 and modified with AGM-28 Hound Dog capability. It was transferred to North American Aviation on 3 April 1959 for testing and redesignated JB-52G. It was redesignated B-52G on 21 February 1962.

B-52G (57-6473) was delivered in place to Boeing on 12 December 1958 and used for AGM-48 SKYBOLT testing. It was redesignated JB-52G during the test program and then returned to B-52G on 22 July 1963. Another B-52G (57-6477) was delivered from the Boeing production line to ARDC at Eglin AFB, FL on 15 April 1959 for AGM-48 SKYBOLT testing. It was redesignated JB-52G during the test program and then returned to B-52G on 26 February 1963.

B-52G (58-0159) was the first B-52G to receive AGM-28 Hound Dog capability during production. It was delivered to ARDC Eglin AFB, FL on 8 October 1959 and used for AGM-28 testing. It was redesignated JB-52G on 1 December 1959 and reverted to B-52G after testing on 2 December 1960.

B-52G (58-0182) was delivered in place to Boeing on 1 August 1959 for test. It was transferred to ARDC at Eglin AFB, FL on 22 August 1961 for additional testing. It was redesignated JB-52G during the test and later reverted to B-52G. It was assigned to AFFTC at Edwards AFB, CA from 2 June 1972 to 26 June 1975 for testing. It was assigned to Boeing from 28 October 1975 to 12 July 1977 for AGM-86A ALCM testing. It was painted gloss white overall during the ALCM test period.

JB-52H. Several B-52H aircraft were used for test programs and redesignated JB-52H. B-52H (60-0003) was delivered in place to Boeing on 3 March 1961 for Category II and ALE-25 testing. It was transferred to AFFTC at Edwards AFB, CA on 4 August 1961 for test and redesignated JB-52H in October 1961. It then returned to Boeing on 20 July 1962 and was redesignated B-52H. B-52H (60-0004) was delivered in place to Boeing on 3 April 1961 where it remained for ASQ-38 and ECM test programs until 11 January 1973. It was redesignated JB-52H in September 1967 and B-52H in September 1968.

B-52H (60-0005) was delivered in place to Boeing on 3 April 1961 for ASQ-38 and ECM testing. It was transferred to AFSC at Wright-Patterson AFB, OH on 25 July 1961 for testing and redesignated JB-52H on 2 October 1961. It returned to Boeing on 23 February 1962 and was redesignated B-52H.

B-52H (60-0006) was delivered in place to Boeing on 26 March 1961 for SKYBOLT testing. It was transferred to AFFTC at Edwards AFB, CA on 1 July 1961 for testing and then returned to Boeing on 15 June 1962. It was also used as the EVS test bed. The aircraft was redesignated JB-52H in September 1961 and then back to B-52H after test.

B-52H (61-0023) was delivered in place to Boeing and used for testing from 18 May 1962 to 17 October 1973. It was redesignated JB-52H during test and later returned to B-52H. Boeing used the aircraft for turbulence tests. On 10 January 1964, during the test flight the entire vertical fin and rudder were ripped off, and the aircraft was able to land safely at Blytheville AFB, AR. The aircraft was repaired and returned to service (RTS).

JB-52H (61-0023) was used for turbulence tests. On 10 January 1964, during the test flight the entire vertical fin and rudder were ripped off. (U.S. Air Force)

RB-52B. Two RB-52B (52-0005, 52-0006) aircraft were delivered to AFFTC, Edwards AFB, CA on 13 and 21 March 1955, respectively, to support ARDC test activities.

B-52D. One B-52D (55-0112) was transferred to Boeing Wichita on 25 June 1973 and underwent destructive testing. It was scrapped on 17 December 1973. B-52D (56-0595) supported Quail flight testing. B-52D (56-0612) was transferred to Eglin AFB, FL on 4 April 1977 for ECM testing. B-52D (56-0616) was transferred to Tinker AFB, OK on 21 May 1971 and underwent destructive testing. It was scrapped on 5 August 1971.

B-52F. One B-52F (57-0071) was initially delivered in place to Boeing on 19 February 1959 and used for various tests before being delivered to Carswell AFB, TX on 2 April.

B-52F (57-0073) was delivered in place to Boeing on 20 March 1959 for use as a static test aircraft and never saw SAC service. It was scrapped in place on 1 October 1964. B-52F (57-0181) was transferred to Boeing on 9 October 1967 and tested to destruction.

B-52G. B-52G (57-6469) was delivered in place to Boeing on 7 November 1958 and used for testing. It was transferred to AFFTC at Edwards AFB, CA for ARDC testing on 15 March 1959 and then to Boeing on 3 August. B-52G (57-6471) was delivered to Boeing (Seattle) on 4 December 1958 and used for TF-33 engine testing. It is sometimes referred to as the YB-52H prototype. B-52G (57-6474) was delivered in place to Boeing on 7 January 1959 and used for ASG-21 DFCS testing. B-52G (57-6475) was delivered in place to Boeing on 9 January 1959 and used for various tests. B-52G (58-0204) was used as the test bed for the fly-off between the Boeing ALCM and the Tomahawk cruise missiles, as well as the Phase VI ECM testing. B-52G (58-0224) was delivered to AFFTC at Edwards AFB, CA on 14 January 1960 for testing before seeing operational service at Loring AFB, ME on 26 August.

B-52H. B-52H (60-0020) was assigned to 6510th TS at Edwards AFB, CA on 13 February 1984 to support AGM-129 ACM testing and other black programs. It was returned to SAC service with the 5th BMW, Minot AFB, ND on 24 August 1991. B-52H (60-0050) was transferred to Boeing for OAS and ALCM modifications on 16 January 1985. It has been assigned to the AFFTC at Edwards AFB, CA since 1 August 1985 and used for CSRL and other tests. It launched the first AGM-136A Tacit Rainbow from a B-52H on 10 January 1989. It launched the first JDAM from a B-52H on 30 April 1997. It was also used as the launch vehicle for the first flight of the WAVERIDER unmanned scramjet on 26 May 2010. B-52H (61-0009) was transferred to Boeing, Tinker AFB, OK on 22 January 2022, after retirement at AMARG on 25 September 2008, to support CERP and radar upgrade integration testing. B-52H (61-0034) became the first B-52 to be completely powered by synthetic fuel when it made a 7-hour test flight from Edwards AFB, CA on 15 December 2006.

B-52H/D-21. Two B-52H (60-0036, 61-0021) aircraft were modified under an ultra-secret project called SENIOR BOWL to launch the D-21 ramjet powered reconnaissance drone. The Lockheed Skunk Works designed the D-21 under the direction of U-2 and SR-71 designer Kelly Johnson. It was intended to allow overflights of the USSR and China after the shoot down of a U-2 on 1 May 1960, which caused the U.S. to discontinue manned flights.

The D-21 was considered a high-speed and high-altitude reconnaissance vehicle and a smaller unmanned version of the A-12. Initially, it was launched from the back of a modified A-12, designated M-21. However, it required a very complicated launch process and separation problems led to the collision of a D-21 with its M-21 launch vehicle on 30 July 1966. Soon after, Johnson suggested dropping the D-21 from a B-52 like the X-15 instead of trying to launch it from an M-21.

B-52H carrying two D-21 high-speed, high-altitude, unmanned reconnaissance drones. (U.S. Air Force)

The Air Force accepted this approach and the first B-52H (61-0021) arrived at Air Force Plant 42 in Palmdale, CA for modification on 12 December 1966. The second aircraft B-52H (60-0036) began modifications in August 1967. The modifications were extensive and included replacing the ECM and Gunner station equipment with D-21 launch control panels, installing high-speed cameras in the wheel wells to capture the D-21 launch, and replacement of the AGM-28 pylons with new X-15 like pylons.

The first launch attempt from a B-52H occurred on 6 November 1967 but was unsuccessful. The D-21 flew 500 nautical miles on 2 December 1967 before it ran out of hydraulic fluid. It flew 2,850 nautical miles on 16 June 1968 reaching an altitude of 90,000 feet. Several other test flights occurred throughout 1968 and 1969 which were mostly successful. The first operational mission occurred on 9 November 1969, and the D-21 crashed somewhere in the Soviet Union. The next two missions were successful flights but the hatch containing film was not recovered. The final mission occurred on 20 March 1971, and the D-21 was shot down over China. The program was subsequently cancelled.

Armament and Weapons

Defensive Armament

The B-52 employed a variety of DFCS equipment which typically controlled a tail gun turret with four .50-caliber M3 machine guns. The turret was hydraulically controlled and gyroscopically stabilized. Each gun could fire 600 rounds per minute. The B-52A was equipped with the A-3A DFCS which was capable of automatic search and tracking operation. In all B-52A through the F models the tail gunner sat in a pressurized compartment just aft of the tail and had direct line-of-sight to the targets. An optical sight was provided to allow the gunner to manually operate the guns if needed.

The gunner's position in the tail of a Boeing B-52D Stratofortress. (National Archives)

Nine of the initial ten RB-52B aircraft (52-0004 to 52-0008; 52-0010 to 52-0013) also retained the tail turret with the four .50-caliber M3 machine guns used on the B-52A controlled by the same A-3A DFCS, but this system was deficient. One of these initial ten RB-52B (52-0009) aircraft was equipped with a tail turret containing two M24A-1 20mm cannons controlled by a new MD-5 DFCS. This system was also installed in the remaining 17 RB-52B (52-8710 to 52-8716; 53-0366 to 53-0372; 53-0377 to 53-0379) aircraft and 16 B-52B (53-0373 to 53-0376; 53-0380 to 53-0391) aircraft. Unfortunately, the MD-5 didn't meet expectations and was replaced in the last seven B-52B (53-0392 to 53-0398) aircraft with an improved version of the A-3A DFCS. These aircraft were also fitted with the original tail turret with four .50-caliber M3 machine guns. This system was also used in most B-52C aircraft but proved unreliable and was replaced with the MD-9 DFCS on the last B-52C (54-2688) and on subsequent B-52D through F models.[1]

The B-52G retained the original tail turret with four .50-caliber M3 machine guns. However, the gunner was moved to the forward crew cabin and now controlled the guns using the new AN/ASG-15 DFCS. This system was capable of automatic search and tracking operation like the previous systems, but the optical sight was replaced by closed-circuit television.

The B-52H featured a single 20mm M61A-1 cannon in the tail gun turret in lieu of the four .50-caliber machine guns. This new six-barrel Gatling-gun type cannon was controlled by the new AN/ASG-21 DFCS and could fire 3,000 to 6,000 rounds per minute. The AN/ASG-21 DFCS was more automated and featured just four modes of operation compared to 20 in the AN/ASG-15 DFCS. Otherwise, the systems operated the same with both automatic radar search and track, and manual operation using closed-circuit television. The AN/ASG-21 DFCS also controlled forward-firing penetration chaff rocket

B-52H 20mm Gatling Gun capable of firing 3,000 to 6,000 rounds per minute. (U.S. Air Force)

launchers mounted on pylons installed between the engine pods.[2] Although it originated as a tube system, it was later upgraded to a mostly digital system.

Air Force Staff Sergeant Samuel O. Turner became the first B-52 gunner to shoot down a MiG-21 on 18 December 1972, the first night of Operation LINEBACKER II. Turner was the gunner flying in a B-52D (56-0676) over North Vietnam when he spotted two MiGs approaching from below and behind the bomber. He fired his four M3 .50-caliber machine guns in a single 6-8 second burst until the MiG exploded. Turner received the Silver Star and was later awarded the Distinguished Flying Cross and several Air Medals. B-52D (56-0676) was the last B-52D retired and is now on display at Fairchild AFB.

A1C Albert Moore became the only other B-52 gunner to shoot down a MiG-21. Moore was flying in a B-52D (55-0083) over North Vietnam during Operation LINEBACKER II. He encountered the MiG during a mission to bomb the railyards near Hanoi, the capital of what was then North Vietnam, on Christmas Eve 1972. Moore said in his combat report, "I observed a target on my radar scope at 8 O'clock low. It stabilized at 4,000 yards. I called for evasive action and when the target reached 2,000 yards, I opened fire." Moore kept firing until the MiG disappeared. Another tail gunner saw the MiG falling from the sky and confirmed the kill. B-52D (55-0083) is now displayed at the United States Air Force Academy, Colorado Springs, CO along with a plague honoring Moore.

Nuclear Munitions

The B-52 was designed as a nuclear bomber from its beginning. As new munitions were developed the latest B-52 models were authorized to carry them. The following table shows the munitions authorized for each model. Most munitions were eventually retired from service leaving only the B-61 and B-83 authorized for carriage on the B-52H. However, it was becoming apparent to most observers as early as 2010 that the B-52 was

Nuclear Munitions (Internal/External)								
Weapon Type	Retired	B-52B	B-52C	B-52D	B-52E	B-52F	B-52G	B-52H
Mk-6 (Kiloton Class)	1962	1/0	1/0	1/0	1/0	1/0	1/0 [1]	
Mk-15 (Megaton Class) [2]	1965				2/0	2/0	2/0	2/0
Mk-21 (Megaton Class) [3]	1957	2/0	2/0	2/0	2/0	2/0	2/0	
Mk-28 (Megaton Class) [6]	1991				4/0	4/0	4/0	4/0
Mk-36 (Megaton Class) [4]	1962					2/0	2/0	2/0
Mk-39 (Megaton Class)	1966					2/0	2/0	2/0
Mk-41 (Megaton Class) [5]	1976				2/0	2/0	2/0	2/0
Mk-43 (Megaton Class)	1991					4/0	4/0	4/0
Mk-53 (Megaton Class) [6]	1997				2/0	2/0	2/0	2/0
Mk-57 (Megaton Class) [6]	1993				4/0	4/0	4/0	4/0
B-61 (Kiloton Class) [8]	N/A						4/0	8/0 [7]
B-83 (Megaton Class) [8]	N/A						4/0	8/0 [7]

(1) First 154 B-52G aircraft only
(2) Replaced by Mk-39
(3) Converted to Mk-36
(4) Replaced by Mk-41
(5) Replaced by Mk-53
(6) Replaced by B-61 and B-83
(7) Increased from four to eight after CSRL installation
(8) No longer authorized for B-52H which now relies on only the AGM-86B ALCM for nuclear capability

no longer capable of penetrating modern Integrated Air Defense Systems (IADS) to drop nuclear gravity weapons on target. A 2018 update to the *Department of Defense 2016 Nuclear Matters Handbook* seemed to acknowledge this fact when it listed removal of the B-61 and B-83 from the B-52H approved weapons. Furthermore, the 1 August 2022 version of *AFI 91-111, Safety Rules for U.S. Air-Launched Nuclear Weapon Systems*, indicates that the B-52H is no longer authorized to carry B-61 and B-83 gravity bombs leaving the AGM-86B ALCM as the only weapon remaining to meet its nuclear mission requirements.

Mk-6 Gravity Bomb. The Mk-6 was an improved version of the Mk-4 "Fat Man" bomb dropped on the Japanese mainland to end WWII. It was 61 inches in diameter and 128 inches long. It weighed 7,600 to 8,500 pounds and produced a yield of 21 kilotons. It could be detonated with either airburst or surface contact. A number of bombs were produced and underwent various modifications from July 1951 to early 1955. It was retired from operational use in 1962. One Mk-6 could be carried in the B-52B through G model bomb bay.

Mk-15 Gravity Bomb. The Mk-15 was the first lightweight thermonuclear bomb. It was 34.4 to 35 inches in diameter and 136 to 140 inches long. It weighed 7,600 pounds and was a megaton class weapon. It could be detonated with either airburst, surface contact, or laydown. A number of bombs were produced and underwent various modifications from April 1955 to February 1957. It was retired from operational use between August 1961 and April 1965 and replaced by the Mk-39. Two Mk-15 bombs could be carried in the B-52E through H model bomb bay.

(Left) B-52B through G models could carry one Mk-6 atomic bomb weighing up to 8,500 pounds with a yield of 21 kilotons. (Right) Two Mk-15 thermonuclear bombs could be carried in the bomb bay of the B-52E through H models. (U.S. Air Force)

Mk-21 Gravity Bomb. The Mk-21 was 56.2 to 58.5 inches in diameter and 149 to 150 inches long. It weighed 15,000 to 17,700 pounds and was a megaton class weapon. It could be detonated with either airburst, surface contact, or laydown. A number of bombs were produced and underwent various modifications from December 1955 to July 1956. It was retired from operational use in 1957 by conversion to Mk-36. Two Mk-21 bombs could be carried in the B-52B through G model bomb bay.

Mk-28 Gravity Bomb. The Mk-28 (a.k.a, B-28) was a multipurpose thermonuclear tactical and strategic bomb. It was 20 to 22 inches in diameter and 96 to 170 inches long. It weighed 1,700 to 2,320 pounds and was a megaton class weapon. It could be dropped free-fall or retarded with parachutes and detonated with either airburst, surface contact, or laydown. A number of bombs were produced in 20 modifications and variants from January 1958 to May 1966. It was retired from operational use between 1961 (early variants) and September 1991. Four Mk-28 bombs could be carried in the B-52E through H model bomb bay. The Mk-28 was replaced by the B-61 and B-83 bombs.

Mk-36 Gravity Bomb. The Mk-36 was a thermonuclear strategic bomb. It was 56.2 to 59 inches in diameter and 149 to 150 inches long. It weighed 17,500 to 17,700 pounds and was a megaton class weapon. It could be dropped free-fall or retarded with parachutes and detonated with either airburst or surface contact. A number of bombs were produced from April 1956 to June 1958. The Mk-36 was retired from operational use between August 1961 and January 1962. Two Mk-36 bombs could be carried in the B-52F through H model bomb bay. The Mk-36 was replaced by the Mk-41.

Mk-39 Gravity Bomb. The Mk-39 was an improved version of the Mk-15 thermonuclear strategic bomb. It was 35 to 44 inches in diameter and 136 to 140 inches long. It weighed 6,650 to 6,750 pounds and was a megaton class weapon. It could be dropped with low-level retarded laydown and detonated with either airburst or surface contact. The Mk-39 was retired from operational use between January 1962 and November 1966. Two Mk-39 bombs could be carried in the B-52F through H model bomb bay.

Two Mk-15 bombs being loaded into the bomb bay of a B-52F Stratofortress. The Mk-39 was an improved version of the Mk-15. (U.S. Air Force)

Mk-41 Gravity Bomb. The Mk-41 replaced the Mk-36. It was 52 inches in diameter and 148 inches long. It weighed 10,500 to 10,670 pounds and was a megaton class weapon.

It could be dropped free-fall or retarded with parachutes and detonated with either airburst, surface contact, or laydown. The Mk-41 was retired from operational use between November 1963 and July 1976. Two Mk-41 bombs could be carried in the B-52E through H model bomb bay. The Mk-41 was replaced by the Mk-53.

Mk-43 Gravity Bomb. The Mk-43 was 18 inches in diameter and 150 to 164 inches long. It weighed 2,060 to 2,125 pounds and was a megaton class weapon. It could be dropped free-fall or retarded with parachutes and detonated with either airburst, surface contact, or laydown. The Mk-43 was retired from operational use by April 1991. Four Mk-43 bombs could be carried in the B-52F through H model bomb bay.

Four Mk-43 bombs could be carried in the B-52F through H models. (U.S. Air Force)

Mk-53 Gravity Bomb. The Mk-53 was 50 inches in diameter and 148 to 150 inches long. It weighed 8,850 to 8,900 pounds and was a megaton class weapon. It could be dropped free-fall or retarded with parachutes and detonated with either airburst, surface contact, or laydown. The Mk-53 was retired from operational use between July 1967 and early 1997. Two Mk-53 bombs could be carried in the B-52E through H model bomb bay. The Mk-53 was replaced by the B-61 and B-83 bombs.

Mk-57 Gravity Bomb. The Mk-57 was a lightweight multipurpose tactical strike weapon. It was 14.75 inches in diameter and 118 inches long. It weighed 490 to 510 pounds and was a megaton class weapon. It could be dropped free-fall or retarded laydown with parachutes and detonated with either airburst or surface contact. The Mk-57 was retired from operational use between June 1975 and June 1993. Four Mk-57 bombs could be carried in the B-52E through H model bomb bay. The Mk-57 was replaced by the B-61 and B-83 bombs.

B-61 Gravity Bomb. The B-61 was a lightweight multipurpose tactical and strategic weapon. It was 13.3 inches in diameter and 141.64 inches long. It weighed 695 to 716 pounds and was a kiloton class weapon. It could be dropped free-fall or retarded and detonated with either airburst or laydown. Early versions of the B-61 were retired from operational use during the 1970s and 1980s. It had the longest production run of any U.S. nuclear weapon and is the oldest design still in service. It is also part of the U.S. enduring stockpile. Four B-61 bombs could be carried in the B-52G and eight in the B-52H bomb bay with CSRL installed. However, the B-61 is no longer authorized for the

Four B-61 bombs could be carried in the B-52G and eight in the H models. (U.S. Air Force)

B-52H, and it now relies on only the AGM-86B ALCM for nuclear stand-off capability.

B-83 Gravity Bomb. The B-83 was a high-yield thermonuclear strategic bomb. It was 18 inches in diameter and 145 inches long. It weighed 2,400 pounds and was a megaton class weapon. It could be dropped free-fall or retarded and detonated with either airburst, contact, or laydown. It remained in service as of 2025. Four B-83 bombs could be carried in the B-52G and eight in the B-52H bomb bay with CSRL installed. However, like the B-61 it is no longer authorized for the B-52H.

Four B-83 bombs could be carried in the B-52G and eight in the H models. (U.S. Air Force)

Nuclear Stand-Off

Early on in its development the Air Force realized the value of launching stand-off weapons from the B-52. Initially, these were nuclear weapons intended to neutralize Soviet air defense systems to allow the bomber to reach its target. Several systems were researched and tested but only the AGM-28 Hound Dog missile was operationally deployed and was the main stand-off weapon until its retirement in 1978. It was replaced by the AGM-69A SRAM which was a much smaller and more capable missile. A new OAS was developed in the late 1970s and installed on about 166 B-52G and all H model aircraft during the 1980s to increase the accuracy of B-52 nuclear weapons.

About 98 OAS modified B-52G and all B-52H aircraft also received the Cruise Missile Integration (CMI) modification to allow launching of the AGM-86B ALCM. The CMI modified B-52G aircraft were also modified with nonfunctional, rounded wing-to-fuselage junctions called "strakelets" that readily identified the aircraft as ALCM capable to overhead Soviet satellites under terms of the Strategic Arms Limitations Treaty (SALT) II. Since all B-52H aircraft received the CMI modification and these aircraft were easily identified by their engine inlets no strakelets were required.

Rotary launcher improvements were the key to expanding the B-52 internal weapons load capacity. When it was first developed the ALCM was planned to be carried on the existing SRAM rotary launchers. However, with extended range improvements, the final ALCM design was too large to fit on the SRAM launcher. Therefore, a compromise was reached that allowed up to six ALCMs to be carried externally under each wing on a newly developed ALCM wing pylon. This was also true of the AGM-129A ACM meant to replace the ALCM. The A/A48K-1 CSRL was developed beginning in 1989 and allowed the B-52H to carry eight ALCMs and ACMs internally, as well as 12 externally, for a total of 20 weapons.

Nuclear Stand-Off (Internal/External)								
Weapon Type	Retired	B-52B	B-52C	B-52D	B-52E	B-52F	B-52G	B-52H
ADM-20 Quail (Decoy)	1978				4/0	4/0	4/0	4/0
AGM-28 Hound Dog	1978	0/2 [1]	0/2	0/2	0/2	0/2	0/2	0/2
AGM-69A SRAM	1993						8/12	8/12 [2]
AGM-86B ALCM	N/A						0/12	8/12 [3]
AGM-129A ACM	2012							8/12 [3]
(1) Only later model B-52B aircraft were equipped								
(2) No longer supported eight in bomb bay after CSRL installation								
(3) Eight in bomb bay after CSRL installation								

Quail. The McDonnell ADM-20 (formerly GAM-72) Quail was a small delta wing drone, equipped with a GE J85-GE-7 turbojet engine. It was not an actual nuclear weapon but was used to enable nuclear missions. It had a range of several hundred miles and could match B-52 performance.[3] Internal electronic devices, including radar jammers, made it look like a B-52 on enemy radar, and it acted as a decoy to confuse enemy SAM radars and saturate enemy defenses with multiple targets. Operational versions carried radar reflectors and a radar repeater or chaff dispenser. An infrared burner in the tail could simulate the bomber's heat signature. The B-52E through H models carried four Quail decoys in the bomb bay and released them as it approached enemy defenses using a retractable arm that lowered the decoy into the slipstream.

The Air Force began to develop a short-range decoy missile requirement in April 1955 and released its operational requirements on 18 January 1956. McDonnell was issued a development contract on 1 February 1956. The Quail was originally intended to support bombers flying high and fast to reduce the time defenders had to respond before it flew out of range. While this worked well for fighter-interceptors it was of little use with SAMs that could reach the bomber in a matter of seconds after detecting it on radar.

ADM-20 Quail decoy missile on display at the National Museum of the United States Air Force. (U.S. Air Force)

SAC began flying low-level attack routes and soon modified the Quails with capability to fly at lower altitudes. However, this significantly reduced its range and flight time and caused Air Force leaders to seriously question its effectiveness.

The Quail included a high-mounted delta wing, no horizontal stabilizer, and two vertical stabilizers. The Quail was 12.75 feet long and 3.25 feet high with wings unfolded. The wingspan was 2.3 feet with wings folded and 5.3 feet unfolded. Total weight was 1,198 pounds. It originally used the GE J85-GE-1 engine with 1,900 to 2,100 pounds of thrust, but production models incorporated the J85-GE-7 with 2,450 pounds of thrust. The guidance system included an autopilot with a rate integrating gyro and could be pre-programmed to execute two turns and one speed change during its typical 45-to-55-minute flight time. It had a maximum speed of Mach 0.9 and initially flew at altitudes between 35,000 to 50,000 feet. It could reach targets 400-500 miles away depending on altitude.

Glide flight testing of the Quail began in 1957 at Holloman AFB, NM, and White Sands Missile range. The first successful powered flight occurred in August 1958. The first Quail assigned to SAC arrived at the 4135th SW at Eglin AFB, FL on 27 February 1960 and operational testing using a 4135th B-52G began on 8 June 1960. The Quail reached

IOC with the 4135th SW on 1 February 1961 and 400 Quails were in SAC inventory by the end of 1961. The first B-52 airborne alert with Quails onboard occurred on 1 January 1962. Full Operational Capability (FOC) was achieved on 15 April 1962 when 14 units were equipped with Quail missiles. Quail testing was also performed on a B-52D (56-0595). A total of 585 were produced and the final delivery occurred on 28 May 1962. SAC had 492 in inventory by 1963. Like the Hound Dog, inventory was slowly reduced but the Quail remained on alert until 30 June 1978. The last Quail retired from service on 15 December 1978.

Hound Dog. The North American AGM-28 Hound Dog (designated GAM-77 until June 1963) was an air-to-surface missile powered by a single P&W J52-P-3 turbojet engine with 7,500 pounds of thrust allowing it to reach supersonic speeds up to Mach 2.1. The missile was huge at 42.5 feet long and 9.3 feet high, 28 inches in diameter with a 12.2-foot wingspan and weighing 10,147 pounds. It was essentially the first ALCM. However, it was not used as a stand-off weapon. Instead, it was intended to take-out Soviet de-

AGM-28 Hound Dog air-to-surface missile mounted on a B-52 wing pylon. (U.S. Air Force)

fenses and clear the way for the bomber to reach its target. It was equipped with an Inertial Navigation System (INS), which was continuously updated before launch by a Kollsman KS-120 Astrotracker mounted in each of the aircraft's wing pylons. It carried a Mk-28 nuclear warhead and was a megaton class weapon. The Hound Dog flew a pre-programmed course and altitude and could reach ground targets such as radar complexes and SAM sites 700 miles away when launched at high altitude and supersonic speed. The distance was reduced to 200 nautical miles at low altitude and subsonic speed.[4]

While the 54 earliest B-52G aircraft were modified post-production, the remaining G models were the first equipped in production to carry two AGM-28 Hound Dog missiles with one carried on each wing pylon. The B-52H was also subsequently equipped during production, while a small number of B-52B and all B-52C through F models were retrofitted to carry the missiles. The missile was connected to the B-52 launch aircraft electrical and fuel systems. Consequently, its engine could be used to assist take-off, and the missile could be refueled in flight from the bomber's fuel system. This gave the B-52 tremendous striking power, but the missile was ultimately highly unreliable.[5]

A B-52F takes off with AGM-28 Hound Dog air-to-surface missiles mounted on the wing pylons. (U.S. Air Force)

The AGM-28 was first tested on a B-52E model in April 1959. The first Hound Dog was formally accepted by General Thomas S. Power, Commander in Chief Strategic Air Command (CINCSAC), on 21 December 1959. It was then immediately delivered to

SAC at the 4135th SW at Eglin AFB, FL on 23 December 1959 for operational testing. The first launch of a Hound Dog from a 4135th B-52G occurred on 29 February 1960. The first operational mission occurred on 12 April 1960 when a 4135th B-52G left Eglin and flew a round trip to the North Pole and back and then launched the missiles over the Cape Canaveral Missile Test Center in Florida.[6] Dubbed Operation BLUE NOSE, the mission completed extensive tests of the B-52 and Hound Dog guidance systems in temperatures as low as -75 degrees.

The AGM-28A Hound Dog missile system reached IOC in July 1960, and 225 missiles were in SAC inventory by the end of 1961. The Hound Dog was used for the first time during airborne alert in January 1962. FOC was achieved in August 1963 when 29 B-52 units were equipped, and SAC had 593 Hound Dogs in service.

Around 452 AGM-28A missiles were upgraded to the AGM-28B configuration with a Kollsman KS-140 Stellar Navigation System integrated into the INS and radar absorbent materials added to improve survivability.[7] The improved KS-140 Astrotracker was now installed inside the missile to allow continuous navigation updates. The AGM-28B also had an integrated radar altimeter for improved low-level flight performance and a larger fuel tank for increased range. An AGM-28C configuration was tested in 1971 but never produced. Most importantly, this configuration tested a newly developed Terrain Contour Matching (TERCOM) navigation system like the one later used on the AGM-86B ALCM.

AGM-28 Hound Dog air-to-surface missile in flight. (U.S. Air Force)

The Hound Dog was intended as a temporary system, which would ultimately be replaced by the AGM-48 Skybolt missile. However, the Skybolt was eventually cancelled, and a number of Hound Dogs were subsequently produced. Although the inventory was slowly reduced, the Hound Dog remained in service on alert aircraft until 30 June 1975. The last Hound Dog missile was retired from the 42nd BMW at Loring AFB, ME on 15 June 1978. It was replaced by the AGM-69A SRAM.

AGM-48 Skybolt ALBM

The Douglas AGM-48 Skybolt (formerly GAM-87) ALBM was a two-stage, solid fuel missile with a range of 1,150 miles. It carried a nuclear warhead and was a megaton class weapon.[8] The warhead was the same W59 lightweight thermonuclear warhead used on the Minuteman Intercontinental Ballistic Missile (ICBM). The missile was 38.25 feet long, had a wingspan of 5.5 feet, a 2.92-inch diameter, and weighed 11,000 pounds. It had eight movable fins on one stage and a gimballed nozzle on the second. The B-52G and H models would have carried four Skybolt missiles with two under each wing. They were mounted on an "inverted Y" dual-launch adapter and projected from the wing leading edge to allow the missile's astrotracker to continually observe the sky.[9]

AGM-48 Skybolt ALBM. (U.S. Air Force)

The Skybolt was the first true stand-off missile planned for the B-52. The concept was that bombers on airborne alert would fly in holding patterns far outside the range of any Soviet missile defenses, before launching their missiles on command. Because an airborne bomber could receive radio instructions in flight, Skybolt could be retargeted before launch giving it potential as a second-strike weapon.[10]

A decision to proceed with the Skybolt was reached in February 1960, and this would have led to initial deployment in 1964. The UK joined the project in June 1960, ordering 100 Skybolts for carriage on their Vulcan bombers, although they planned to use their own "Red Snow" warhead on safety grounds.[11] President Kennedy called for additional Skybolt development funds on 28 March 1961 to extend the life of SAC's heavy bombers into the missile age. Congress approved the request.

However, as the Navy's Polaris Submarine Launched Ballistic Missile (SLBM) and the Air Force's Minuteman ICBM entered service the need for the Skybolt became less apparent. Poor test results, including five failed attempts, were the nail in the coffin. President Kennedy confirmed his intention to curtail Skybolt development on 7 December 1962, much to the chagrin of the British who had seen the Skybolt as a key to their nuclear deterrence strategy. The Skybolt finally completed its first successful test flight on 19 December 1962. Despite this late success, Kennedy announced cancellation of the program on 21 December in a joint statement with British Prime Minister Macmillan. SECDEF McNamara confirmed the cancellation was due to higher-than-expected development and production costs, and the successful development of other weapons that could carry out the task of suppressing enemy defenses at substantially lower costs. McNamara had always opposed the Skybolt and believed that manned bombers were obsolete and unnecessary in the age of the SLBM and ICBM.[12]

SRAM. The Boeing AGM-69A SRAM was a solid fuel ballistic missile that had a maximum range of about 127 miles. Like the Hound Dog, it was intended to clear the way for the bomber to reach its target. It was fitted with a Lockheed SR75-LP-1 two-pulse solid fuel rocket motor. It carried a nuclear warhead and was a megaton class weapon. The SRAM was 15.8 feet long with the tail fairing and 14 feet without the fairing. It had a wingspan of 2.5 feet, a 17.5-inch diameter, and weighed 2,230 pounds. The missile flew at a maximum speed of Mach 3.5 and could be launched outside the approximately 20-mile range of Soviet SA-2 SAMs. At this speed and range, it could take out the SA-2 site before the bomber reached it. Six SRAMs could be carried externally on each wing of the B-52G and H model on specially designed pylons and standard MAU-12 ejector racks. Eight SRAMs could also be carried on a rotary launcher in the bomb bay, for a total of 20 missiles, plus the standard load nuclear gravity bombs normally carried.

AGM-69A SRAMs and Mk-28 (B-28) nuclear bombs loaded in the B-52 bomb bay. (U.S. Air Force)

The SRAM had a General Precision/Kearfott KT-76 Inertial Measurement Unit (IMU) and a Stewart-Warner radar altimeter which enabled the missile to be launched in either a semi-ballistic or terrain-following flight path. The SRAM was also capable of performing one "major maneuver" during its flight which gave the missile the capability of reversing its course and attacking targets that were behind it, sometimes called an "over-the-shoulder" launch. The missile was completely coated with 0.8 inches of soft rubber, used to absorb radar energy, and dissipate heat during flight. The three fins on the tail were made of a phenolic material designed to minimize any reflected radar energy.[13]

SAC accepted the first SRAM on 1 March 1972, and it was delivered to the 42nd BMW at Loring AFB, ME on 4 March for test and evaluation. The first SRAM launch by a 42nd BMW B-52G occurred on 15 June at the White Sands Missile Range, NM. The missile officially entered service with the 42nd BMW on 15 September 1972. A number of missiles were produced by the time the last delivery occurred at the 320th BMW at Mather AFB, CA on 20 August 1975.

A new AGM-69B version was proposed in the late 1970s that included an improved engine and warhead. However, President Carter cancelled it in 1978. A new AGM-131 SRAM II was also planned to be carried on the B-1B and began development in 1981, but it was cancelled in 1991 by President George H.W. Bush citing the end of the Cold War. The AGM-69A SRAM was ultimately phased out in 1993 and replaced by the AGM-86B ALCM. There were growing safety concerns over its warhead and rocket motor when several suffered cracking of the propellant which could have caused catastrophic failure.

ALCM. The ALCM was the first true stand-off weapon used on the B-52 and designed to take out strategic targets. It was originally intended as a replacement to the Quail drone decoy. Designated the AGM-86A Subsonic Cruise Aircraft Decoy (SCAD) it was designed to fit in the same launcher as the SRAM. Like the Quail, it was designed to fly a pre-programmed course and act as a decoy to fool enemy radars. However, as nuclear warheads became lighter and smaller, the Air Force modified its specifications with the intent of attacking missile and radar sites at the end of the decoy's flight. Eventually, the requirement evolved into a B-52 nuclear stand-off weapon, now called the Subsonic Cruise Armed Decoy (SCAD), allowing the B-52 force to launch hundreds of small, low-flying targets to saturate Soviet defenses and attack SAM sites.

However, William P. Clements, Jr., Deputy Secretary of Defense, cancelled the SCAD on 30 June 1973 in favor of a stand-off system like the SRAM but dedicated to the long-range attack mission. On 20 July 1973, Dr. Malcolm R. Currie, Director of Defense and Engineering, directed the Air Force to develop a long-range missile based on SCAD development efforts. Although it retained the AGM-86A designation, the new missile was renamed as the ALCM and flew for the first time in March 1976. The AGM-86A ALCM was given the production go-ahead in January 1977. The Air Force was simultaneously considering other options including an Extended Range Version (ERV) of the ALCM that would have increased range to about 1,700 miles. However, it would need to be much larger than the AGM-86A to hold the required fuel load and would be unable to fit on the B-52 SRAM launchers. Worse, it would be too large to fit in the bomb bay of the B-1A.

Meanwhile, the Navy was working on a cruise missile program of its own called the Sea-Launched Cruise Missile (SLCM) program. This program resulted in the development of the General Dynamics BGM-109 Tomahawk SLCM which had several features in common with the Boeing AGM-86A ALCM. Consequently, in 1977 the Air Force and Navy were ordered to collaborate under the Joint Cruise Missile Project (JCMP) to use as many common parts as possible. By 1978, President Carter had cancelled the B-1A program. The JCMP was no longer constrained by the size of the B-1A bomb bay, and the Air Force elected to incorporate the ERV specifications. A fly-off was conducted that proved both missiles were capable of being launched from a B-52.

Tomahawk ALCM dropping from a B-52G during testing in 1979. (National Archives)

In the end, the Air Force selected the Boeing AGM-86B ALCM incorporating the ERV design but agreed to use the McDonnell-Douglas AN/DPW-23 TERCOM and the Williams F107 engines developed for the Tomahawk. The AGM-86B ALCM was much larger than the AGM-86A. It also featured a more rounded nose compared to the "sharklike" nose on the AGM-86A. The AGM-86A was intended to be carried on the SRAM rotary launcher in the bomb bay, but the AGM-86B was too large to fit on the launcher. Therefore, a total of 12 ALCMs could be carried externally on OAS modified B-52G aircraft with six on each newly developed SUU-67/A ALCM wing pylon and MAU-12D/A ejector rack and none could be carried internally. The aircraft could also still carry SRAMs and nuclear bombs internally. The B-52H can carry 12 externally as well as eight internally, for a total of 20 ALCMs, on aircraft with the A/A48K-1 CSRL installed.

The AGM-86B first flew in June 1979. The first two missiles were delivered to the 416th BMW, a B-52G unit at Griffiss AFB, NY on 11 January 1981 for environmental and maintenance training. The 416th BMW became the first wing fully equipped and operational with ALCM missiles. It received the first OAS modified B-52G with the CMI package capable of launching ALCM missiles on 15 August 1981.

B-52G releasing an AGM-86B ALCM during flight testing near Edwards AFB, CA. (U.S. Air Force)

The Air Force announced in 2007 that it intended to phase out the AGM-86B at the end of its service life in 2020. However, the Air Force ultimately decided to implement a Service Life Extension Program (SLEP) to extend the AGM-86B useful life to at least 2030. It planned to reduce the total inventory of missiles and consolidate them at Minot AFB, ND. The excess missiles would be destroyed to comply with the Strategic Offensive Reductions Treaty (SORT), which required total deployed nuclear weapons to be reduced below 2,200 by 2012. The Air Force plans to replace the AGM-86B ALCM with the Raytheon AGM-181 LRSO missile in the late 2020s.

The Boeing-developed AGM-86B ALCM is 20.75 inches long and 24.5 inches in diameter with a 12-foot wingspan and weighs 3,150 pounds. It is powered by a Williams F107-WR-100 turbofan engine with 600 pounds of thrust and top speed over 550 mph (Mach 0.73). It has a range of over 1,500 miles and carries a W80-1 nuclear warhead and is a megaton class weapon. The ALCM uses the TERCOM guidance system like the one originally tested in the AGM-28C Hound Dog. The TERCOM uses INS, barometric, and radar altimeters to measure elevation above the terrain. The ALCM computer then compares the elevation to preloaded terrain elevation maps and provides position updates to the INS. Together, they provide a highly accurate guidance system.

ACM. The General Dynamics AGM-129A ACM was intended as an improved replacement to the AGM-86B ALCM. The Air Force began studies in 1982 to replace the ALCM with a low-observable stealth missile capable of penetrating future air defense systems. AGM-129A design began in 1983 and resulted in a missile with forward swept wings and reduced radar cross section. It was 20.8 feet long, 2.4 feet in diameter, had a 10.2-foot wingspan, and weighed over 3,700 pounds. It incorporated the Williams International F112-WR-100 turbofan engine with 600 to 730 pounds of thrust. It had a maximum speed of 500 mph and range of 2,148 to 2,304 miles. It had an inertial navigation guidance system with TERCOM and laser doppler velocimeter updates for highly accurate speed updates. It carried a W80-1 nuclear warhead and was a megaton class weapon like the AGM-86B ALCM.

AGM-129A Advanced Cruise Missile in flight. (U.S. Air Force)

The AGM-129A was only employed on B-52H aircraft, which could carry a total of 20 missiles with six on each SUU-72/A wing pylon and eight on the CSRL. It first flew in 1985 and, after various funding and production issues, began operational service in June 1990. A total of 1,460 production units were planned to replace all AGM-86B

ALCMs. However, President Bush reduced the total procurement in January 1992, and this total was eventually capped with final deliveries in 1993. The Air Force decided in 2007 to reduce its total nuclear-capable cruise missile inventory to comply with the SORT. This inventory would consist of only AGM-86B ALCMs. All AGM-129A ACMs were destroyed by April 2012 due to poor reliability and higher maintenance costs.[14]

Conventional Munitions

The B-52 was intended as a nuclear delivery platform and through the early years of its service life all models remained dedicated to that role. The B-52B, C, and E models could carry up to 27 general purpose or cluster bombs internally but were never used in the conventional role. However, beginning during the Vietnam War the B-52D, F, and G models took on the conventional role in addition to their nuclear role. Beginning in

AGM-28 pylon with I-Beam adapter and two MERs capable of carrying 12 weapons on each wing. (U.S. Air Force)

June 1964, 27 B-52F aircraft were modified under project SOUTH BAY to carry twenty-four 500- or 750-pound bombs externally on their wing pylons, along with 27 internally in the bomb bay, for a total of 51 bombs.

The SOUTH BAY modification included an "I-Beam Adapter" that could be mounted to the existing AGM-28 Hound Dog wing pylons. Two Multiple Ejector Racks (MER), capable of carrying six weapons each, were then mounted to each adapter to control the release of the conventional weapons. One year later, another 46 aircraft were modified under project SUN BATH. These modifications nearly doubled the B-52F conventional bombload to 38,250 pounds. Consequently, 27 B-52F aircraft from the 7th BMW at Carswell AFB, TX and the 320th BMW at Mather AFB, CA were the first to deploy to Guam for service over SEA. The B-52F quickly proved its versatility as a reliable weapon.

Meanwhile, SAC began implementing a new modification called "BIG BELLY" on its B-52D fleet in December 1965. These modified B-52D aircraft were intended to replace the B-52F as the primary conventional bomber for SEA operations. The B-52D fleet had over twice the number of aircraft as the B-52F fleet, and after the modification, could carry twice the bombload.

Contrary to popular belief, the "BIG BELLY" modification did not increase the size of the bomb bay. Instead, it strengthened the bomb bay structure and modified the bomb rack attachment points to allow carriage of three newly developed high density bomb racks called "clips" that could hold twenty-eight 500-pound or fourteen 750-pound bombs each. These clips increased efficiency and allowed loading in one-third the time previously required.

The "BIG BELLY" modification also incorporated the AGM-28 wing pylons with I-Beam Adapters and MERs for conventional weapons like the B-52F. The AGM-28 pylons were eventually supplemented with a new "Stub Pylon" to allow the limited number of AGM-28 pylons to support the nuclear alert mission. The Stub Pylons could accommodate the I-Beam adapters used with the AGM-28 pylons, as well as a new Heavy Stores Adapter Beam (HSAB). The HSABs used nine MAU-12 adapters in lieu of the two MERs to carry larger and heavier weapons but were consequently limited to nine on each wing.

Total bombload increased to 60,000 pounds after the modification consisting of eighty-four 500-pound bombs in the bomb bay and twenty-four 750-pound bombs on the wing pylons. The bomb bay could also accommodate forty-two 750-pound bombs in addition to twenty-four 500-pound or 750-pound bombs on the wing pylons. Alternatively, the B-52D could accommodate two Hayes Industries SUU-24 cluster bomb dispensers in the bomb bay. Each dispenser consisted of 24 vertical cells which could each hold three bomblet adapter units. They typically carried a total of 10,656 BLU-3 "Pineapple" bomblets.

The modified B-52D bombers, now adorning the infamous SEA camouflage paint scheme, began replacing B-52F aircraft for Operation ARC LIGHT missions beginning in April 1966. These operations required significant numbers of aircraft to support the planned operational missions. The "BIG BELLY" modifications were completed on 13

Stub pylon with HSAB and nine MAU-12 adapters capable of carrying nine weapons on each wing. (U.S. Air Force)

September 1967. These aircraft were also upgraded with all-weather bombing capability. The D model became the primary bomber used in SEA through most of the war. However, B-52G aircraft were finally deployed in 1972 to support Operation LINEBACKER I and LINEBACKER II. While the D models were deployed to both Guam and U-Tapao Royal Thai Navy Airfield (RTNAF), Thailand, the B-52G aircraft were deployed to Guam only and were not modified to carry weapons externally. They were limited to carrying twenty-seven 500- or 750-pound bombs internally. The B-52G could also accommodate one Hayes Industries SUU-24 cluster bomb dispenser with a total of 5,328 bomblets in the bomb bay.[15]

As the B-52D fleet was retired in the early to mid-1980s, about 46 B-52G aircraft were removed from the CMI modification schedule and set aside to support the conventional mission. These aircraft were modified with the Integrated Conventional Stores Management System (ICSMS) and the same external weapons capability as the B-52F and D models using the AGM-28 pylon or Stub Pylon and the I-Beam Adapter with two MERs to carry 12 weapons on each wing.[16] The G model also used the Stub Pylon with the HSAB limited to nine weapons on each wing. While all three pylon configurations could be used on the B-52G and H models, the AGM-28 pylons did not fit well on the new hard points designed for the ALCM pylons. Therefore, only the Stub Pylons with HSABs are now used on the B-52H limiting the external carriage to 18 weapons total.

Most of the bombs dropped in the Vietnam War were standard General-Purpose bombs. These were essentially "dumb bombs" with no corrections applied. However, in the Gulf War B-52G aircraft began dropping cluster bombs using a Tactical Munitions Despenser (TMD). These bombs were highly successful and in the years after the Gulf War B-52H aircraft were equipped with more capable bombs. A new Wind Corrected Munitions Despenser (WCMD) guidance kit was developed to correct the TMD for winds when dropped from high altitudes. The WCMD was loaded with an aimpoint from the B-52H OAS before launch and corrected the TMD for winds and other variants.

B-52H internal weapons loaded using CRL. (U.S. Air Force)

All B-52H aircraft were also modified with the Internal Weapons Bay Upgrade (IWBU). This upgrade modified the bomb bay and the CSRL with a MIL-STD-1760 digital databus giving the aircraft the capability to carry up to eight Joint Direct Attack Missiles (JDAM) and other smart weapons internally. It also reconfigured the external Stub Pylons to carry up to 16 Laser JDAMs (LJDAM).[17] The CSRL was modified and renamed the Conventional Rotary Launcher (CRL) after this modification. These upgrades were conducted under the Smart Weapons Integration Next Generation (SWING) program, which continues to introduce new smart weapons capability.

The B-52H is now capable of carrying the largest array of conventional weapons of any Air Force platform including gravity bombs, sea mines, cluster bombs, and laser guided

Conventional Munitions (Internal/External)								
Weapon Type	Retired	B-52B	B-52C	B-52D [1]	B-52E	B-52F	B-52G [3]	B-52H [4]
Mk-81 Bomb (250 pounds)	1970s			74/24				
Mk-82 Bomb (500 pounds)	N/A			84/24		27/24	27/24	27/18
M117 Bomb (750 pounds)	2015			42/24		27/24	27/24	27/18
Mk-83 Bomb (1,000 pounds)	Unknown			28/12				
Mk-84 Bomb (2,000 pounds)	N/A			8/10 [2]			8/10 [2]	8/10
Mk-36 Mine (500 pounds)	1970s			60/24		27/24	27/24	27/18
Mk-52 Mine (1,000 pounds)	Unknown			18/0		12/0	12/0	12/0
Mk-53 Mine (500 pounds)	Unknown			48/0				
Mk-55 Mine (2,000 pounds)	Unknown			8/12 [2]			8/12 [2]	8/12
Mk-56 Mine (2,000 pounds)	N/A			8/12 [2]			8/12 [2]	8/12
Mk-60 Mine (2,000 pounds)	N/A						8/10 [2]	8/10
Mk-62 Mine (500 pounds)	N/A							27/18
Mk-63 Mine (1,000 pounds)	N/A							0/18
Mk-64 Mine (2,000 pounds)	N/A							0/10
Mk-65 Mine (2,000 pounds)	N/A							0/10
M-36 Cluster (750 pounds)	Unknown			42/24		27/24	27/24	
CBU-24 (750 pounds)	Unknown			42/24		27/24	27/24	
CBU-49 (750 pounds)	Unknown			42/24		27/24	27/24	
CBU-52 (750 pounds)	N/A			42/24		27/24	27/24	27/18
CBU-58 (750 pounds)	N/A			42/24		27/24	27/24	27/18
CBU-71 (750 pounds)	N/A			42/24		27/24	27/24	27/18
CBU-87 (1,000 pounds)	N/A						6/22	6/18
CBU-89 (750 pounds)	N/A						6/24	6/18
CBU-97 (1,000 pounds)	N/A							6/18
CBU-103 (1,000 pounds)	N/A							0/16
CBU-104 (750 pounds)	N/A							0/16
CBU-105 (1,000 pounds)	N/A							0/16
CBU-107 (1,000 pounds)	N/A							0/16
GBU-10 LGB (2,000 pounds)	N/A							0/10
GBU-12 LGB (500 pounds)	N/A							0/10
GBU-28 LGB (5,000 pounds)	N/A							0/4
GBU-31 JDAM (2,000 pounds)	N/A							8/12 [6]
GBU-38 JDAM (500 pounds)	N/A							8/16 [7]
GBU-39 SDB (250 pounds)	N/A							32/0
GBU-54 LJDAM (500 pounds)	N/A							8/16

(1) "BIG BELLY" modified aircraft with AGM-28 pylon or Stub Pylon, I-Beam Adapter, and MERs
(2) Requires Stub Pylon with HSAB installed
(3) Only 18 externally with Stub Pylon and HSAB installed
(4) B-52H now uses only Stub Pylon with HSAB installed
(5) Eight internally (including LJDAM) after IWBU Increment 1.1 and CRL installation
(6) Eight internally (including LJDAM) and 16 externally after IWBU Increment 1.1 and CRL installation

bombs. It can perform strategic attack, Close Air Support (CAS), interdiction, counter-air, maritime surveillance, mine laying, anti-surface, and anti-submarine operations using the appropriate weapons. Total payload is more than 70,000 pounds of various munitions as shown in the above table. Even the GBU-57A/B Massive Ordnance Penetrator, the largest conventional bomb in the Air Force inventory, was tested.

B-52H with Stub Pylon and HSAB testing release of PDU-5B leaflet bombs. (U.S. Air Force)

Mk-81 Gravity Bomb. The Mk-81 is a general-purpose 250-pound conventional bomb. It is 9 inches in diameter and 74 inches long. It weighs 262 pounds and contains 96 pounds of Tritonal, Minol, or Composition H6 high explosive. It is the lightest of all Mach 80 series bombs and was developed in the 1950s. It was first used in the Vietnam War dropped from B-52D bombers and tactical aircraft. However, it was determined ineffective and quickly discontinued from U.S. service. Licensed versions remain in service with other nations. It can be dropped free-fall or retarded with high drag "Snake Eye" tail fins for low level delivery. The B-52D "BIG BELLY" modified aircraft could carry a total of 98 with 74 in the bomb bay and 12 on each wing pylon.

Mk-82 Gravity Bomb. The Mk-82 is a general-purpose 500-pound conventional bomb. It is 10.75 inches in diameter and 66.15 inches long. It weighs 510 to 570 pounds and contains 192 pounds of Tritonal high explosive. It can also be filled with PBXN-109 thermally insensitive explosive and is designated BLU-111 in this configuration. It can be dropped free-fall or retarded with high drag "Snake Eye" tail fins for low level delivery. It was developed in the late 1940s and is one of the most used conventional bombs in the world. Thousands were dropped from multiple aircraft types, including the B-52, in

Vietnam, Iraq, and Afghanistan. It remained in service as of 2025. The B-52D "BIG BELLY" modified aircraft could carry a total of 108 with 84 in the bomb bay and 12 on each wing pylon. A total of 51 Mk-82 could be carried on the B-52F and G models using the AGM-28 Pylon and I-Beam Adapter with 27 in the bomb bay and 12 on each wing pylon. A total of 45 Mk-82 could be carried on the B-52G and H models using the Stub Pylon and HSAB with 27 in the bomb bay and nine on each wing pylon. The Mk-82 is the core warhead for the Guided Bomb Unit (GBU)-12 PAVEWAY II Laser-Guided Bomb (LGB), Mk-62 QUICKSTRIKE Sea Mines, and the GBU-38 JDAM.

M117 Gravity Bomb. The M117 was a general-purpose 750-pound conventional and demolition bomb. It was 16.1 inches in diameter and 89.43 inches long. It weighed 823 pounds and contained 386 pounds of Tritonal high explosive. It could be dropped free-fall or retarded with MAU-19A/B high drag tail fins for low-level delivery. It used a delay fuze that allowed the weapon to penetrate buildings or structures before detonation. It was developed during the Korean War and was carried by several fighter-bombers, attack, and bomber aircraft. Thousands were dropped from the B-52D, F, and G in Vietnam and from the B-52G in Iraq. The M117 was phased out of service and the last one in PACAF inventory was dropped from a B-52H during a practice bomb run on an island near Guam on 26 June 2015. The B-52D "BIG BELLY" modified aircraft could carry a total of 66 with 42 in the bomb bay and 12 on each wing pylon. A total of 51 M117 could be carried on the B-52F and G models using the AGM-28 Pylon and I-Beam Adapter with 27 in the bomb bay and 12 on each wing pylon. A total of 45 M117 could be carried on the B-52G and H models using the Stub Pylon and HSAB with 27 in the bomb bay and nine on each wing pylon.

Mk-83 Gravity Bomb. The Mk-83 is a general-purpose 1,000-pound conventional bomb. It replaced the Mk-65 that was previously used as the standard 1,000-pound bomb. It remained in service as of 2025 primarily used by the U.S. Navy. The B-52D "BIG BELLY" modified aircraft could carry a total of 40 with 28 in the bomb bay and six on each wing pylon. The Mk-83 is the core warhead for the GBU-16 PAVEWAY II LGB, Mk-65 QUICKSTRIKE Sea Mines, and the GBU-32 JDAM.

Mk-84 Gravity Bomb. The Mk-84 is a general-purpose 2,000-pound conventional hard-target penetrating bomb. It was designed in 1970 and was introduced during the Vietnam War. It became a commonly used heavy unguided bomb. It can be dropped free-fall or with retarding devices for precision guided delivery. A total of 18 Mk-84 could be carried on the B-52D "BIG BELLY" modified aircraft, B-52G and H models with eight in the bomb bay and five on each wing using the Stub Pylon and HSAB. The Mk-84 is the core warhead for the GBU-10/24/27 PAVEWAY LGB, GBU-15 Electro-Optical Bomb, Mk-65 QUICKSTRIKE Sea Mines, and the GBU-31 JDAM.

Mk-36 Sea Mine. The Mk-36 Sea Mine was a 500-pound air launched mine based on the Mk-82 gravity bomb and fitted with the Mk-75 Destructor (DST). Development began in 1966 and ended in 1970. It was widely used during the Vietnam War but was compromised when the North Vietnamese were given the fusing details to disable the mines at the end of the war. They were replaced by the new QUICKSTRIKE mines in the late 1970s. The B-52D "BIG BELLY" modified aircraft could carry a total of 84 with 60 in the bomb bay and 12 on each wing pylon. A total of 51 could be carried on the B-52F and G models using the AGM-28 Pylon and I-Beam Adapter with 27 in the bomb bay and 12 on each wing pylon. A total of 45 could be carried on the B-52G and H models using the Stub Pylon and HSAB with 27 in the bomb bay and nine on each wing pylon.

Mk-52 Sea Mine. The Mk-52 Sea Mine was a 1,000-pound air launched mine. It weighed about 1,200 pounds. A total of 18 Mk-52 could be carried internally on the B-52D "BIG BELLY" modified aircraft with none on the wing pylons. B-52F, G, and H models could carry 12 internally.

Mk-53 Sea Mine. The Mk-53 Sea Mine was a 500-pound air launched mine. It was designed as an anti-sweep device to hinder sweeping operations by damaging the sweep gear when actuated. A total of 48 Mk-53 could be carried internally on the B-52D "BIG BELLY" modified aircraft with none on the wing pylons.

Mk-55 Sea Mine. The Mk-55 Sea Mine was a 2,000-pound air launched mine. It weighed an average of 2,105 pounds. A total of 20 Mk-55 could be carried on the B-52D "BIG BELLY" modified aircraft, as well as B-52G and H models, with eight internally and six on each wing with Stub Pylons and HSABs installed.

Mk-56 Sea Mine. The Mk-56 Sea Mine was a 2,000-pound air launched mine first deployed in 1966. It weighed 2,059 pounds. A total of 20 Mk-56 could be carried on the B-52D "BIG BELLY" modified aircraft, as well as B-52G and H models, with eight in the bomb bay and six on each wing using the Stub Pylon and HSAB.

Mk-60 Sea Mine. The Mk-60 is an Encapsulated Torpedo (CAPTOR) mine that uses a Mk-46 torpedo with an aluminum case. It can be launched by aircraft, surface ships, or submarines. It was first deployed in 1979. A total of 18 Mk-60 could be carried on the B-52G and H models with eight in the bomb bay and five on each wing using the Stub Pylon and HSAB.

Mk-62 Sea Mine. The Mk-62 QUICKSTRIKE Sea Mine is a general-purpose Mk-82 500-pound conventional bomb with one of three available sensor packages attached that detonate the mine. A total of 45 Mk-62 can be carried on the B-52H models with 27 in the bomb bay and nine on each wing using the Stub Pylon and HSAB.

Mk-63 Sea Mine. The Mk-63 QUICKSTRIKE Sea Mine is a general-purpose Mk-83

Mk-60 CAPTOR mine being readied for loading on a B-52G in November 1989. (U.S. Air Force)

1,000-pound conventional bomb with one of three available sensor packages attached that detonate the mine. A total of 18 Mk-63 can be carried on the B-52H with nine on each wing using the Stub Pylon and HSAB.

Mk-64 Sea Mine. The Mk-64 QUICKSTRIKE Sea Mine is a 2,000-pound DST mine first deployed during the Vietnam War. A total of 10 Mk-64 can be carried on the B-52H with five on each wing using the Stub Pylon and HSAB.

Mk-65 Sea Mine. The Mk-65 QUICKSTRIKE Sea Mine is a new development 2,000-pound mine first deployed in 1983. It weighs 2,390 pounds. A total of 10 Mk-65 can be carried on the B-52H with five on each wing using the Stub Pylon and HSAB.

M-36 Cluster Bomb. The M-36 Cluster Bomb was filled with 182 incendiary bomblets and fitted with a burster, fin assembly, two fuses, and an arming wire. The B-52D "BIG BELLY" modified aircraft could carry a total of 66 with 42 in the bomb bay and 12 on each wing pylon. A total of 51 CBU-52 could be carried on the B-52F and G models using the AGM-28 Pylon and I-Beam Adapter with 27 in the bomb bay and 12 on each wing pylon.

CBU-24 Cluster Bomb. The Cluster Bomb Unit (CBU)-24 was an unguided anti-personnel and anti-material weapon. The B-52D "BIG BELLY" modified aircraft could carry a total of 66 with 42 in the bomb bay and 12 on each wing pylon. A total of 51 CBU-52 could be carried on the B-52F and G models using the AGM-28 Pylon and I-Beam Adapter with 27 in the bomb bay and 12 on each wing pylon.

CBU-49 Cluster Bomb. The CBU-49 was an unguided 750-pound cluster bomb. The B-52D "BIG BELLY" modified aircraft could carry a total of 66 with 42 in the bomb bay and 12 on each wing pylon. A total of 51 CBU-52 could be carried on the B-52F and G models using the AGM-28 Pylon and I-Beam Adapter with 27 in the bomb bay and 12 on each wing pylon.

CBU-52 Cluster Bomb. The CBU-52 is a cluster bomb designed to be used against personnel and light-skinned vehicles. It weighs about 750 pounds (766 pounds). The B-52D "BIG BELLY" modified aircraft could carry a total of 66 with 42 in the bomb bay and 12 on each wing pylon. A total of 51 CBU-52 could be carried on the B-52F and G models using the AGM-28 Pylon and I-Beam Adapter with 27 in the bomb bay and 12 on each wing pylon. A total of 45 could be carried on the B-52G and H models using the Stub Pylon and HSAB with 27 in the bomb bay and nine on each wing pylon.

CBU-58 Cluster Bomb. The CBU-58 is a cluster bomb designed to be used against personnel and light-skinned vehicles. It weighs about 750 pounds (800 pounds). The B-52D "BIG BELLY" modified aircraft could carry a total of 66 with 42 in the bomb bay and 12 on each wing pylon. A total of 51 CBU-58 could be carried on the B-52F and G models using the AGM-28 Pylon and I-Beam Adapter with 27 in the bomb bay and 12 on each wing pylon. A total of 45 could be carried on the B-52G and H models using the Stub Pylon and HSAB with 27 in the bomb bay and nine on each wing pylon.

CBU-71 Cluster Bomb. The CBU-71 is an unguided cluster bomb. It holds 650 BLU-68 incendiary submunitions. The B-52D "BIG BELLY" modified aircraft could carry a total of 66 with 42 in the bomb bay and 12 on each wing pylon. A total of 51 CBU-71 could be carried on the B-52F and G models using the AGM-28 Pylon and I-Beam Adapter with 27 in the bomb bay and 12 on each wing pylon. A total of 45 could be carried on the B-52G and H models using the Stub Pylon and HSAB with 27 in the bomb bay and nine on each wing pylon.

CBU-87/103 Cluster Bomb. The CBU-87 is a cluster bomb designed to be used against armor, personnel, and soft targets. It consists of a TMD that contains 202 BLU-97 bomblets. It spins when dropped and releases several bomblets, or clusters, which cover large or small areas depending on the size selected. It is 16 inches in diameter and 91 inches long. It weighs about 1,000 pounds (951 pounds). It can be dropped free-fall and is aimed by the aircraft before release. It was introduced in 1986 to replace Vietnam era cluster bombs. It can also be used with the WCMD guidance tail kit for precision guidance and is designated CBU-103 in that configuration. A total of 28 CBU-87 could be carried on the B-52G using the AGM-28 Pylon and I-Beam Adapter with six in the bomb bay and 11 on each wing pylon. A total of 24 CBU-87 could be carried on the B-52G and H models using the Stub Pylon and HSAB with six in the bomb bay and nine on each wing pylon. The CBU-103 load is limited to 16 weapons with eight on each wing pylon and none internally.

CBU-89/104 Cluster Bomb. The CBU-89 is a cluster bomb containing 72 antitank and 22 antipersonnel mines. It releases and arms the mines when the dispenser opens after being dropped from the aircraft. The mines can be preset to self-destruct after between four hours and 15 days. It is 16 inches in diameter and 92 inches long. It weighs about 1,000 pounds. It was introduced in the 1980s and was used extensively in the Gulf War. It can also be used with the WCMD guidance tail kit for precision guidance and is designated CBU-104 in that configuration. A total of 30 CBU-89 could be carried on the B-52G using the AGM-28 Pylon and I-Beam Adapter with six in the bomb bay and 12 on each wing pylon. A total of 24 CBU-89 could be carried on the B-52G and H models using the Stub Pylon and HSAB with six in the bomb bay and nine on each wing pylon. The CBU-104 load is limited to 16 weapons with eight on each wing pylon and none internally.

CBU-97/105 Cluster Bomb. The CBU-97 is a 1,000-pound amour piercing cluster bomb designed to be used against tanks, armored personnel carriers, trucks, and other vehicles. It releases ten BLU-108 submunitions, or clusters, that each contain four hockey-puck shaped sensor-fused projectiles called skeets. The skeets scan a 1,500 by 500 feet area using infrared and laser sensors and fire a penetrator to destroy any target they find. The CBU-97 is 15.6 inches in diameter and 92 inches long. It weighs 927 pounds. It was introduced in 1992 and was first used in combat in Iraq in 2003. It can also be used with the WCMD guidance tail kit for precision guidance and is designated CBU-105/B Sensor Fused Weapon (SFW) in that configuration. A total of 24 CBU-97 can be carried on the B-52H using the Stub Pylon and HSAB with six in the bomb bay and nine on each wing pylon. The CBU-105 load is limited to 16 weapons with eight on each wing pylon and none internally.

CBU-107 Cluster Bomb. The CBU-107 is a 1,000-pound Passive Attack Weapon (PAW) designed to be used against targets where explosive effects may be undesirable such as fuel storage tanks and chemical weapons stockpiles. It uses a WCMD guidance tail kit for precision guidance. It consists of a SUU-66 TMD containing 3,750 penetrating rods that are released at a high enough altitude terminal velocity. The inner chamber begins to rotate before impact, and the projectiles are released in rapid succession by centrifugal force. The CBU-107 is 15.6 inches in diameter and 92 inches long. It weighs

927 pounds. It first flew in 2002 and achieved IOC in December 2002. It was first used in combat in Iraq on 28 March 2003. A total of 16 CBU-107 can be carried on the B-52H using the Stub Pylon and HSAB with eight on each wing pylon.

GBU-10 LGB. The GBU-10 is laser-guided version of the 2,000-pound Mk-84 general purpose bomb. It also uses the BLU-109 penetrating warhead to attack bunkers, aircraft shelters, and reinforced concrete structures. The GBU-10 is controlled using the AN/AAQ-28 Litening II Targeting Pod laser target designator, AN/AAQ-33 Sniper Advanced Targeting Pod, or a laser designator from another aircraft or ground personnel that illuminates the target. Enhanced versions use INS and GPS to augment the laser guidance. The bomb steers itself toward the illuminated target using the MXU-651 fin assembly kit including adjustable tail fins and forward canards, or control surfaces, coupled with a PAVEWAY II Computer Control Group. It is highly accurate, but its accuracy can be easily compromised by rain, dust, smoke, and cloud cover. The GBU-10 is 18 inches in diameter and 14.8 feet long with a 5.5 feet span. It first flew in 2002 and achieved IOC in December 2002. A total of 10 GBU-10 can be carried externally on the B-52H.

GBU-12 LGB. The GBU-12 is laser-guided version of the 500-pound Mk-82 general purpose bomb. The GBU-12 is controlled using a laser target designator that illuminates the target like the GBU-10 LGB. Enhanced versions use INS and GPS to augment the laser guidance. It is modified with an MXU-650 fin assembly kit and PAVEWAY II Computer Control Group. The GBU-12 is 11 to 18 inches in diameter and 10.8 feet long with a 4.4 feet span. A total of 10 GBU-12 can be carried externally on the B-52H.

GBU-28 LGB. The GBU-28 is PAVEWAY III laser-guided version of the 4,700-pound BLU-113 bunker buster bomb. The GBU-28 is controlled using a laser target designator that illuminates the target like the GBU-10 LGB. The PAVEWAY III enables greater accuracy over the PAVEWAY II. It can be dropped from all altitudes and is effective against high-value targets. Enhanced GBU-28B versions use INS and GPS to augment the laser guidance for all weather targeting. GBU-28C versions use a more powerful BLU-122 warhead. The GBU-28 is 15 inches in diameter and approximately 20 feet long. Two can be carried on each HSAB of the B-52H.

GBU-31 JDAM. The GBU-31 is a guided version of the 2,000-pound Mk-84 general purpose bomb. The Mk-84 is modified with a Guidance Kit that converts it to a precision guided munition. The Guidance Kit includes a Guidance and Control Unit (GCU) which is mounted in the tail assembly. The GCU uses both INS and GPS to significantly increase precision and accuracy of the bomb. It can be pre-programmed with targeting information or updated in flight using the aircraft datalink. A laser kit can be easily installed in the field converting the weapon to a LJDAM and giving it the flexibility to hit targets of opportunity including mobile and maritime targets. Alternatively, the BLU-109/B 2,000-pound penetrating bomb can be used as the GBU-31 warhead allowing the bomb to penetrate buried and fortified targets. The first JDAM test on a B-52H (60-0050) occurred on 30 April 1997 at the U.S. Navy China Lake range. It was dropped from 20,000 feet at Mach 0.8. The GBU-31 is 25 inches in diameter and approximately 12 feet long. The JDAM achieved IOC on the B-52H on 5 December 2000, the first aircraft to do so.[18] Six GBU-31 JDAMs can be carried on each wing pylon of the B-52H. Eight can also be carried in the bomb bay after IWBU Increment 1.2 and CRL installation.

GBU-38 JDAM. The GBU-38 is a guided version of the 500-pound Mk-82 general purpose bomb. The Mk-82 is modified with the same guidance kit used for the GBU-31

B-52H carrying a load of JDAM smart bombs. (U.S. Air Force)

JDAM that converts it to a precision guided munition. A laser kit can be easily installed in the field converting the weapon to a LJDAM and giving it the flexibility to hit targets of opportunity including mobile and maritime targets. The GBU-38 is 14 inches in diameter and 7.8 feet long. Eight GBU-38 JDAMs can be carried on each wing pylon of the

B-52H. Eight can also be carried in the bomb bay after IWBU Increment 1.2 and CRL installation.

GBU-39 SDB. The GBU-39 SDB is a miniature JDAM capable of penetrating more than three feet of reinforced concrete. It is 70.8 inches long, 7.5 inches in diameter, and weighs 285 pounds with a range of about 69 miles. The BRU-61A Smart Bomb Rack holds four BGU-39 SDBs each. It first flew on 23 May 2003 and reached IOC on 2 October 2006. Eight BRU-61A bomb racks, for a total of 32 SDBs, can be carried in the B-52H bomb bay on the CSRL.

GBU-54 LJDAM. The LJDAM is a GPS/INS guided, autonomous, all-weather attack weapon for use against fixed and moving targets. It is a joint USAF-Navy development that combines a laser guidance kit with the GPS/INS-based navigation of the existing GBU-38 JDAM. The current LJDAM is a dual-mode 500-lb guided weapon capable of attacking moving targets with precision. It was developed as an urgent operational need, and testing was completed in less than 17 months. It was first delivered in May 2008 and deployed in combat in Iraq three months later. A total of 24 can be carried on the B-52H with eight in the bomb bay and eight on each wing pylon.

Conventional Stand-off

The original B-52 air-launched weapon was the AGM-28 Hound Dog nuclear capable missile. By the 1970s, the Hound Dog was beyond its operational service and obsolete.

B-52H with display of potential weapons. (U.S. Air Force)

The Air Force began developing more modern and capable missiles with the AGM-69A SRAM being deployed in the late 1970s. Development of nuclear capable stand-off weapons for the B-52 continued in the 1980s with the deployment of the AGM-86B ALCM. Meanwhile, the B-52's value as a conventional stand-off weapon launch platform was being realized. In the mid-1980s the Air Force began equipping 30 ICSMS modified B-52G aircraft with AGM-84D Harpoon missiles in cooperation with the Navy. Another eight ICSMS modified B-52G aircraft were modified to carry the AGM-142A HAVE NAP air-to-surface stand-off missile. By 1991, the CALCM had completed testing and was launched from B-52G aircraft in the opening salvo of Operation DESERT STORM.

Two other conventional missile programs were planned for the B-52 in the 1980s. The Northrop AGM-136A TACIT RAINBOW was a low-cost missile designed to autonomously locate and neutralize enemy air defense radars. It would fly along a programmed flight path until it detected a radar that matched its target data files and then destroy the radar. The B-52G would have carried 30 missiles in the bomb bay. It first flew from a B-52H (60-0050) on 10 January 1989. However, it was cancelled in 1992 with the end of the Cold War. The Northrop AGM-137A Tri-Service Surface Attack Missile (TSSAM) was intended to provide the B-52G with a low-observable strike weapon. It would have been used in conjunction with the TACIT RAINBOW to destroy enemy targets. The B-52G would have carried 12 missiles with six on each wing pylon. It too was cancelled due to cost and technical problems in 1994.[19]

By the end of 1994 SAC was long since deactivated and all B-52G aircraft were retired from service to comply with force reductions and arms control treaty limitations. ACC elected to keep the B-52H in service for as long as possible and decided to convert some of these aircraft to the conventional role. Four B-52H (60-0013, 61-0013, 61-0019, and 61-0024) aircraft were immediately modified to carry the AGM-84D Harpoon missiles and another four (60-0014, 60-0025, 60-0062, and 61-004) received AGM-142A HAVE NAP capability under the RAPID EIGHT program.

By the end of 1999, 66 B-52H aircraft had received the new Conventional Enhancement Modification (CEM) including the stub pylons and HSAB used on the B-52G with the ability to carry up to nine weapons on each wing. All CEM upgraded aircraft were equipped to carry Harpoon missiles and eight aircraft were given AGM-142 HAVE NAP II capability. Perhaps the most important aspect of the CEM upgrade that eventually allowed the B-52H to carry more different weapon types than any other Air Force platform, was the introduction of Block II OAS software incorporating SMO modules. This improvement allowed new weapons to be quickly added to the B-52H arsenal without regression testing the entire OAS. The CEM upgrades also included integrating GPS into the OAS.

In 2012, as the Air Force continued to comply with SORT obligations, the B-52H fleet was reduced to 76 operational aircraft of which only 46 were allowed to be nuclear capable. The Air Force was thus forced to decide if it would retire the remaining aircraft or convert them to a conventional only role. In 2015, Air Force Global Strike Command (AFGSC), now in control of the Air Force bomber force, announced that 30 active aircraft craft and 11 aircraft then in storage would be converted to a conventional-only con-

Conventional Stand-Off (Internal/External)								
Weapon Type	Retired	B-52B	B-52C	B-52D	B-52E	B-52F	B-52G	B-52H
AGM-84D Harpoon	N/A						0/8 [1]	0/8 [1]
AGM-86C/D CALCM	2019						0/12	8/12 [2]
AGM-142A HAVE NAP	N/A						0/4 [3]	0/4 [3]
AGM-154A JSOW	2004							0/12 [4]
AGM-158A JASSM	N/A							8/12 [5]
AGM-158B JASSM-ER	N/A							8/12 [6]
ADM-160B/C MALD/MALD-J	N/A							8/16 [7]

(1) Typically carried four on each wing pylon although six could be accommodated
(2) Eight internally after CSRL installation
(3) Two on each wing pylon or two on one pylon and one on the other with datalink pod
(4) Integrated on B-52H but not used in favor of AGM-158 JASSM
(5) Eight internally with IWBU Increment 1.2 and CRL installation
(6) Eight internally (increased from four) and 12 externally with IWBU Increment 1.2 and CRL installation
(7) Eight internally with IWBU Increment 1.2 and CRL installation

figuration. B-52H (61-0021) assigned to the Air Force Reserve Command's (AFRC) 307th BW at Barksdale AFB, LA was the first aircraft converted in September 2015.

As with nuclear weapons, the A/A48K-1 CSRL was key to expanding the B-52H internal weapons load capacity. The CSRL allowed the B-52H to carry missiles and other conventional ordnance internally. For example, the CSRL gave the B-52 the ability to carry a total of 20 AGM-86C CALCMs with eight internally and 12 externally. However, the CSRL could not accommodate the new smart weapons fielded after the CEM upgrade. These weapons, such as the JDAM, were carried on the wing pods only.

The IWBU, which achieved IOC in March 2016, converted 44 CSRLs to the new CRL configuration. The CRL allowed internal carriage of up to eight JDAMs, LJDAMs, Joint Air-to-Surface Standoff Missiles (JASSM), JASSM-Extended Range (JASSM-ER), Miniature Air-Launched Decoys (MALD), or MALD-Jammer (MALD-J). Although the modification now allowed all weapons to be carried internally to reduce fuel costs and conceal the weapons when desired, it also added capability to carry up to 16 LJDAMs or 12 JASSM-ERs externally.

Harpoon. The McDonnell-Douglas AGM-84 Harpoon missile is an anti-ship weapon developed for the U.S. Navy. It can be launched from land, ships, submarines, and various types of aircraft including the P-3, P-8, AV-8, F-111, F/A-18, and B-52. It flies at low-level "sea-skimming" altitude using active radar guidance and employed a WDU-18/B blast fragmentation target penetrating warhead with 488 pounds of high explosive to ensure survivability and effectiveness. The Harpoon was originally developed in the early 1970s but was upgraded several times. The latest development for anti-ship missions, designated AGM-84D Block 1C, was developed in 1982 and was operational with the U.S. Navy and Air Force as well as more than 30 allied nations.

The Soviet Union had expanded its Naval fleet to 1,764 active vessels by the late 1970s and deployed long-range anti-ship missiles on its land-based bomber fleet. In response the U.S. Air Force began training crews in ocean surveillance, maritime strike, and aerial mine laying operations in cooperation with the Navy. SAC also began equipping B-52G aircraft with AGM-84D Harpoon anti-ship missiles in the early 1980s. One B-52G (58-0202) from the 320th BMW at Mather AFB, CA was modified with the AN/AWG-19 Har-

B-52G (58-0202) takes from Mather AFB, CA for Harpoon missile tests. (National Archives)

poon Launch Control System (HALCS) and missiles in March 1983. After three successful live launches at Point Mugu, CA from 15-28 March 1983, a total of 30 ICSMS-equipped conventional B-52G aircraft were modified to carry the HALCS and missiles by 30 June 1985.[20] These aircraft were assigned to the 42nd BMW at Loring AFB, ME and the 43rd SW at Andersen AFB, Guam as the only operational Harpoon capable wings. This mission was ultimately transferred to B-52H aircraft from the 2nd BMW at

Barksdale, LA in March 1994 when the B-52G fleet was retired. The mission was also shared by the 5th BMW at Minot AFB, ND. Initially, four B-52H aircraft were modified with the HALCS. The launch control system was subsequently integrated into the B-52H OAS and the HALCS was eliminated allowing the entire H model fleet to carry Harpoons by 1997.

The AGM-84D Harpoon is 12.6 feet long and 13.5 inches in diameter. It has a three-foot wingspan and weighs 1,145 pounds. The Harpoon is powered by a Teledyne CAE J402 turbojet engine with 660 pounds of thrust and has a maximum speed of 537 mph. Operational range is about 86 miles. It flies at sea-skimming altitude using a GPS/INS

B-52G (58-0202) flew three successful AGM-84D Harpoon test flights at Point Mugu, CA in 1983 proving the capability for B-52G aircraft. (U.S. Air Force)

guidance system monitored by an onboard radar altimeter and active radar terminal homing allowing it to find and track its target autonomously. Production began in 1977 and a total of 7,000 Harpoons in various configurations were delivered by 2004. A total of 12 Harpoons could be carried on HALCS modified B-52G and all B-52H with six on each wing pylon. However, they typically carry only four on each wing. Funding was provided in 2019 to begin integrating the AGM-158C Long Range Anti-Ship Missile (LRASM), currently carried by the B-1B Lancer, into the B-52H to replace the Harpoon.

CALCM. A total of 239 AGM-86B ALCMs were converted to replace the nuclear warhead with a 1,000-pound conventional fragmentation warhead beginning in January 1991. Its TERCOM system was tied to the bomber's GPS and the onboard map system, had all hills, valleys, streets, buildings, and rivers in its memory allowing it to literally fly down a designated street until it found the correct address.[21] This version was designated AGM-86C CALCM, Block 0. In 1996-97 a Block 1 version was developed which included a 3,000-pound conventional warhead, and a multi-channel GPS along with upgraded software that increased navigation accuracy. This GPS upgrade was eventually incorporated into all Block 0 CALCMs giving them all the same avionics package. A new eight-channel GPS and precision navigation capability was integrated in 2001 and designated Block 1A. Another 50 AGM-86B ALCMs were converted beginning in November 2001 to replace the nuclear warhead with an AUP-3(M) "Bunker Buster" conventional warhead capable of destroying hardened and deeply buried targets. This version was designated AGM-86D and considered CALCM Block 2. It first flew in November 2002.

The AGM-86C CALCM was first used in combat on 17 January 1991 when seven B-52G aircraft, launched from Barksdale AFB, LA, fired 35 missiles on the opening morning of Operation DESERT STORM. These aircraft were the first aircraft to launch for the

Boeing-designed AGM-86C CALCM in flight. (U.S Air Force)

operation and conducted a "round robin" mission of over 14,000 miles and 35.4 flight hours to attack high-priority targets in Iraq. This was considered the longest bombing mission in history at the time. The missiles also saw service being launched for the first time from a B-52H with the CSRL installed in September 1996 during Operation DESERT STRIKE, as well as Operation DESERT FOX in 1998, Operation ALLIED FORCE in 1999, and the AGM-86D which incorporated a 1,000-pound penetration warhead was first used during Operation IRAQI FREEDOM in 2003.

The CALCM was retired on 20 November 2019, and all remaining AGM-86C/D missiles were in storage at Barksdale AFB, LA awaiting disposal as of 2022. Its mission was replaced by the Lockheed Martin AGM-158B JASSM-ER. This replacement left a significant range gap as the JASSM-ER has a range of only 575 miles compared to the 1,500-mile range of the CALCM. The Air Force planned to close this gap with a new development AGM-158D JASSM-Extreme Range (JASSM-XR) beginning in 2024 if it meets its 1,200-mile range requirement.

HAVE NAP. The AGM-142A HAVE NAP was originally developed by the Israeli company Rafael as the Popeye air-to-surface stand-off missile. It is electro-optically (television) guided and controlled using an AN/ASW-55 datalink pod. The television camera

The HAVE NAP was fielded on eight B-52G aircraft and later fitted on B-52H aircraft. (American Aviation Historical Society)

is used to acquire the target and the datalink locks-on and guides the missile to the target. The Air Force selected the missile for the HAVE NAP program and integration with the B-52G in 1988. It is 15.8-feet long, 21-inches in diameter, has a 5.7-foot wingspan, and weighs 3,000 pounds. It is powered by a solid-fueled rocket engine that produces supersonic speeds and has a range of 50 miles. It carries a 750-pound conventional blast-fragmentation warhead. The HAVE NAP was fielded on eight B-52G aircraft. It was later fitted on the B-52H aircraft after the G models were retired. The HAVE NAP load is configured with two missiles on one wing pylon and one missile and a datalink pod on the other.

JSOW. The AGM-154 JSOW is a long-range, precision-guided glide bomb developed by Raytheon for the U.S. Navy and Air Force. Its primary purpose is to enable aircraft to engage targets from beyond enemy air defenses. The JSOW is a glide bomb that lacks a rocket motor, relying instead on aerodynamic lift and gravity for propulsion. Its modular airframe incorporates pop-out wings and tail fins that deploy after release. It is 160 inches long, has a wingspan of 106 inches, and weighs approximately 1,065 pounds. Its box-shaped body has a 13-inch cross-section. It uses a combination of INS and GPS guidance. The AGM-154C variant also includes a terminal Imaging Infrared (IIR) seeker, which provides enhanced accuracy during the final phase of flight. The JSOW C-1 variant further integrates a Link-16 data link, allowing for real-time target updates during flight. The B-52H can carry 12 JSOWs externally with six on each wing.

JASSM/JASSM-ER. The Lockheed Martin AGM-158A JASSM is a stealthy long range guided cruise missile designed to attack well-defended targets. The AGM-158B JASSM-ER, like its predecessor, is 14 feet long with a 7.8-foot wingspan and weighs 2,251 pounds. It incorporates a 1,000-pound WDU-42/B blast-fragmentation penetrator warhead. However, it has a more efficient Williams International F107-WR-105 turbofan engine and larger fuel capacity than the AGM-158A giving it a range of more than 575 miles. It has a GPS aided INS guidance system with an infrared seeker for infrared homing automatic target recognition. It was initially deployed on Air Force B-1B bombers in 2014 and later integrated on the B-52H, which can carry a total of 20 missiles with 12 external and eight internal with IWBU Increment 1.2 and CRL installation.

MALD/MALD-J. The ADM-160B MALD is a programmable, low cost, modular, autonomous flight vehicle that mimics aircraft to confuse enemy radars and defensive systems. It is a decoy like the ADM-20 Quail using state-of-the-art technologies. The ADM-160B MALD-J jammer version uses radar jamming to degrade or deny the enemy's ability to establish a track or strike an incoming aircraft. The MALD is 9.3 feet long with a 5.6-foot wingspan and weighs less than 300 pounds. It uses a Hamilton Sundstrand TJ-50 turbojet engine with 337 pounds thrust and range up to 575 miles. The first flight of a MALD occurred in 1999 and the MALD-J in 2009. The first MALD-J was delivered on 6 September 2012 and achieved IOC in 2015. The F-16 and B-52H were the lead aircraft for MALD employment. The B-52H can carry a total of 24 decoys with 16 external and eight internal on aircraft incorporating IWBU Increment 1.2 and CRL installation.

Organization and Basing

At the time of the Japanese surrender from World War II, the AAF had 218 operational groups. By January 1946, the United States was rapidly demobilizing its military forces and the AAF was down to 30 percent of its wartime strength. The War Department's postwar planning called for a "bedrock minimum" air force of 70 groups and a personnel strength of 400,000. General Carl A. "Tooey" Spaatz took over as Chief of Staff in February. He held to the objective of 70 air groups and was supported in that position by the Army Chief of Staff, General of the Army Dwight D. Eisenhower. When Spaatz established SAC on 21 March 1946, it inherited a force of approximately 1,300 bomber, fighter, reconnaissance, and support aircraft. The initial SAC force included 100,000 personnel with 22 major installations and 30 minor bases.

However, the 70-group plan evaporated in August 1946 when President Truman ordered "economy commitments" to reduce the budget by $2.2 billion, with 75 percent of that to come from the armed forces. By summer, the inventory of aircraft had been reduced by half, most of them cut up for scrap.[1] At the end of 1946, the AAF had 55 groups remaining, only two of which could be counted as combat ready. "We are not ready to fight a war if one came today—and we won't be for quite a long time," Spaatz said. The Air Force had not yet reached bottom. It would sink to 48 groups and a personnel strength of 304,000 before the buildup for the Korean War began.[2]

SAC was reduced after the demobilization to a force of 279 aircraft including 148 B-29, 85 P-51, 31 F-2, and five C-54. Personnel were reduced to 37,092 with only 18 bases remaining. Its bomber force was organized into nine Bomb Groups (BG), designated Very Heavy (VH), comprised of B-29 bombers. Three of these groups had no aircraft assigned. Each bomb group was an operational combat unit typically comprised of up to three operational squadrons assigned directly to the group headquarters. A base commander, often a non-flying administrator, was the highest echelon of command on the base and controlled all support functions. The base commander was also the immediate superior of the group commander.

Beginning on 17 November 1947, the 7th BMW was formed at Carswell AFB, TX as a test unit for the newly developed "Wing-Base" plan, also known as the "Hobson Plan", in which the wing became the predominant operational unit on each base. The wing commander, a highly experienced aviator, became the operational commander and the combat group reported directly to the wing. The combat group was still comprised of up to three operational squadrons and a group headquarters responsible for all flying activities. The wing commander also controlled three support groups, consisting of maintenance and supply, airdrome, and medical, which gave the wing complete control of all flying and support functions. The base commander now had responsibility only for base housekeeping and administrative functions. The test was successful, and the 7th BMW was formally activated as the first B-36 wing on 1 August 1948. Both the 7th and 11th BG located at Carswell operated as components of the 7th BMW until 1 February 1951 when the 11th BG was reassigned to the newly formed 11th BMW.

This designation of the 11th BMW at Carswell in 1951 was part of the reorganization of SAC's combat command structure, based on a plan devised by General LeMay, using experience gained from Korean War operations. LeMay wanted an organization designed for peacetime that required minimum change to rapidly switch to war footing. Previously, despite having a base commander assigned, the wing commander was still occupied with too many base housekeeping and administrative tasks that distracted from his ability to focus on combat operations. Also, the wing commander relied heavily on the combat group commanders to conduct daily flying operations. With the reorganization, the Air Base Group commander assumed all base housekeeping and administrative duties. Although the combat groups continued to exist, they became subordinates to the wing and the wing commander served as the group commander. This gave full control and focus of combat operations to the wing commander.

Both the 7th and 11th BG located at Carswell operated B-36 aircraft as components of the 7th BMW until 1 February 1951 when the 11th BG was reassigned to the newly formed 11th BMW. (American Aviation Historical Society)

In conjunction with this reorganization, SAC received authority from Headquarters Air Force to organize Air Division (AD) headquarters on double wing bases and to operate only one Air Base Group on these installations. The AD commander, typically a 1- or 2-star general officer, exercised direct control over the two wing commanders and the Air Base Group commander. For example, on 16 February 1951 the 19th AD was established at Carswell to oversee operation of the two B-36 wings assigned to the base.[3] SAC fully implemented this change across the command over the following year. On 16 June 1952, all SAC combat groups were eliminated, and the combat squadrons reported directly to the wing.

All SAC wings and ADs reported directly to a Numbered Air Force (NAF) commander. SAC employed up to three NAFs during its history, the 2nd, 8th, and 15th. Initially, medium and heavy bomber units were assigned to the 8th AF while the 15th AF had mostly medium bombers and 2nd AF concentrated on reconnaissance. This arrangement resulted in an unbalanced organization with wings in the same geographical region being assigned to different NAFs. On 1 April 1950, SAC reorganized the NAFs on a geographical basis with 2nd AF in the eastern, 8th AF in the central, and 15th AF in the western United States.

The B-36 was not capable of aerial refueling but it was capable of flights of over 10,000 miles and could stay aloft for over 51 hours, making it the first true intercontinental

In 1947, SAC began construction of a new base in Limestone, ME built from the ground up to accommodate 60 B-36 bombers. (American Aviation Historical Society)

bomber. Still, aircraft operating from Carswell and other "Southern Tier" bases had to forward deploy to Thule AB, Greenland, Goose Bay, Labrador, and Alaska to allow them to fly over the north pole and hit targets in the Soviet Union. As the B-36 role as a nuclear bomb delivery platform and the growing Cold War with the Soviet Union became more apparent, the need for "Northern Tier" basing became clear. In 1947, SAC began construction of a new base in Limestone, ME built from the ground up to accommodate 60 B-36 bombers.[4] In 1950, SAC established the 28th Strategic Reconnaissance Wing (SRW) at Rapid City AFB, SD. In 1951, SAC established the 92nd BMW and 99th SRW at Fairchild AFB, WA and in 1953 the 42nd BMW at Limestone AFB, ME. SAC bombers could now reach targets in the Soviet Union from its United States bases. These bases also allowed additional forward operating locations for the southern based wings.

SAC also continued to rely on forward operating bases in Alaska, Goose Bay, and Thule to shorten the distance to targets in the Soviet Union. However, these bases presented significant challenges to B-36 operations due to the extreme cold conditions. The B-36

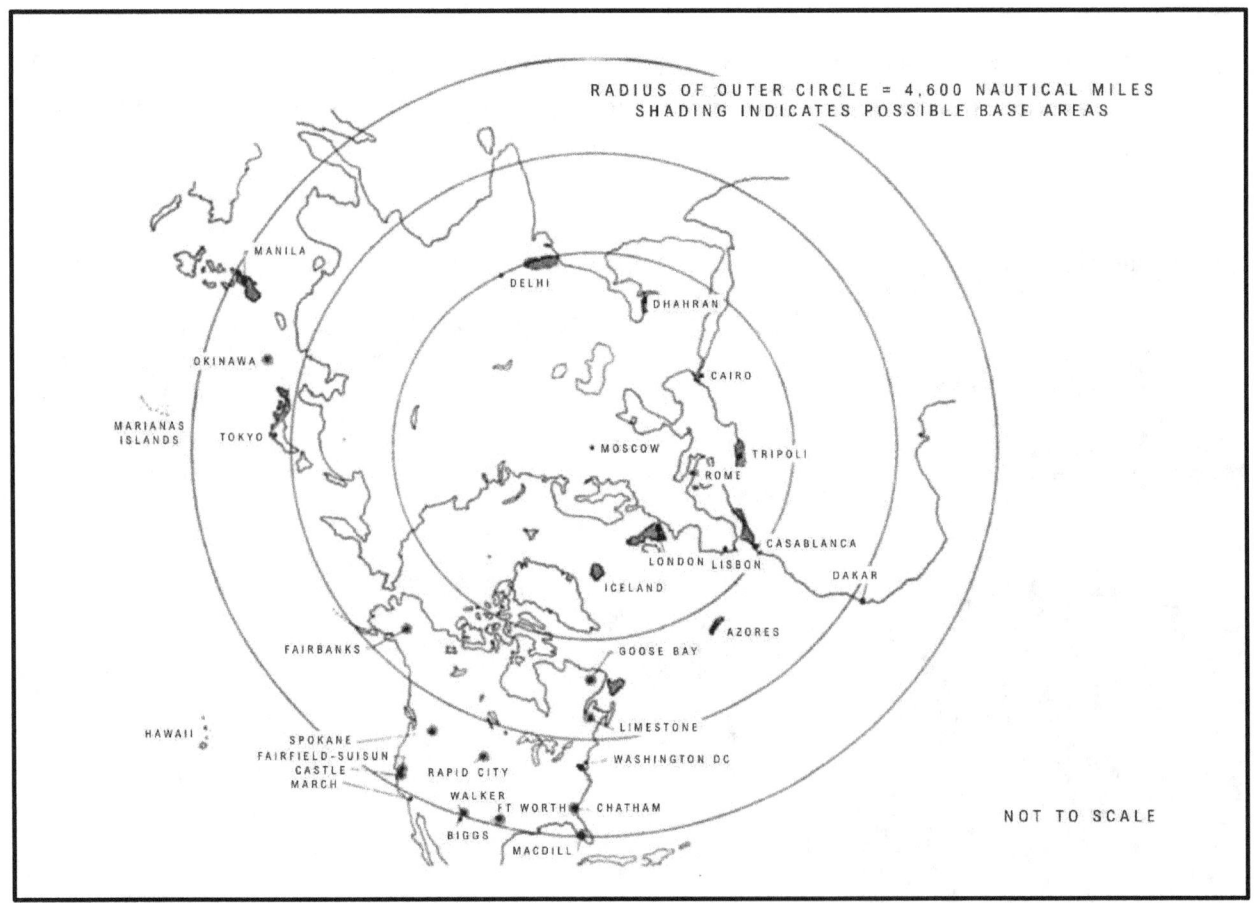

SAC B-36 basing plan allowed southern tier units to forward deploy to Alaska, Labrador, and Greenland to put them within reach of Soviet targets (History of AMC Support to Strategic Air Command 1946-1952).

did not respond well to cold weather and the aircraft had to be preheated before each flight. Furthermore, the facilities at these locations and all SAC northern tier bases were inadequate for the B-36 due to its size.

SAC addressed this issue by equipping their northern tier bases with new "nose dock" hangars which allowed the aircraft, except the tail, to be indoors out of the harsh weather for maintenance. Both Limestone and Rapid City also got new "Arch Hangars" which fully enclosed two B-36 aircraft. These hangars were built in 1947-48 and used a thin-shell concrete design in which huge concrete arches created a huge two and half acre open floor space. SAC intended to build up to 10 of these hangars at Limestone and similar numbers at other Northern Tier bases. However, this plan was downscaled and eventually abandoned due to a shortage of concrete and other materials in the wake of the post-WWII building boom.

Instead, beginning in the early 1950s, SAC settled on a new Double-Cantilever (DC) hangar design capable of fully enclosing, except the tail, two B-36 aircraft in each bay. The DC hangar could be constructed with up to three bays capable of housing five B-36 aircraft in total. Three-bay DC hangars were constructed to accommodate B-36 maintenance at Biggs, Carswell, Limestone, and Walker and at several other bases to accommodate other SAC aircraft. Two-bay DC hangars were built at other SAC bases and supplemented with nose-dock hangars.

(Left) Loring AFB Arch Hangar capable of fully enclosing two B-36 bombers with the DC Hangar capable of holding five partially enclosed B-36 aircraft in the background. (Right) Nose dock hangars built at Limestone AFB, ME and most other SAC bases allowed maintenance crews to work on B-36 Bombers with most of the aircraft sheltered from severe weather. (JC Garbinski)

All of these hangars built for the B-36 would prove very useful when B-52 aircraft replaced the B-36 fleet. For example, the Arch Hangar became Limestone's (renamed Loring AFB) phase dock and housed one B-52 and two KC-135 aircraft during inspections. Nose dock hangars were modified, and the DC hangar continued service to support B-52 aircraft.

SAC also began to develop forward operating locations in warmer climates such as the United Kingdom (UK), Morocco, and Guam. These locations put B-36 aircraft within range of the Soviet Union and were much more operationally suitable. B-36 aircraft flying from Guam were only 6,100 miles from Moscow and 2,200 miles from Vladivostok putting most targets in the Soviet Union in range. All of these locations were converted to host B-52 aircraft when they replaced the B-36.

When former General of the Army Dwight D. Eisenhower became President of the United States in 1953, he assembled his advisors for discussions regarding defense strategy. The U.S. had just finished fighting a war in Korea that ultimately became a stalemate. Most observers believed the Soviets instigated it in their quest to spread Communism across the world. Eisenhower believed the Soviets could provoke more confrontations when and where in the world they wanted. The U.S. was unable to match the Soviets in conventional forces due to its draw down of forces after WWII. The Soviets could draw on at least 175 regular and 125 reserve divisions, compared to only 29 U.S. regular and seven reserve divisions.

Eisenhower soon decided on a "New Look" defense strategy that was outlined in NSC 162/2, dated October 1953. The new strategy relied on nuclear weapons to provide a deterrence against Communist expansion. It was designed to provide a massive retaliatory strike in the event of unprovoked attacks. Eisenhower saw this strategy as the most economical way to deter Communist aggression and protect the U.S. homeland. While the U.S. was considerably behind the Soviets in conventional might, it was significantly ahead in atomic stockpiles with 841 weapons compared to only 120 for the Soviets at the end of 1952.

This new strategy had a significant impact on the Air Force, which was given the primary nuclear delivery role for the U.S. military. The B-36 was in service and capable of carrying considerable nuclear payloads. The B-47 was entering service, and the B-52 was planned to begin operational service in 1955. The Air Force budget was about 47 percent of the DoD appropriation compared to 29 percent for the Navy and 22 percent for the Army. Air Force combat strength grew from 95 wings in the last Truman budget to 137 wings in the first Eisenhower budget. SAC was assigned a total of 92 wings and quickly equipped them with hundreds of B-47 bombers.

In 1955, SAC began converting some B-47 wings to B-52 aircraft. The Air Force Council recommended a B-52 fleet of 576 aircraft, and Secretary Talbott approved between 399 and 576 aircraft depending on available funding. SAC plans called for 11 B-52 wings assuming funding for 576 aircraft would ultimately be approved. SAC's B-52 basing plan followed the same strategy as its B-47 and B-36 wings with 45 bombers[5] and 15-20

93 BW B-52F 57-0161 at Castle AFB on 16 October 1967 by Stephen Miller

The 93rd BMW at Castle AFB, CA became the first operational B-52 wing when it traded-in its B-47 aircraft for B/RB-52B models on 29 June 1955. Castle eventually hosted all B-52 models except the C model. (American Aviation Historical Society)

tankers assigned to one overcrowded base. Many of these wings were planned for Northeastern bases and SAC again reorganized its NAFs on 13 June 1955 to better support bomber operations. Headquarters 8th AF moved from Carswell AFB, TX to Westover AFB, MA. The 2nd AF was now responsible for the southeast (including Texas), 8th AF was responsible for the northeast and central United States, while the 15th AF had the southwest and west.

The 93rd BMW at Castle AFB, CA became the first B-52 wing when it traded in its B-47 aircraft. It received the first operational RB-52B (52-8711) on 29 June 1955. This aircraft and the subsequent RB-52 aircraft were B-52B bombers that could be converted to the reconnaissance mission by uploading a reconnaissance pod in the bomb bay. However, they were used primarily in the bomber role, and most were redesignated B-52B upon assignment to SAC. Deliveries continued through March 1956, and 45 B/RB-52B aircraft were assigned to the 93rd BMW, which became the primary B-52 training wing with activation of the 4017th Combat Crew Training Squadron (CCTS) on 8 January 1955.

Initial problems delayed full operational capability including uncertain aircraft delivery schedules, lack of spare parts, shortages of ground support equipment, dual bomb racks, crew kits, ECM components, and training items. Delayed construction and shortage of operational facilities were serious handicaps. Facilities were upgraded and constructed beginning in 1959, including warehouses, nose docks, hangars and alert facilities, as well as operations and engineering buildings, but funding issues delayed some much-needed work. In addition, the failure of ramps and taxiways together with runway deterioration interfered with operations. In mid-1958, paving projects were started at nine of 13 bases needing immediate attention. Parking ramps and runways were strengthened to handle the 450,000-pound loads expected for later B-52 models and taxiway widths were increased to 175 feet to accommodate the B-52 wingspan and outrigger wheels.[6]

SAC also began converting all B-36 wings and bases to B-52 aircraft in 1956. The 42nd BMW at Loring AFB, ME became the second B-52 wing when it received the first operational B-52C (53-0400) on 16 June 1956. However, phasing out the B-36 proved more complicated than expected since SAC needed to maintain its global nuclear delivery capability. Initial shortages of B-52 aircraft forced the withdrawal of B-36 aircraft from reclamation contracts. B-36 aircraft remaining in service were supported with parts from out of service aircraft until the B-36 could be completely phased out.[7]

By the fall of 1956 SAC began receiving B-52D models. The first ones began arriving at Castle in June-August and at Loring in December. By the end of the year SAC had nearly 100 B-52 aircraft (40 B-52B, 32 B-52C, and 25 B-52D), but combat-ready crews lagged with only 26 in the 93rd BMW and 16 in the 42nd BMW. SAC responded quickly and increased the training throughput at Castle. By 1958, the command had 402 combat-ready crews for 380 aircraft.[8]

By March 1959, over 400 B-52 aircraft had been delivered, and all B-36 aircraft were phased out giving SAC an all-jet bomber force. SAC continued to accept B-52D, E, F, and G models throughout 1958-59 and bed down issues were a constant concern. By the late 1950s, SAC's B-52 basing strategy with 45 aircraft in each wing presented rich targets for more accurate and prevalent Soviet missiles. SAC planners were concerned that the Soviets could take out most of the bomber fleet with just a few missiles. It was also impossible to launch so many aircraft from one runway within a survivable response time. Therefore, SAC began breaking up the large concentrations of aircraft in 1958 to complicate the Soviet targeting problem and increase available runways. Each B-52 wing was divided into three smaller wings of 15 bombers and 10-15 tankers each. These new wings were designated as four-digit identified SWs and relocated to other existing bases, most owned by other commands.

SAC also reorganized its B-47 and B-52 wings beginning 1958 to support its new one-third ground alert program. The first two wings to convert on 1 September were the 4123rd SW, a B-52 unit at Little Rock AFB, AR and the 307th BMW, a B-47 unit at Lincoln AFB, NE. SAC plans dictated that one wing in each NAF would convert each month until complete. The reorganization called for a new Deputy Commander for Operations (DCO) and Deputy Commander for Maintenance (DCM) to assist the Wing Commander to provide combat-ready crews and aircraft to support ground alert requirements. Maintenance was removed from the operational combat squadrons and periodic

SAC continued to accept B-52D, E, F, and G models throughout 1958-59 including B-52G (59-2575) pictured above at Travis AFB, CA on 17 October 1967. (American Aviation Historical Society)

maintenance squadron, and centralized into a newly formed Organizational Maintenance Squadron (OMS). Finally, the Air Base Group was redesignated as the Combat Support Group (CSG) to emphasize its role in supporting the ground alert mission.

In 1959, SAC again realigned its NAF responsibilities. However, it was a relatively minor change moving one unit from the 8th AF to the 15th AF, but it gave the 8th AF responsibility for the Eastern United States, while the 2nd AF gained responsibility for the Central section, and the 15th AF retained the Western section. Along with this change, SAC realigned its ADs and gave them broader responsibilities. Before this change ADs were primarily assigned to double-wing bases and had responsibility of the wings and support groups assigned to the base. After the realignment, which included an increase in the headquarters' staff from 17 to 25 personnel, each AD became responsible for three to five dispersed wings. Initially the realignment applied to existing ADs, but by the end of 1959 six new ADs were activated.

On 15 February 1962, SAC redesignated ADs and wings having two or more aircraft, missiles, or space systems using the new Air Force policy that directed use of "Aerospace" in organization titles. ADs and bomb wings that directed both aircraft and missiles became Strategic Aerospace Wings (SAW). The first redesignations were the 17th AD, 18th AD, 21st AD, and 821st AD that became Strategic Aerospace Divisions, as well as 92nd BMW that became the 92nd SAW.

By 1962, SAC had 639 B-52 aircraft assigned to 11 BMWs (one 45 UE, two 30 UE, and eight 15 UE), three SAWs (one 45 UE and two 15 UE), and 22 SWs (15 UE) with four-digit identifiers. The four-digit SWs were MAJCOM controlled, and their lineage ended when they were inactivated and could never be revived. SAC discontinued these four-digit designations during January to September 1963 and activated Air Force controlled (one, two, or three digit identified) units in their place.

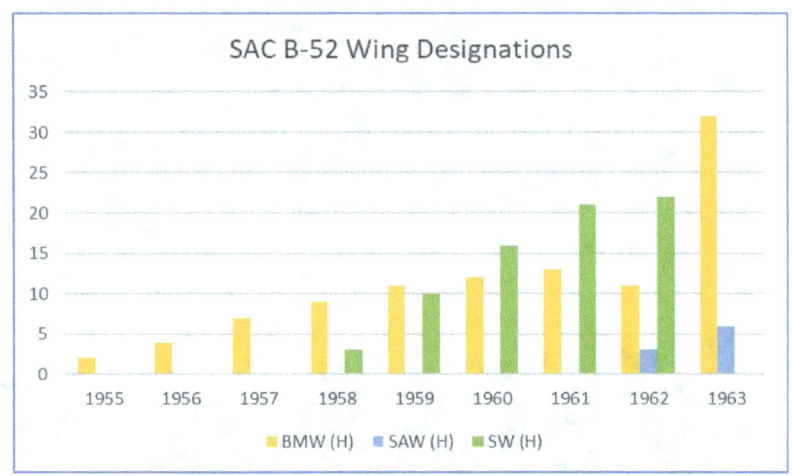

SAC redesignated all dispersed B-52 SWs and BMWs in 1963 to ensure all wings would retain their lineage going forward.

This was intended to ensure the wings would retain their lineage going forward and to perpetuate the lineage of illustrious WWII units no longer active. Air Force headquarters agreed to this organizational change. However, the SWs were inactivated and the change was not considered a redesignation so the SW records, awards, and achievements could not be inherited by the newly activated wings.

On 8 December 1965, SECDEF Robert McNamara announced the planned phase-out of all B-58 and early B-52 (C, D, E, and F) aircraft by end of June 1971. This phase-out would also include base closures and unit relocations. All B-52B aircraft were retired in 1966 and the 22nd BMW at March AFB, CA was converted to B-52D aircraft while the 95th BMW at Biggs AFB, TX was inactivated.

B-52 aircraft began supporting SEA operations from Andersen AFB, Guam in 1965. Normally the aircraft, crews, and combat support elements of two B-52 wings with augmentee aircraft and crews from other wings were maintained on Guam. They were assigned to the 4133rd BMW (Provisional), which was activated under the 3rd AD on 1

February 1966, during their TDY to Guam. Each wing typically supported combat operations for six months before being replaced by another wing from the CONUS.

Three B-52 squadrons were inactivated in 1967 to comply with McNamara's directive, but their aircraft did not immediately retire. The B-52D aircraft with "BIG BELLY" modifications were used to supplement wings supporting SEA combat operations. Excess B-52F aircraft were retained with the 4017th CCTS at Castle AFB, CA through 1974 to support combat crew training requirements. SAC continued to phase-out some of its bomber force in 1968. Six squadrons, accounting for 90 aircraft, were inactivated. All remaining B-52E and all B-52F aircraft that had reached the end of their service life were retired by 1971.

As the war in Vietnam dragged on SAC was tasked to provide B-52 sorties in support of Operation ARC LIGHT missions while still maintaining about 40 percent of its bomber force on ground alert. SAC established a Replacement Training Unit (RTU) on 15 April 1968 within the 4017th CCTS to better balance its SEA commitment and alter requirements. The unit cross-trained every B-52 crew, from B-52E through the B-52H model, in the operation of B-52D aircraft. After two weeks of training, the crews were used to augment the cadre of units in SEA. This spread-out combat duties more equitably among the entire B-52 force and provided the crews needed to meet the increased bombing effort.[9]

SAC again realigned its NAFs on 31 March 1970 as part of Air Force directed manpower reductions.[10] Headquarters 8th AF was closed at Westover, and its forces were transferred to the 2nd and 15th AF. The 2nd AF became an all-manned aircraft command consisting of B-52, FB-111, and KC-135 aircraft. The 15th AF got all missiles, strategic reconnaissance, and a few B-52 and KC-135 units. The 8th AF was reactivated the next day at Andersen AFB, Guam, replacing the 3rd AD, and received responsibility for all SEA operations. The 43rd BMW, previously a B-58 unit, became the 43rd SW and replaced the 3960th SW at Andersen. At the same time the 307th SW replaced the 4258th SW at U-Tapao RTNAF, Thailand[11] and 376th SW replaced 4252nd SW at Kadena AB, Okinawa to retain the illustrious histories of each wing.

SAC created several new 8th AF units to effectively control the additional B-52 and KC-135 aircraft and crews deployed to SEA. Completed on 1 July 1972, the reorganization included activating provisional units at seven Western Pacific and SEA bases. B-52 operational changes included establishing the 57th AD (Provisional) at Andersen and 17th AD (Provisional) at U-Tapao. In addition to the 43rd SW, the 57th AD directed the 72nd SW (Provisional) and the 303rd Consolidated Aircraft Maintenance Wing (CAMW)

B-52D aircraft on the ramp at Andersen AFB, Guam. (Ronny Young Collection)

(Provisional) at Andersen. The 17th AD directed the 307th SW, along with the 310th SW (Provisional) (KC-135 unit) and the 340th CAMW (Provisional) at U-Tapao.

With the conclusion of SEA operations, SAC moved the 8th AF to Barksdale AFB, LA effective 1 January 1975 and deactivated the 2nd AF.[12] At the same time the 3rd AD was reactivated on Guam. Effective 1 July 1976, SAC implemented the "Tri-Deputy" concept in its wing organizations to comply with Air Force directives. This reorganization added a Deputy Commander for Resources to the existing DCO and DCM. This new commander reported directly to the Wing Commander and received the functions of Comptroller, Procurement, Supply, and Transportation from the CSG. This freed up the Base Commander to focus on people programs.

In 1982, SAC made a major change in the assignment of its OCONUS ADs, which had been assigned directly to SAC Headquarters since the early 1950s. Effective 31 January, the 3rd AD at Andersen AFB, Guam, was assigned directly to 8th AF Headquarters, and the 7th AD was assigned to the 15th AF. The CONUS ADs were also realigned to better distribute units and balance division strengths and distances between the division headquarters and subordinate units.

Throughout the 1980s SAC began to place more emphasis on its conventional capabilities and responsibilities. This became codified in a new concept called the twin triad which embodied a nuclear capability and an overlapping conventional capability. The nuclear triad included SAC bombers, ICBMs, and Navy SLBMs. The conventional triad integrated SAC's bombers, tankers, and battle management assets to conduct conventional operations in support of combat commanders. SAC reorganized its NAFs along functional lines on 1 September 1991 to support the Twin Triad concept. The 2nd AF was reactivated and assigned all reconnaissance and battle management assets. The 20th AF was activated to manage the ICBM force. SAC's two existing NAFs were reorganized to gradually place all bomber operations under 8th AF and tanker operations under 15th AF.

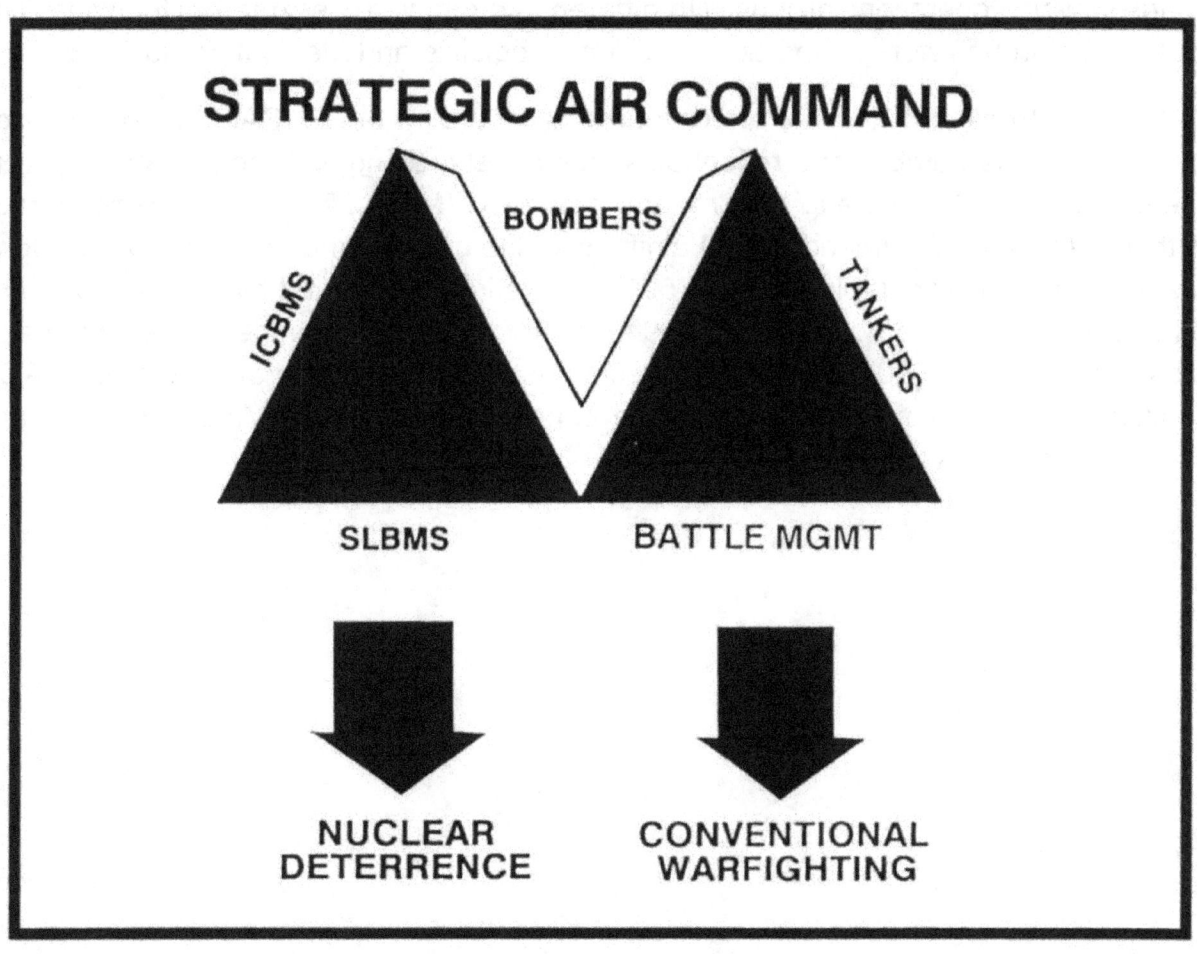

THE TWIN TRIAD CONCEPT

SAC placed more emphasis on its conventional capabilities in the 1980s using the Twin Triad concept. (U.S. Air Force)

The Air Force implemented its new Objective Wing concept in 1991 under the direction of the new Chief of Staff, General Merrill McPeak. Perhaps the most visible change was the redesignation of the "Bombardment Wing" to the simple term "Wing". For example, the 2nd Bombardment Wing was redesignated as the 2nd Wing. This change was completed by 1 September 1991, although it was short-lived, and designations were changed to "Bomb Wing" (BW) beginning on 1 June 1992. The Objective Wing also reintroduced the Bomb Group, now simply called the "Operations Group", assigned directly to the Wing as the operational combat unit. All Bomb Squadrons reported to the group as they had before the wing became the predominant unit under LeMay's 1951-52 reorganization. Each Bomb Squadron also had its own maintenance directly assigned to allow easier deployment of the group for combat operations. This wing organization supported McPeak's "One Base, One Boss" philosophy and created composite wings in which operations groups with different aircraft were assigned to the wing. The operations groups were supported by common logistics and medical groups.

In 1992, McPeak also directed a reorganization of the Air Force. SAC was deactivated by June and its bomber and reconnaissance assets, along with some tankers, were merged with TAC fighters to become the new ACC. Most SAC tankers were merged with Military Airlift Command (MAC) airlift assets to become the new Air Mobility Command. ACC quickly completed the B-52G fleet retirement started by SAC in 1989 and all aircraft were in the boneyard by 1994. Many former SAC bomber bases were also closed by the mid-1990s due to BRAC recommendations made in the mid-1980s. ACC realigned the remaining B-52H fleet and concentrated them at Minot AFB, ND and Barksdale AFB, LA where they remain.

Firsts and Records

On 17 January 1961, the first and only B-52B (53-0397) to deploy to the UK arrived at Royal Air Force (RAF) Upper Heyford from the 93rd BMW at Castle AFB, CA. It spent the weekend there before returning to California.

On 15 September 1982, the crew onboard a B-52G from the 19th BMW at Robins AFB, GA won the Mackay Trophy for outstanding feat of airmanship and bravery. While returning to base from a training mission, the aircraft lost both rudder-elevator hydraulic systems. The pilot assessed the situation and decided to attempt landing at Robins, the first time this operation had never been accomplished by a SAC crew. The entire crew worked in concert to ensure a safe and uneventful landing.

On 19 February 1985, a B-52G from the 319th BMW at Grand Forks AFB, ND launched an AGM-86B ALCM for a flight test over the Beaufort Sea, north of Canada, for the first time. The missile flew down a 1,500 by 50-mile pre-programmed course and made a parachute-assisted landing in the target area near Cold Lake, Alberta, after a 4-hour and 30-minute flight.

On 17 January 1991, seven B-52G (57-6475, 58-0177, 58-0183, 58-0185, 58-0238, 59-2564, 59-2582) aircraft from the 2nd BMW at Barksdale AFB, LA launched 35 AGM-86C CALCMs against Iraqi targets at the beginning of Operation DESERT STORM. These aircraft were the first aircraft to launch for the operation having taken off from Barksdale on the morning of 16 January. The aircraft all then returned non-stop to Barksdale 35.4 hours later, completing the longest bombing mission in history at that time. This was also the first time the CALCM was used in combat operations.

From 1-2 August 1994, two B-52H (60-0008, 60-0059) aircraft from the 2nd BW at Barksdale AFB, LA flew a 20,062-mile, 47.2-hour flight during the GLOBAL POWER 94-7 exercise. The flight marked the fourth anniversary of the 1990 Iraqi invasion of Kuwait. Both bombers hit targets in the Middle East with 27 Mk-82 bombs and returned Barksdale by circling the earth, proving the B-52 can hit targets anywhere in the world. This was the first time an aircraft dropped bombs while circumnavigating the globe.

Record Flights

Quick Kick. Four B-52B aircraft from the 93rd BMW participated in Operation QUICK KICK from 24-25 November 1956. They joined up with four B-52C aircraft from the 42nd BMW. The group completed a non-stop flight around the perimeter of North America in a flight lasting 31.5 hours and covering 13,500 nautical miles. The flight required four in-flight refueling hook-ups using KC-97 tankers. SAC quickly pointed out that the flight could have been completed five or six hours faster using the jet-powered KC-135 tankers then being developed by Boeing.

A Boeing B-52 Stratofortress refuels in flight from a Boeing KC-97 Stratotanker. The KC-97 had to enter a shallow dive to increase its speed, while the B-52 flew in landing configuration at reduced speed to stay with the tanker. (U.S. Air Force)

Power Flite. The 93rd BMW won the MacKay Trophy in 1957 for an around-the-world flight called Operation POWER FLITE which was recognized by the National Aeronautic Association as the Outstanding Flight of 1957. Five B-52B aircraft from the 93rd BMW including B-52B (53-0397) Lucky Lady I, B-52B (53-0398) Lucky Lady II, and the lead aircraft B-52B (53-0394) Lucky Lady III, took-off from their home station of Castle AFB, CA on 16 January 1957. Two aircraft were designated as spares with one replacing a mission aircraft that experienced ice build-up in its refueling receptacle and diverted to Goose Bay, Labrador. The second spare also left the flight, as planned, over the Atlantic Ocean in the first leg of the journey and landed near Casablanca, Morocco. The remaining three aircraft, including the lead aircraft piloted by Major General Archie J. Old, Jr., 15th AF commander, made the flight non-stop with five KC-97 refueling hookups. They landed at March AFB, CA on 18 January completing the 24,325-mile flight at an average speed of 534 mph in 45 hours and 19 minutes, which was less than half

The three Boeing B-52B Stratofortress aircraft at March AFB, CA on 18 January 1957 after completing a 24,325-mile around-the-world flight. (U.S. Air Force)

time required by the previous record holder, a B-50A named Lucky Lady II, which made the flight in February 1949. LeMay proudly announced that the operation was a demonstration of SAC's ability to strike any target on the face of the earth.[1]

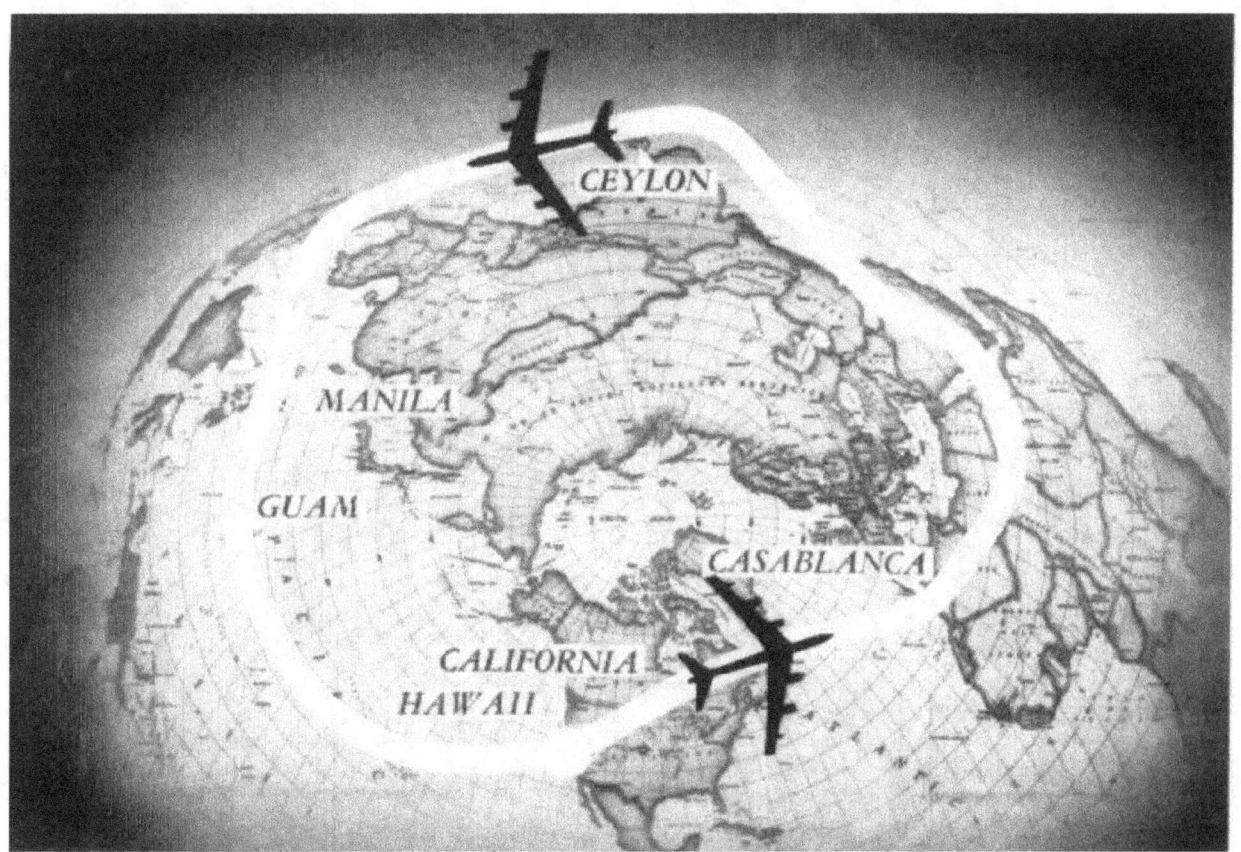

Three B-52B aircraft completed an around-the-world non-stop flight during Operation POWER FLITE from 16-18 January 1957 and demonstrated SAC's ability to strike any target on the face of the earth. (U.S. Air Force)

Long Legs. SAC completed Operation LONG LEGS on 16-17 November 1957 when six B-52 aircraft from the 42nd BMW at Loring AFB, ME flew a 10,600-mile non-stop flight from Homestead AFB, FL to Buenos Aires and back to Plattsburgh AFB, NY. It took two refueling hook-ups by KC-97 tankers and one by a KC-135 tanker.

Long Jump. A B-52G from the 5th BMW at Travis AFB, CA completed a world record flight on 14 December 1960 covering 10,078.84 miles without refueling. The flight lasted 19 hours and 44 minutes on a closed course breaking the previous record of 8,854 miles set by a B-29 in 1947.

Persian Rug. A B-52H (60-0040) from the 4136th SW at Minot AFB, ND completed a record "distance in a straight line" from 10-11 January 1962 under Operation PERSIAN RUG. The aircraft flew 12,532.28 miles unrefueled in 21 hours and 52 minutes from Kadena AB, Okinawa to Torrejon AB, Spain at altitudes between 40,000 to 50,000 feet

Colonel Clyde P. Evely, U. S. Air Force, with the crew of the record-setting B-52H (60-0040) Stratofortress. (U.S. Air Force)

with an average speed of 604.44 mph and top speed of 662 mph. The flight beat the previous record of 11,235.6 miles held by the propeller-driven U.S. Navy "Truculent Turtle".[2] The flight set 11 Fédération Aéronautique Internationale (FAI) records, several of which still stand, and seven National Aeronautic Association records for speed over a recognized course are also current.

Other Flights. Two B-52D aircraft from the 28th BMW at Ellsworth AFB, SD established two world speed records on 26 September 1958. They flew two different closed-circuit routes simultaneously. One aircraft (55-0049) completed 10,000 kilometers without payload at 560.705 mph (902.369 kilometers per hour) in 11 hours and 9 minutes. The actual distance covered was 6,233.6 miles which set a world distance record on the same mission. The other (56-0694) flew a 5,000-kilometer route without payload at 597.676 mph (961.867 kilometers per hour) in 5 hours, 11 minutes, and 49 seconds.

On 15 December 1958, a B-52G with a Boeing flight crew set a world distance record by flying unrefueled for 10,078.84 miles. The flight lasted 19 hours and 44 minutes with an average speed of 510.75 mph. This record was broken two years later by B-52H (60-0040) during Operation PERSIAN RUG.

In August 1959, a B-52G made a 28-hour flight covering 12,942 miles non-stop flying over the capitals of every continental U.S. state including Alaska. Another B-52G flew a non-stop flight in April 1960 under Operation BLUE NOSE taking off from Eglin AFB, FL and flying over the North Pole before returning to Florida. It fired two AGM-28 Hound Dog missiles over the coastal missile range before landing back at Eglin. The aircraft covered 10,800 miles in 22 hours.

On 7 June 1962, a B-52H from the 19th BMW at Homestead AFB, FL broke the world record for a closed course of 11,336.92 miles from Seymour-Johnson AFB, NC, and return. The old record of 10,078.84 miles was held by a B-52G of the 5th BMW at Travis AFB, CA since 14 December 1960.[3]

On 10 January 1964, a B-52H (61-0023) landed at Blytheville AFB, AR with the vertical tail assembly broken off. The Boeing flight crew was conducting low-level dynamic loads flight tests with two AGM-28 Hound Dog missiles in the Spanish Peaks Mountains in Colorado. The aircraft was instrumented to investigate the effects on the airframe from high-speed, low level, flight. It was flying at 14,300 feet and 397 mph indicated airspeed. Severe turbulence suddenly broke off the tail, but the aircraft remained flight worthy. It landed safely after consulting with the authorities to determine a suitable landing site. The aircraft was eventually repaired and returned to service. It remained active until 24 July 2008 when it was placed in storage at Tinker AFB, OK. This incident occurred nearly one year after a B-52C (53-0406) was lost on 24 January 1963 near Greenville, ME when its tail broke off due to turbulence during a low-level flight exercise. Three

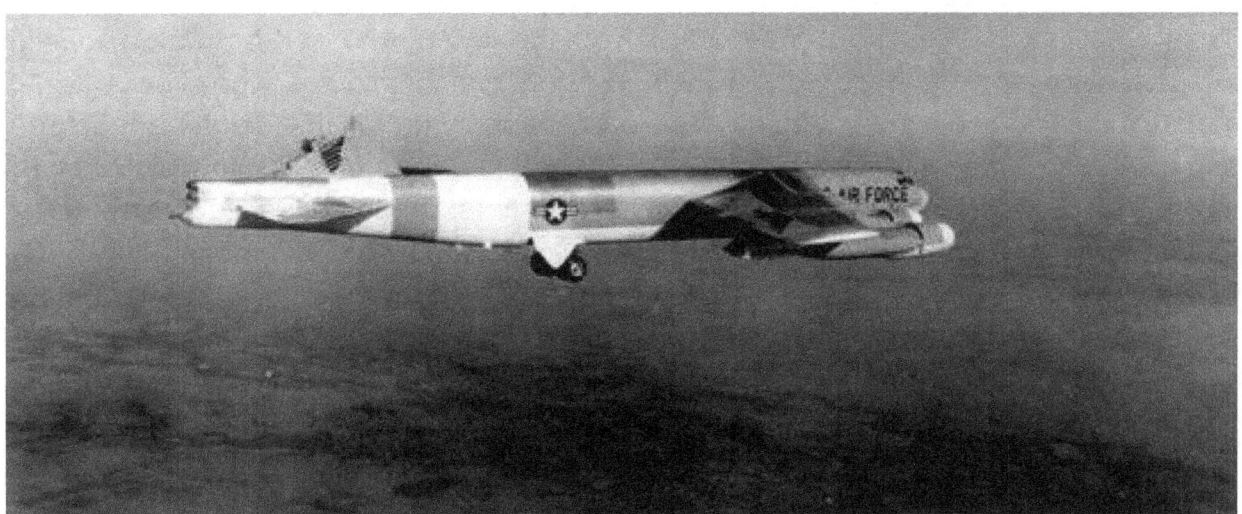

Boeing B-52H (61-0023) "Ten-Twenty-Three" after losing the vertical fin on 10 January 1964. (U.S. Air Force)

days later, a B-52D (55-0060) was lost near Barton, MD when its tail fin and rudder were ripped off during turbulence in a blizzard. These incidents led to the implementation of ECP 1124-2 to reinforce the tail section, and ECP 1128 to strengthen the upper fuselage and vertical fin of all B-52B through H models.

Brigadier General Jimmy Stewart flew as an observer on the last combat mission of his career on 20 February 1966. He flew aboard a B-52F (57-0149) from Andersen AFB, Guam on a 12 hour and 50-minute Operation ARC LIGHT mission over Vietnam. Jimmy Stewart was a famous American actor who starred in several popular movies including *Strategic Air Command* and *The Spirit of St. Louis*. He

Brigadier General Jimmy Stewart, USAFR (center) with the crew of B-52F (57-0149) at Anderson AFB, Guam on 20 February 1966. (U.S. Air Force)

flew B-24 Liberators during WWII and joined the reserves after the war. He flew B-36, B-47, and B-52 aircraft during his years of service. He retired on 1 June 1968.

On 12-14 March 1980, two B-52H aircraft from the 410th BMW at K.I. Sawyer AFB, MI made a non-stop around the world flight in 42 hours and 30 minutes. They flew a 19,353 nautical-mile course across Canada, the North Atlantic, Europe, the Mediterranean Sea, Indian Ocean, Strait of Malacca, South China Sea, and the North Pacific before returning to K.I. Sawyer. The aircraft refueled several times and received almost 600,000 pounds of fuel. General Richard H. Ellis, CINCSAC, said the flight demonstrated SAC's "ability to rapidly project military power to any point in the world in a matter of hours." The crew was awarded the 1980 Mackay Trophy. On 15 May, the aircraft commanders received the Distinguished Flying Cross (DFC) and the crews received Air Medals.

On 25 August 1995, a B-52H from the 2nd BW at Barksdale AFB, LA set a FAI world record for top speed over a 10,000-kilometer course on an airplane weighing 440,000 to 550,000 pounds. The aircraft flew nonstop from Edwards AFB, CA to Alaska and return without refueling. It averaged 556 mph during the 11-hour and 23-minute flight.

On 3 September 1996, three B-52H (60-0014, 60-0025, and 60-0054) aircraft, each loaded with a CSRL and eight CALCMs, took off from Guam. The spare aircraft (60-0025) returned to Guam after the first refueling and the remaining two continued to the target. They fired three CALCMs each against targets in Southern Iraq. They then pulled back to reload mission tapes and reboot their systems before turning back to the targets. They ultimately fired a total of 13 missiles hitting their targets, while three malfunctioned and could not be launched. This 34-hour mission covered 13,600 miles, making it the second longest bombing mission in history at the time (slightly less than the 1991 Operation DESERT STORM flight) resulting in the award of the 1996 McKay Trophy.

In February 2018, CRL equipped B-52H bombers dropped unparalleled numbers of Precision Guided Munitions (PGM) to destroy nearly 70 Taliban targets. One aircraft dropped 24 guided munitions, including 19 JDAMs, on Taliban positions and vehicles setting a record for the most precision weapons dropped on a single B-52 sortie.

On 12 January 2021, two B-52H bombers from the 5th BW at Minot AFB, ND flew a 36-hour non-stop show-of-force flight aimed at Iran from Minot to the Arabian Gulf and back. The bombers were escorted by Saudi F-15 aircraft when they entered Saudi airspace. This was the fourth show of force in 60 days. Another 5th BW B-52H made a similar flight on 7 March. This time the bomber was escorted by F-15 aircraft from Israel, Qatar, and Saudi Arabia as it flew through their respected airspace.

Operations

During 1949 SAC had only about 40 B-36 aircraft and only 5-8 were considered operational.[1] The Air Force placed top priority on manning and equipping SAC to support the emphasis on the strategic air warfare mission by the JCS. In his first full year as SAC Commanding General[2], Lieutenant General Curtis LeMay began to make his lasting imprint on the command. "We didn't have one crew, not one crew, in the entire command who could do a professional job," LeMay wrote of the SAC he inherited. He challenged his crews to stage a practice bomb raid on Dayton, OH, from 30,000 feet, using photographs taken in 1941 – the best they would have for the Soviet Union. After the fiasco that ensued, LeMay whipped the crews into shape. He moved the best people from other groups to make the nuclear capable 509th BG combat-ready, then did the same for the next most promising group.[3]

Training of bomber crews was intensified and accuracy of high altitude bombing substantially improved. Combat crew proficiency was raised through a system of lead crew training that proved successful during WWII. Units were deployed on a rotational schedule at overseas bases for limited periods to familiarize personnel with operating conditions outside the United States. SAC continued to play an important role in the field of atomic energy through the development of strategies, tactics, techniques, and logistics to assure the most effective combat employment of atomic weapons in the national interest. SECAF W. Stuart Symington pointed out at the end of 1949: "Existence of this strategic atomic striking force is the greatest deterrent in the world today to the start of another global war."

At 4:00 am local time on 25 June 1950, North Korean troops stormed across the 38th parallel. In November, Chinese "volunteers" joined them. These developments marked the end of President Truman's defense economy drive. First Germany, then Japan, then Russia, and now events in Korea had succeeded in advancing the cause of the B-36. Suddenly plenty of money was available for megabombers, and for supercarriers as well. The Korean war produced another milestone for SAC: Truman released atomic bombs to the military. They probably didn't leave the country, but the B-36 did, flying from Texas to airfields in Britain and Morocco in the spring and fall of 1951. Only six airplanes were involved, and their visits were short, but the message could not have escaped Moscow's attention. However briefly, the capital and most of the territory of the Soviet Union had come within the combat radius of the B-36.[4] This set the stage for the Cold War, and from this point forward SAC maintained a strong and lethal bomber force including the B-52 to keep the Soviets at bay.

Cold War Operations

Ground Alert Program. By 1957, the Soviet Union was developing much improved intercontinental ballistic missiles and SAC planners began exploring concepts for getting its force airborne 15 minutes after detecting a missile attack. They settled on a ground alert program in which at least one-third of the bomber and tanker force would be loaded with crews standing by for immediate take-off. This concept was initially applied to B-47 and KC-97 aircraft and eventually expanded to the B-52 and KC-135 force.

A B-52B (52-8711) alert crew assigned to 22nd BMW runs to their aircraft during ground alert in 1965 at March AFB, CA. (U.S. Air Force)

SAC conducted three test programs at B-47 units to verify the operational viability of the program and to identify improvements. The first, Operation TRY OUT, was conducted from November 1956 through March 1957 at Hunter AFB, GA. It proved the concept viable and identified numerous changes needed. SAC conducted two additional tests to perfect these changes. Operation WATCH TOWER was conducted at Little Rock AFB, AR from April to November 1957 and Operation FRESH APPROACH was conducted at Mountain Home AFB, ID in September. These tests all proved very successful and, although there were still some details to work out, General Thomas Power, CINCSAC, directed that ground alert operations would start at several CONUS and OCONUS bases on 1 October 1957.

General Power penned a memorandum to all SAC alert crew members on 9 November 1957 to mark the occasion. He wrote, "As a member of SAC's Alert Force you are contributing to an operation which is of the utmost importance to the security and welfare of this nation and its allies in the free world... the only way of insuring the survival of some of SAC's combat capability, even in the case of the most unexpected and massive attack, is our Alert Force. As long as the Soviets know that, no matter what means they may employ to stop it, a sizable percentage of SAC's strike force will be in the air for the counterattack within minutes after they have initiated aggression, they will think twice before undertaking such aggression. For this reason, it is my considered opinion that a combat-ready Alert Force of adequate size is the very backbone of our deterrent posture."

The ground alert force was soon called into action during the LEBANON CRISIS in July and August 1958. Lebanon's president was fearful of an imminent Soviet attack and asked the United States for help. President Eisenhower sent ground, naval, and air forces to the area and ordered SAC to place its bombers on alert. SAC had not yet achieved its one-third alert goal but immediately generated additional aircraft for ground alert. Over 1,100 aircraft were poised and ready for take-off within a few hours, and they were maintained on alert for several days until it was clear the Soviets did not intend to invade.

The SAC alert force was again poised and ready during the TAIWAN CRISIS in August and September 1958. This time China began artillery bombardment of islands off the coast of China. SAC immediately increased its ground alert strength at Andersen AFB, Guam and alerted several bomb wings for possible deployment to the Pacific. Since the Commander-in-Chief Pacific Command (CINCPAC) did not anticipate having to use the SAC forces they were soon returned to normal configuration.

SAC achieved its one-third ground alert goal in May 1960. Alert bombers (B-47 and B-52) and tankers (KC-97 and KC-135) were loaded and ready to take off upon detection of incoming Soviet missiles. SAC forces were on alert at 46 bases across the United States. This greatly increased the number of runways available to support alert take-offs. SAC's reaction capability was further enhanced through a new Minimum Interval Take-Off (MITO) procedure implemented beginning in mid-1960. This capability was tested in the spring of 1960 at several bases under Project OPEN ROAD and allowed the alert force to become airborne within 15 minutes, giving the United States the ability to conduct massive retaliatory attacks on the Soviet Union.

SAC began operating a new underground command post at SAC headquarters along with EC-135 Airborne Command Post aircraft, called LOOKING GLASS, to provide around the clock control of all SAC nuclear weapons. SAC also implemented a new HF communication system, called SHORT ORDER, that allowed positive control of the alert bombers enroute to the target. Using positive control procedures, the bombers flew to a designated point outside enemy territory and awaited a Go-Code before proceeding to their targets. If no code was received, they returned to home station.

On 28 March 1961, President Kennedy requested funds to strengthen and protect the strategic deterrent force and strengthen the ability to wage limited war. He ordered 50 percent of SAC's B-52 and B-47 bombers, along with KC-97 and KC-135 tankers, to be placed on ground alert. He also directed accelerated B-47 phase out to provide the crews for expanded ground alert. SAC achieved the 50 percent target in July with up to 519 bombers on alert. This number increased to 625 by the end of 1962.

President Kennedy ordered 50 percent of SAC's B-52 and B-47 bombers, along with KC-97 and KC-135 tankers, to be placed on ground alert beginning in 1961. (American Aviation Historical Society)

SAC began reducing its alert force in 1963 with the phase-out of B-47 bombers directed by President Kennedy. In 1964, SAC had 464 bombers on alert and replaced its B-47 ground alert force at Andersen AFB, Guam with B-52 aircraft. B-47 ground alert across the command terminated on 11 February 1966 and only 301 B-52 aircraft remained on alert. By 1967, SAC had reduced its alert force to 40 percent with 223 B-52 bombers on alert due to Vietnam War commitments.

SAC implemented a new B-52 dispersal program in 1968 based on its earlier success with the B-47 force. Aircraft were dispersed during times of crisis to various military and civilian airfields and placed on ground alert to spread out the force and complicate the enemy's targeting problem. On 20 February 1969, SAC began testing a new B-52 satellite basing program at Homestead AFB, FL (TAC base) using aircraft relocated from Ramey AFB, Puerto Rico. A small maintenance and support contingent deployed with the aircraft, while the crews rotated back and forth between the satellite base and their home base. The test was completed by 20 May, and several wings were brought into the program on 1 July. These programs were similar except the aircraft were regularly deployed to satellite bases whereas the dispersal program was used only during times of crisis. Both programs multiplied the available runways and increased the probability of launching the entire alert force before Soviet missiles could strike.

President Ronald Reagan recognized the SAC Alert Force contributions to the national security on 9 October 1987 after thirty years of alert service. He wrote, "In October 1957, aircrews of the Strategic Air Command went on alert for the first time. From that historic day forward, SAC's demonstrated readiness has been a cornerstone for peace and security for the free world. Today, strategic deterrence is still the foundation on which rests the peace of the world and the protection of freedom." This recognition was followed by General John T. "Jack" Chain (CINCSAC) declaring 1988 the "Year of the SAC Alert Force". Despite this renewed emphasis, the alert force had been reduced over the years to only 69 B-52 aircraft on alert.

But the world was quickly changing. By the late 1980s the Soviet economy was in shambles, and the Berlin Wall fell in November 1989. The Soviet Union soon began withdrawing its forces from Eastern Europe and the Treaty on Conventional Forces was signed between the Soviets and the U.S. in November 1990. SAC's EC-135 LOOKING GLASS flew its last continuous airborne mission on 24 July 1990 and began ground alert operations. The bomber alert force was now down to 53 aircraft.

The USSR announced the dissolution of the Warsaw Pact on 1 July 1991 and thirty days later signed the Strategic Arms Reduction Treaty (START). Soviet President Mikhail Gorbachev survived a poorly organized coup attempt in September 1991 and recognized the independence of the Baltic Republics on 6 September 1991. President George H. W. Bush recognized the move to democracy in the Soviet Union and ordered an end to B-52 ground alert on 17 September 1991. General George "Lee" Butler

(CINCSAC) and his staff immediately oversaw the standdown of the alert force which was completed at 2:59 PM on 18 September.

Butler promptly called his wing and unit commanders from the SAC Command Center and passed on his thanks and congratulations, concluding "…rest secure in the knowledge that for the first time in over 40 years we can truly promise our children and our grandchildren a world drained from the tension of superpower confrontation. God bless you all for what you have accomplished. CINCSAC out."

Airborne Alert Program. SAC began testing airborne alert operations under Operation HEAD START I with the 42nd BMW at Loring AFB, ME from 15 September to 15 December 1958. SAC conducted Operation HEAD START II from 2 March to 30 June 1959 with five B-52 bombers from the 92nd BMW at Fairchild AFB, WA constantly airborne. The results proved satisfactory and General Power testified before Congress in February 1959, "We in Strategic Air Command have developed a system known as airborne alert where we maintain airplanes in the air 24 hours a day, loaded with bombs, on station, ready to go to the target… I feel strongly that we must get on with this airborne alert… We must impress Mr. Khrushchev that we have it, and he cannot strike this country with impunity."

B-52G (57-6471) refueling from KC-135A. (U.S. Air Force)

The command conducted more than 6,000 airborne alert sorties over the next two years and proved the feasibility of the concept. General Power announced on 18 January 1961 that indoctrination training was completed, and B-52 bombers were conducting airborne alert training under realistic conditions. Up to 12 B-52 aircraft from 24 bases were constantly airborne under the program called Operation CHROME DOME. Four aircraft flew the southern route and six flew the northern route. Additionally, two SAC bombers flew Operation HARD HEAD (Thule Monitoring) missions to maintain constant surveillance of the Ballistic Missile Early Warning System (BMEWS) at Thule which provided early warning of Soviet missile launches.

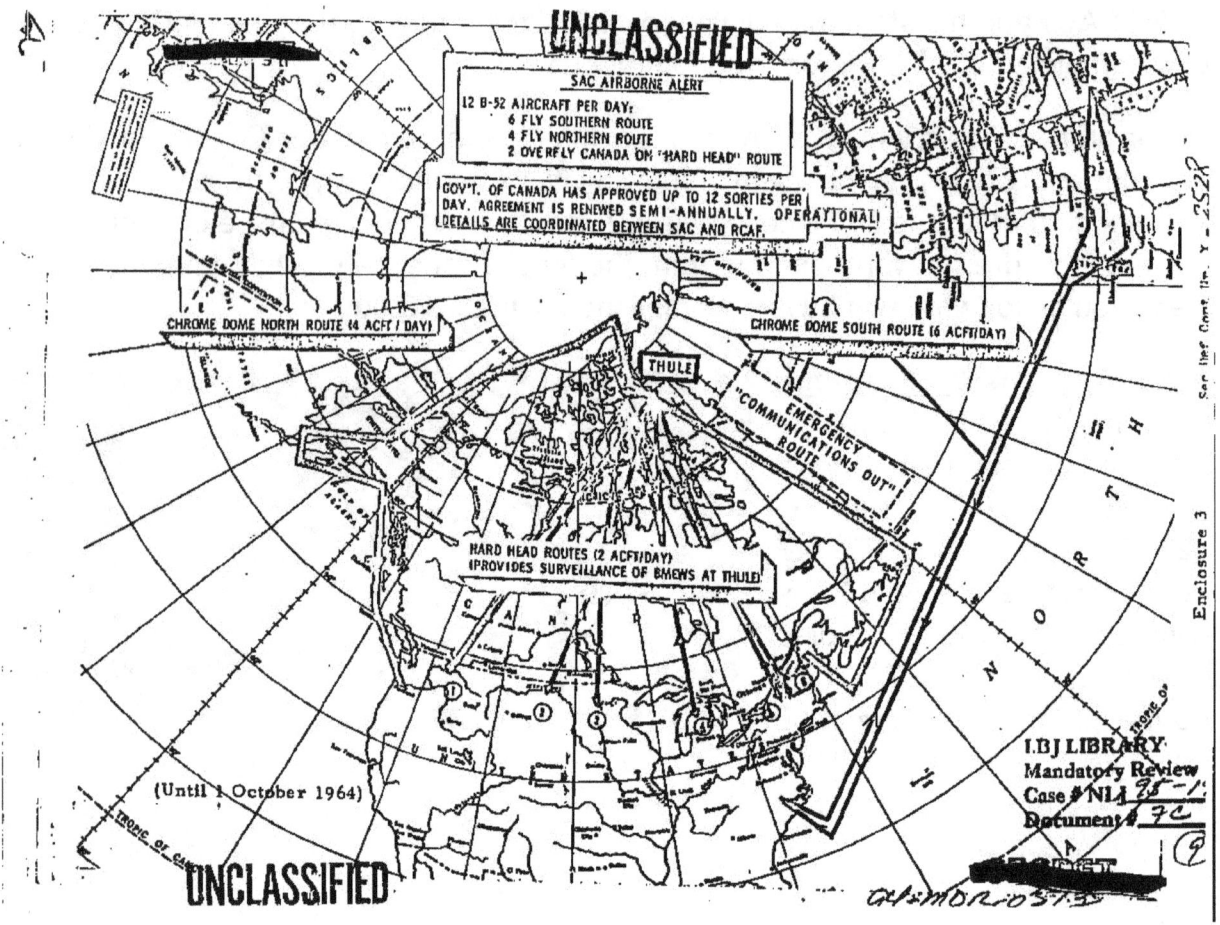

Overview of daily B-52 airborne alert flights conducted under Operation CHROME DOME in 1962. (U.S. Air Force)

The airborne alert program played a key role during the CUBAN MISSILE CRISIS from 22 October to 20 November 1962. SAC U-2 aircraft had confirmed the presence of Soviet missiles on Cuba after several flyovers beginning on 14 October. On 22 October President Kennedy announced the quarantine of all ships bound for Cuba and demanded that the Soviets remove all missiles currently on the island. SAC responded by placing its battle staff on 24-hour alert, dispersing B-47 aircraft for ground alert at several military and civilian airfields in the area and designating the B-52 airborne alert indoctrination program as an operational airborne alert mission. Operation CHROME DOME flights were increased within 24 hours from the typical 12 flights per day to as many as 66 nuclear armed flights per day. This included 28 on the northern route, 36 on the southern route, and the two Thule monitor flights.

SAC remained engaged throughout the crisis with U-2 surveillance, RB-47 reconnaissance, and KC-135 refueling flights. On 28 October, the Soviets agreed to remove all missiles from Cuba and over the next several days SAC surveillance aircraft verified that the missiles were dismantled and sent back to the Soviet Union.

By 5 November, as more B-52H aircraft became available, SAC increased its airborne alert to 75 flights per day including 42 on the southern route, 31 on the northern route, and the two Thule monitor flights. An additional flight was added on the last three days bringing the total to 76 flights per day.

On 20 November, the Soviets agreed to remove all bombers from Cuba, and the next day SAC reverted to its normal 50 percent ground alert and returned its B-52 airborne alert to a crew indoctrination mission. During less than one month of operations SAC had launched 2,088 B-52 airborne alert sorties and crews flew a total of 47,168 hours. The safety record was outstanding, and maintenance was 97 percent effective.

President Kennedy visited SAC Headquarters on 7 December and presented a plaque that read, "For outstanding record in flight safety during airborne alert in the Cuban emergency, 22 Oct – 21 Nov 62." The citation accompanying the plaque read in part, "the airborne alert provided a strategic posture under which every United States force could operate with relative freedom of action."

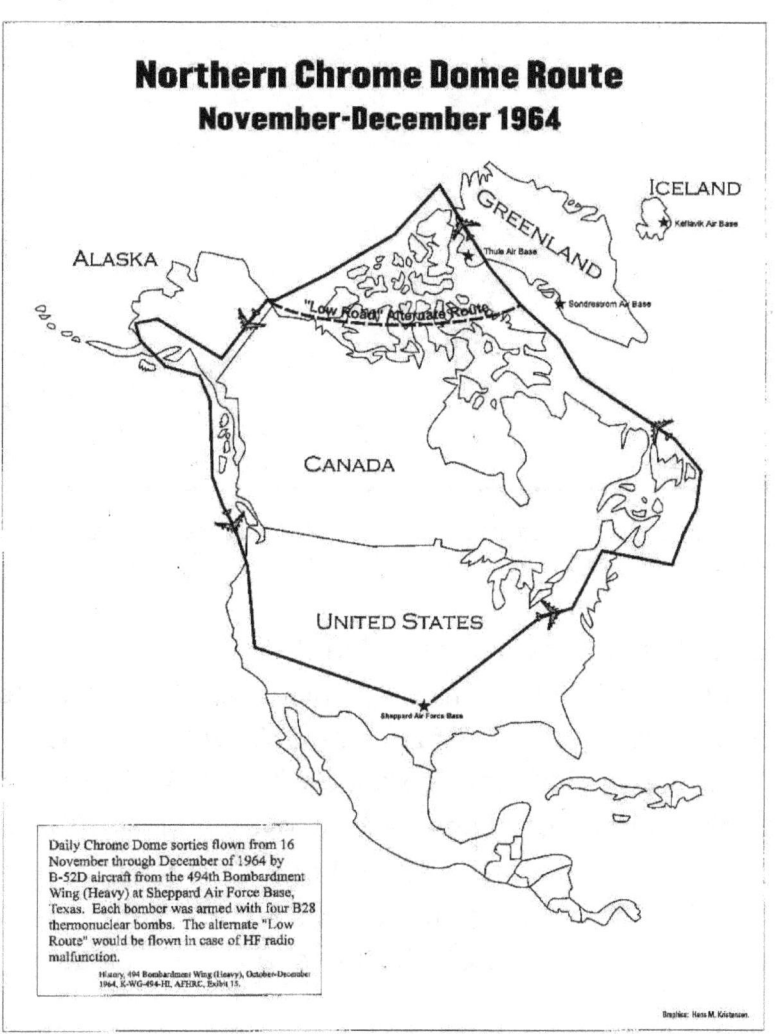

Northern route for B-52 airborne alert flights conducted under Operation CHROME DOME in 1964. (U.S. Air Force)

By 1966, SECDEF McNamara proposed cutting CHROME DOME flights because the BMEWS was fully operational, and bombers were becoming redundant to missiles. SAC and the JCS opposed the plan and reached a compromise to reduce the airborne alert to only four bombers per day. SAC also committed one aircraft to support the Thule

Monitoring requirement without the knowledge of civilian authorities in the U.S. who SAC determined did not have a need to know about secret operational issues.

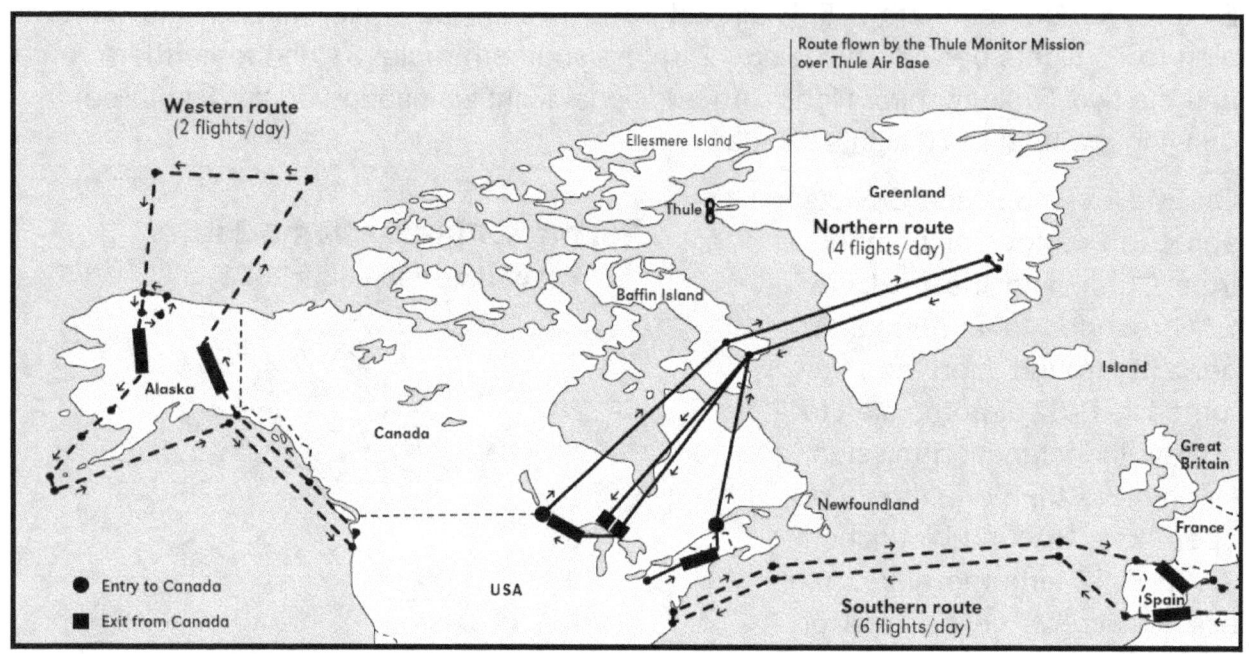

Overview of daily B-52 airborne alert flights conducted under Operation CHROME DOME in 1966 before the JCS agreement with SECDEF McNamara to reduce the force to only four flights per day. (U.S. Air Force)

Despite its overall success, the SAC airborne alert program suffered significant losses resulting in "Broken Arrows", which was SAC's term to describe an incident involving a nuclear warhead. A B-52F (57-0036) with two nuclear weapons on board crashed on 15 October 1959 after colliding with a KC-135 during refueling over Hardinsburg, KY Both unarmed weapons were recovered intact. A B-52F (57-0166) with two Mk-39 nuclear bombs onboard crashed approximately 15 miles west of Yuba City, CA on 14 March 1961. Although the bombs were severely damaged the high explosives did not detonate and no radioactive materials were released.

A B-52G (58-0187) with two Mk-39 nuclear weapons onboard crashed on 24 January 1961 during landing at Seymour-Johnson AFB, NC. The two nuclear weapons broke loose. One parachuted, undamaged, to the ground and one sank to at least 50 feet in a marsh. A B-52D (55-0060) from Turner AFB, GA crashed into Savage Mountain near Barton, MD on 13 January 1964 during a ferry flight enroute to its home base from Westover AFB, MA, where it had landed after an Operation CHROME DOME patrol over Europe. Two nuclear weapons were onboard, and both were in the "Tactical Ferry Configuration" with no electrical connections and safing switches in the SAFE position.

A B-52G (58-0256) from Seymour-Johnson AFB, NC carrying four Mk-28 nuclear weapons crashed near Palomares, Spain on 17 January 1966 during an airborne alert indoctrination mission after colliding with its KC-135 tanker. The four nuclear weapons fell free from the bomb bay. Three were found on land and two of these released radioactive material as the result of non-nuclear explosions. One of the bombs fell into the Mediterranean Sea and was found by a U.S. Navy submarine on 15 March after an extensive search. This crash spelled the immediate end for missions along the southern route.

A B-52G (58-0188) from Plattsburg AFB, NY crashed on the ice on 22 January 1968 at North Star Bay after attempting an emergency landing at Thule AB, Greenland. Four nuclear weapons detonated (non-nuclear) and caused widespread contamination. The crashes continued to mar the program, but the final nail in its coffin was the rising operating costs and advent of the ICBM force. McNamara ultimately ordered suspension of all CHROME DOME missions immediately following the Thule crash.

Despite cancellation of Operation CHROME DOME, President Nixon authorized another temporary airborne alert called Operation GIANT LANCE on 27 October 1969. The purpose of the top-secret operation was to intimidate both the Soviet Union and North Vietnam and achieve a favorable outcome to negotiations and end the Vietnam War. A squadron of 18 B-52 bombers loaded with nuclear weapons were authorized to patrol the Arctic ice caps as a show of force. It was intended to be easily detected by the Soviet Union and initiate a positive response in negotiations. However, the Soviets did not respond (even though they later revealed that they understood Nixon's "madman" intentions). The alert ended on 30 October, and aircraft returned to home station.

Single Integrated Operational Plan (SIOP). SAC controlled most of the United States nuclear striking power in 1960, but significant strength provided by the Navy's SLBM and tactical nuclear weapons required closer coordination between the services. On 16 August 1960, SECDEF Thomas S. Gates announced the creation of the Joint Strategic Target Planning Staff (JSTPS) composed of representatives from all service branches. The JSTPS was charged with maintaining a National Strategic Target List and a SIOP to commit specific weapons to the various targets in case of war. The CINCSAC was designated as the JSTPS Director, and a Navy Vice Admiral was designated as the Deputy Director and oversaw day-to-day activities. The JSPTS was collocated in SAC Headquarters at Offutt AFB, NE to fully use the strategic planning experience and facilities available in SAC. The first SIOP, called SIOP 62, went into effect on 1 April 1961.

In 1980, the JCS formed the Rapid Deployment Joint Task Force (RDJTF) to respond immediately with conventional forces in a crisis. SAC developed the Strategic Projection Force (SPF) as its contribution to the RDJTF. The SPF included various SAC bomber, refueling, and reconnaissance forces including B-52H aircraft from the 5th BMW at Minot AFB, ND and the 319th BMW at Grand Forks AFB, ND.

Low-Level Operations. By 1959, the Soviets had made significant improvements in detecting and defending against aircraft flying at high altitudes. SAC strategists determined that B-52 aircraft would need to fly at extremely low altitudes to penetrate Soviet airspace. Working with the FAA, they established several low-level training routes in November. These routes were 20 miles wide and 500 miles long and were established in remote areas of the United States.

Initial testing turned disastrous on 23 June 1959 when a B-52D (56-0591) crashed during a low-level experimental flight test. Five Boeing employees were on board. The aircraft had been making test runs over Eastern Washington, Oregon, and Idaho since April 10, 1959, and was loaded with special electronic equipment for measuring stress on the airframe and flight surfaces. The bomber was scheduled to fly at lower than 500 feet above the ground on an elliptical course at near maximum speed of 638 mph. SAC needed to know if the B-52, specifically designed to fly at high altitudes, could survive the secondary structural stresses caused by violent air turbulence found at very low altitudes. According to eyewitnesses, the aircraft was flying at extremely low-level below the canyon walls when it began to break apart, suddenly pitched straight up, burst into flames, and then crashed into a surrounding hill. Air Force and Boeing experts concluded the accident was caused by the catastrophic failure of the horizontal stabilizer, affecting the B-52's longitudinal stability. The plane was not designed for the excessive turbulence of high-speed, low-level flight and began to disintegrate.[5]

Despite this tragedy, the tests proved that the B-52 could evade Soviet radar using low-level tactics. Boeing immediately developed structural improvements, and the Air Force implemented them under the HI STRESS and BIG FOUR modification programs. All operational B-52C through H model, and late model B-52B, bombers were modified by 1969 with the required structural improvements and systems to support low-level flying including improved BNS and doppler radar, and a new TA radar. This allowed the aircraft to fly at high speed at 200 feet and lower using hills and valleys to help hide them from enemy SAM radars.

Additional realism was introduced in late 1979 with low-level training flights using TA radar over mountainous areas at night. Before 1979, SAC limited such night training. Yet skillful execution of this tactic could be required during a combat mission. Night training was beneficial because the lack of visual cues forced crews to use information presented in the cockpit and rely less on external references. Thus, increased use of TA presentations led to better instrument interpretation and improved TA performance both day and night. Night training also increased crew confidence in their ability to accomplish the wartime mission under other conditions that restrict visibility, such as adverse weather and thermal curtains. Encouraged by this nighttime TA success, SAC lowered minimum altitudes in low-level operations both day and night.[6]

Operational Readiness Inspections (ORI). ORIs determined the readiness of SAC Units to accomplish their wartime mission. Normally, units "generated" all available aircraft to full alert status, and crews then flew simulated wartime missions without nuclear weapons. These flights involved low-altitude penetration of a predetermined target area and electronic scoring of simulated releases of nuclear weapons. But by the late 1970s SAC began putting more realism in the ORI including testing a wing's ability to conduct conventional operations and deployments.

In December 1979, a no-notice deployment of fourteen B-52H aircraft from Ellsworth AFB, SD to Guam, reflected a dramatic departure from other ORIs. Previous inspections tested a unit's nuclear mission and were conducted in CONUS, but the Ellsworth inspection tested the capability of an entire unit to respond rapidly over great distances. Early in 1980, the remaining B-52H units subsequently flew similar no-notice deployments to Guam as part of their ORIs.[7]

Maritime Mission Support. The Air Force directed SAC on 15 January 1964 to support Air Defense Command in antisubmarine defense of the United States. In September, a B-52 dropped eight different types of aerial mines during tests at the Air Proving Ground on Eglin AFB, FL. B-52 bombers, along with KC-135 tankers, flew their first visual and photographic sea search exercise called Operation WATER GAP during March to April 1965. Vietnam War operations took priority from 1965-73, but SAC resumed maritime mission support operations in 1974. B-52D aircraft supported Navy mine tests on 18 June and on 9 October two B-52H aircraft flew a mission off Puerto Rico.

The long range of the B-52 allowed it to drop mines at almost any oceanic chokepoint in the world. Some U.S. Navy brass were a bit irritated with the Air Force for encroaching on their turf, and in the 1980s the Navy successfully lobbied for limitations on B-52 maritime activities.[8]

Nonetheless, in 1983, B-52D and G models from the 43rd SW at Andersen AFB, Guam, 2nd BMW at Barksdale AFB, LA, and 19th BMW at Robins AFB, GA conducted the largest B-52 aerial mining exercise on 7 March 1983. The bombers dropped mines off the coast of South Korea as a part of TEAM SPIRIT '83. They were joined by U.S. Navy and Marine minelayers and fighters, and South Korean fighters.

B-52D mine loading at Andersen AFB, Guam during TEAM SPIRIT '83 on 7 March 1983. (U.S. Air Force)

SAC also began supporting sea interdiction missions in 1983 when B-52G aircraft from the 42nd BMW at Loring AFB, ME were fitted with AGM-84D Harpoon missiles and achieved limited operational capability on 6 October. The B-52 was highly effective, and two aircraft could monitor 140,000 square miles of ocean surface in two hours. SAC B-52G aircraft from the 43rd SW participated in the TEAM SPIRIT '84/MINEX exercise in March 1984. They dropped mines off the South Korean coast.

Vietnam War

Arc Light. SAC leadership initially resisted deploying B-52 aircraft to support Vietnam War operations. They believed the aircraft should be dedicated to the strategic nuclear role and did not want to divert SAC forces to support tactical

A B-52D drops a load of Mk-82 bombs during the Vietnam War. (U.S. Air Force)

operations in a "brushfire in the Far East." However, Army General William Westmoreland, South Vietnam Theater Commander, wanted to use B-52s in support of ground forces and interdiction of infiltration of troops and supplies by hitting Viet Cong enclaves in South Vietnam. He believed the B-52 bombers were better for this mission than fighter-bombers, because they could deliver more munitions over a wider area. Despite SAC misgivings, the Air Force directed modification of B-52F aircraft to carry heavier bombloads under the SOUTH BAY and SUN BATH programs in 1964-65. A decision was made on 9 February 1965 to transfer 31 B-52F aircraft to Andersen AFB, Guam along with 32 tankers to Kadena AB, Okinawa to support combat operations.[9]

The first B-52 aircraft to enter combat in Vietnam were a total of 27 F models from the 7th BMW at Carswell AFB, TX and the 320th BMW at Mather AFB, CA on 18 June 1965. The wings were deployed to Andersen AFB, Guam and used to provide Operation ARC LIGHT saturation bombing of Viet Cong base camps in the jungles of South Vietnam with conventional 750- and 1,000-pound bombs. These attacks were largely viewed as

Boeing B-52F (57-0144) drops bombs during an ARC LIGHT strike. (U.S. Air force)

ineffective. Two aircraft were lost in a mid-air collision on the way to the target and the Viet Cong had already left the area before the remaining aircraft dropped their bombs. The press tagged it a fiasco akin to "swatting flies with a sledgehammer" and quickly dubbed the missions "carpet bombing". Despite this criticism, General Westmoreland considered them a huge success and requested more B-52 support.

There were no more Operation ARC LIGHT missions in June, but five more totaling 140 B-52F sorties were conducted in July and another five totaling 165 sorties were conducted in August.[10] SAC B-52F crews from the 7th, the 320th, the 2nd BMW from Barksdale AFB, LA, and the 454th BMW from Columbus AFB, MS completed 163 missions consisting of 2,949 sorties from June 1965 to 31 January 1966. B-52F aircraft continued to fly ARC LIGHT missions through April 1966. After less than a year they dropped over 100,000 tons of bombs on Viet Cong targets in South Vietnam.[11]

B-52 attacks also included secret attacks on the Ho Chi Minh trail in South Vietnam, Laos, and Cambodia.[12] These attacks began on 12 December 1965 when 24 B-52F bombers hit Laotian targets with 750-pound gravity bombs and BLU-3 CBUs. This attack was not communicated beforehand to the Laotian Prime Minister. This became the normal approach for maintaining secrecy with some missions designated as Category One, with notification to the Laotian government, and Category Two conducted without notification. These attacks eventually increased to about 300 sorties per month.

Most missions were continued bombing of base camps and supply lines, but many late in the year provided direct tactical support of U.S. Marine Corps Operation HARVEST MOON and the 1st Air Cav operations at La Drang Valley. The Marines had discovered a North Vietnamese Army (NVA) base camp near the Cambodian border on 14 November 1965 and called in air strikes. SAC launched 18 B-52F bombers from Guam and dropped 344 tons of bombs on 16 November. They flew a total of 96 sorties and dropped 1,795 tons of bombs before the missions were completed at the end of November. While most observers considered these missions largely ineffective, Major General L.W. Walt, 3rd Marine Amphibious Force Commander, was impressed with the results. He said the timing was precise, bombing was accurate, and overall effects were awesome. He also stated, "The enemy has abandoned his prepared positions and much of his equipment in great confusion, and this is making our part of the job easier."

While the B-52F that supported initial operations in Vietnam were modified with the SOUTH BAY and SUN BATH modifications to increase their bombloads to 38,250 pounds, the B-52D fleet began receiving the new "BIG BELLY" modification in December 1965. This modification, along with the addition of wing pylons, increased the total bombload to 60,000 pounds. The first modified B-52D aircraft from the 28th BMW at Ellsworth AFB, SD and the 484th BMW at Turner AFB, GA arrived on Guam in April

1966, and the D model soon replaced all F models, which were returned to the U.S. to resume their nuclear alert role.

In 1966, SAC bombers flew 5,217 Operation ARC LIGHT sorties and continued to attack Viet Cong base camps to prevent force build-up. The B-52D effectiveness was enhanced by the introduction of the COMBAT SKYSPOT bombing system, which greatly increased the accuracy of strikes and allowed all-weather operations. The B-52s own BNS was not very useful in Vietnam since the targets were generally featureless jungle with little distinctive terrain or structures to mark them. COMBAT SKYSPOT involved the siting of a network of ground stations with AN/MSQ-77 radar across the country. The ground stations would track the bombers, guide them to the precise target, and tell the crews when to release their bombs. COMBAT SKYSPOT allowed the B-52D models to be used in battle areas where friendly forces were present, greatly reducing the risk of friendly-fire casualties.[13] The bombers flew so high and fast they could not be seen or heard on the ground making the bomb explosions a complete surprise to enemy troops. They were very effective against troop concentrations and supply lines. They also began providing CAS for ground troops in direct contact with the enemy thanks to COMBAT SKYSPOT.

A B-52D takes-off from Andersen AFB, Guam during Operation ARC LIGHT. (National Archives)

On 11 and 26 April 1966, 30 B-52D bombers hit approaches to the Mu Gia Pass in North Vietnam to stop infiltration of enemy troops into Loas and down the Ho Chi Minh Trail. General Westmoreland commented, "... we know from talking to many prisoners and defectors, that the enemy troops fear B-52s, tactical air, artillery and armor, in that order." He also said SAC attacks thwarted enemy plans and prevented massing and carrying out planned maneuvers. However, within two weeks he requested additional strikes because traffic flow had returned to pre-strike levels. CINCPAC denied the request due to the build-up of SA-2 SAM sites and the fact that previous strikes had only closed the Mu Gia Pass for relatively short periods. Consequently, these attacks were followed by a five-month pause against North Vietnamese targets.

B-52D bombers conducted ARC LIGHT attacks in April and May 1966 in the Tay Ninh province destroying 14 base camps with 162 sorties. Several missions were also conducted over the next several months against the Ho Chi Minh trail in North Vietnam. Although the bombers caused heavy damage, it was quickly repaired by North Vietnamese forces. A B-52D from the 454th BMW was nearly hit by two SA-2 SAMs. The EWO detected the inbound missiles and quickly told the pilots to turn sharply while he jammed the missiles causing them to explode just 3,000 feet from the aircraft. This near miss provided the justification Washington needed to eliminate any additional bomber missions above the Demilitarized Zone (DMZ).

SAC B-52 bombers flew 9,686 Operation ARC LIGHT sorties in 1967, nearly double the amount from 1966. On 6 May 1967, SAC flew its 10,000th sortie in SEA and had dropped more than 190,000 tons of bombs in less than two years of combat operations. Weapons included Mk-82 500-pound, M117 750-pound, and 1,000-pound semi-armor piercing bombs. Despite the objections of senior Air Force leaders, including General William W. "Spike" Momyer, 7th AF Commander and later Military Assistance Command Vietnam (MACV) Deputy for Air, who believed Westmoreland's control over the bombers violated the basic doctrine of a separate Air Force run by airmen, and was the product of Army thinking, the bombers continued to support ground operations. Most provided CAS and hit enemy troop concentrations and supply lines near the Cambodian Border and in the DMZ.

The bombers flew in cells of three aircraft each and could saturate an area in the jungle 3,000 yards long and 1,000 yards wide. They were also eventually used to hit Viet Cong tunnel complexes with heavy bombs using delayed action fuses that detonated deep in the ground and caved in the tunnels.[14]

SAC crews typically flew several dozens of ARC LIGHT missions[15] during six months on TDY to SEA. Aircraft operating from Andersen AFB, Guam, faced 12-hour missions consisting of a 5,000-mile round trip to the targets in SEA, including prestrike aerial refueling, and return to Guam almost daily. This was on top of 2-3 hours for crew briefing and aircraft pre-flight, and another two hours for debriefing after landing.

SAC began flying B-52 sorties from U-Tapao RTNAF as a Forward Operating Base (FOB) on 10 April 1967 to reduce mission times and eliminate the need for aerial refueling. The bombers would fly a bombing mission over SEA from Andersen and land at U-Tapao. They would then fly eight more missions out of Thailand and then land back at Andersen. These missions originating from U-Tapao were controlled by the 4258th SW, which had been operating KC-135 tankers for Tactical Air (TACAIR) support since June 1966.

Unfortunately, July 1967 was a sad month for SAC due to the combat losses of three B-52 aircraft. On 7 July, two B-52D (56-0595 and 56-0627) aircraft collided in mid-air and crashed in the South China Sea. Six of the 13 crew members perished including Major General William J. Crumm, 3rd AD Commander. On 8 July, a B-52D (56-0601) diverted to Da Nang AB, South Vietnam due to electrical problems. The pilot approached the runway for a no-flaps landing. The aircraft touched down and then bounced about 6,000 feet down the runway before it touched down again. It overran the runway and crashed in a drainage ditch where it broke up and caught fire. Five of the six crew members perished. Only the tail gunner survived.

SAC committed over 70 B-52 bombers to SEA in 1968 and flew 20,658 Operation ARC LIGHT sorties to support increased bombing efforts in response to NVA offensives. Tensions increased in the early part of the year as the NVA launched the Tet Offensive. The attacks finally drove the consolidation of all aviation forces under the control of the MACV Deputy for Air. On 8 March, Momyer was given overall mission direction and the Air Force was given overall command.

B-52D bombers from Andersen, U-Tapao, and Kadena AB, Okinawa[16] flew over 2,548 around the clock bombing missions and dropped 59,542 tons of bombs during Operation NIAGARA to destroy enemy supply concentrations during the siege of Khe Sanh, which was at its height during the monsoon season on 21 January. TAC fighter bombers were unable to attack during inclement weather, but the B-52s relied on COMBAT SKYSPOT ground radar to see through the weather and hit their targets. These strikes were instrumental in breaking the siege and causing the enemy to withdraw. Bombing continued throughout the year and concentrated in the A Shau Valley, tri-border area, and the assault corridor running from the Cambodian Border to Saigon. By November 1968 the bombers had flown a total of 35,680 sorties and dropped 886,490 tons of bombs.

SAC redesignated U-Tapao RTNAF, Thailand from an FOB to a Main Operating Base (MOB) in January 1969 and it subsequently supported most B-52 bombing operations

A B-52D (55-0110) being loaded with Mk-82 and M117 bombs during Operation ARC LIGHT. (National Archives)

in SEA. B-52D bombers continued Operation ARC LIGHT missions including harassment and disruption of enemy operations, and bombing of troop concentrations, base areas, and supply lines. Most B-52 strikes were concentrated on the assault corridor.

The bombers also performed CAS on targets in Cambodia beginning in March under a series of top-secret missions that became known as "Menu" operations. The first, called Operation BREAKFAST, launched on 18 March 1969, and hit a logistical supply and storage area called Base Area 353 three miles inside Cambodia. These top-secret missions continued in April and May under the code names Operation SUPPER, LUNCH, DESERT, and SNACK. The bombers flew a total of 804 sorties and hit 140 targets.

The Menu operations continued through 26 May 1970 until they were halted after being exposed by a report in the *New York Times*. During 14 months of operations, the bombers dropped 100,000 tons on Cambodia in 3,630 sorties. These missions were supported by 763 overt ARC LIGHT sorties that crossed the border into Cambodia in May

1970. Although exposed, the attacks on targets in Cambodia continued in the open from June 1970 until August 1973.

By January 1970 B-52D sorties from Andersen were no longer needed to support the current operational commitments. B-52D operations from Kadena were discontinued in September 1970. SAC continued Operation ARC LIGHT missions in Vietnam during 1971 using the 42 B-52D aircraft operating out of the 307th SW at U-Tapao. They continued providing CAS for American forces in combat and destroyed enemy troop concentrations and supply lines along the DMZ and Laotian border. The U-Tapao bombers were joined once again by B-52D bombers operating out of Andersen in 1972 and 1973.

Commando Hunt. B-52s began supporting COMMANDO HUNT I operations in Laos on 15 November 1968. These operations evolved into a series of operations (Commando Hunt II, III, etc.) lasting six months each. They were conducted primarily using TACAIR, AC-119 and AC-130 Gunships, and B-52 bombers. They concentrated on the Ho Chi Minh Trail, road networks, terrain including jungle and mountain passes, and AAA/SAM sites. The bombers were most effectively applied against passes and stationary targets. One square mile boxes were established around the Ban Kari, Mu Gia, Ban Raving, and Nape passes. About 27 B-52 sorties per day attacked these boxes resulting in 3,377 sorties over Laos by the end of 1968.

COMMANDO HUNT II began in May 1969 at the height of the monsoon season. While other air assets were grounded due to bad weather, B-52s were used to drop 500-pound bombs on wet mountain passes which caused huge landslides and quickly closed the roads. But the enemy used the pause in bombing of North Vietnam to build up supplies near the border. B-52 bombers also flew interdiction missions in Laos supporting COMMANDO HUNT III from November 1969 to April 1970. The attacks seemed to be working, and the infiltration rate was down more than 50 percent. Consequently, Nixon ordered a reduction in COMMANDO HUNT sorties on 26 February 1970. But this turned out to be a gain for the North Vietnamese, who spent the next two years building up their forces.

During COMMANDO HUNT IV B-52 bombers supported Operation BARRELL ROLL in northern Laos. COMMANDO HUNT V began in late 1970 when intelligence discovered a supply depot on the northern end of the Ho Chi Minh trail. On 30 October, Phase I of the operation began as the U.S. set up a logistics base on the Laotian border and brought in more than 12,000 South Vietnamese Army of the Republic of Vietnam (ARVN) troops to prepare a ground attack on the depot. On 8 February 1971, Phase II began and by the 23rd more than 17,000 ARVN troops entered Laos supported by gunships, TACAIR, and 399 B-52 sorties. But on the 25th the NVA counterattacked and by 3 March the ARVN were bogged down. A week later the ARVN commander ordered withdrawal, which soon turned into a disaster. B-52 strikes provided cover as helicopters airlifted survivors and extracted most of the ARVN by 24 March. The bombers flew 1,358 sorties and dropped 32,000 tons of bombs, while TACAIR flew more than 8,000 sorties.[17]

COMMANDO HUNT V officially ended on 6 April 1971. U.S. helicopter crews saved thousands of ARVN soldiers, but the cost was high, with 107 helicopters lost and 600 damaged. The U.S. lost 176 killed, 1,042 wounded, and 42 missing—many dying while saving their South Vietnamese allies. Enemy losses, mostly due to air strikes, included 14,000 killed and 4,800 wounded as well as 20,000 tons of food and ammunition, 156,000 gallons of fuel, 1,530 trucks, 74 tanks, and 6,000 individual weapons captured or destroyed. The ARVN lost 1,519 killed; 5,423 wounded; 651 missing; 75 tanks; dozens of personnel carriers; 198 crew-served weapons; and 3,000 individual weapons. Perhaps worst of all, during the ARVN retreat they abandoned large quantities of undamaged weapons and supplies, later salvaged, and used by the enemy.[18]

Given these losses, COMMANDO HUNT VI from May to October 1971 was a highly diminished operation and enemy traffic flow on the Ho Chi Minh trail grew markedly. The Communists added 140 miles of new roads; and by October they had more than 2,170 miles of single-lane roads, multi-lane roads, parallel routes, bypasses, and spur roads in Laos. They also added 344 AAA batteries, new MiG bases in southern North Vietnam, and dozens of SA-2 SAM sites, most of which were along the Laotian–North Vietnam border. One estimate placed 96,000 NVA in Laos, 63,000 in Cambodia, and 200,000 in South Vietnam.[19]

COMMANDO HUNT VII, the last of the series, became a response to this massive build up. Planners hoped to use B-52 strikes to close passes, destroy roadways, and force enemy vehicles to congregate in truck parks where they would be attacked and destroyed. But things did not go as planned. The North Vietnamese had already constructed 310 miles of additional roads unknown to U.S. planners. When roads became impassable, they sent supplies down rivers and constructed pipelines to transport petroleum products.

The operation rolled out in three phases. Phase I began when U.S. aircraft, primarily B-52s, struck target boxes near the Mu Gia, Ban Karai, and Ban Raving passes as well as areas in the western DMZ. During the first three weeks tactical aircraft and B-52s dropped 14,400 instantaneously fused 500-pound bombs, 17,100 750-pound bombs, as well as a few dozen 2,000-pound laser-guided bombs, Mk-36 magnetic-influence mines, and cluster bomb unit antipersonnel mines. The initial bombing appraisal determined that enemy traffic had been slowed some of the time in some places. But by 4

November intelligence indicated that the Mu Gia pass roads were already being repaired, and that the traffic flow was near normal levels.[20]

During phase two, which began in late November, U.S. aircraft struck the enemy units as they moved south. Roads were cut by B-52s, which left large craters and created choke points and blocking belts. As enemy truck traffic backed up, Air Force fighters attacked with laser-guided bombs, using data gathered from Task Force Alpha sensors. They also seeded the area with air-dropped mines. As enemy units attempted to clear the mines or repair the roads, further attacks caught them exposed, causing great destruction.[21] Despite heavy bombing, the North Vietnamese were able to quickly repair the damage and keep the routes open.

The third phase began in early 1972 and shifted attacks to exit points from Laos into South Vietnam and Cambodia as well as against enemy AAA batteries. U.S. TACAIR flew 31,500 sorties, half by the Air Force, while B-52 bombers flew 3,176 more. Official reports claimed that large numbers of enemy vehicles were destroyed or damaged, thousands of NVA killed, and tens of thousands of tons of supplies destroyed. U.S. officials declared the operation a success that prevented another Tet-style uprising. However, the operation ended abruptly on 31 March 1972 only one day after the North Vietnamese launched the Easter Offensive.

Freedom Train. President Johnson had agreed to end Operation ROLLING THUNDER bombing of North Vietnam by TACAIR forces on 31 October 1968 to achieve a peace agreement. The bombing halt came with the agreement that the North Vietnamese would not use the area near the DMZ to attack U.S. forces, would not strike cities in South Vietnam, and the U.S. could continue reconnaissance flights over the DMZ to verify compliance.[22] For the next four years SAC supported ARC LIGHT bombing operations in South Vietnam. But the bombing halt failed to produce fruitful peace talks and instead resulted in a diplomatic stalemate. It also allowed the NVA to peacefully buildup its forces in the north and prepare new offensives. Without warning, on 30 March 1972, twelve divisions of NVA forces comprised of 120,000 soldiers and more than 200 tanks swept across the DMZ and overran targets throughout the south in an operation known as the Easter Offensive.

President Nixon ordered an immediate response to include the renewed bombing of strategic targets in North Vietnam. In early April, SAC sent B-52 bombers to Andersen AFB, Guam under Operation BULLET SHOT[23] to support bombing operations from the island for the first time since 1970. B-52D bombers from Andersen and U-Tapao successfully hit targets in North Vietnam for the first time since 1968 under a new more concentrated effort called Operation FREEDOM TRAIN.

On 9 April, 15 B-52D models hit targets at the Vinh railroad yard and Petroleum, Oil, and Lubricants (POL) supply areas. Another 18 aircraft struck Bai Thuong airfield on 12 April. B-52 bombers and TACAIR struck rail and shipping targets in the Haiphong and Thang Hua areas on 16 April 1972. Similar targets at Hamm Rong and Thanh Hoa were struck one week later.[24]

During April to June, U.S. forces flew 27,745 attack and support sorties, 1,000 of which were flown with B-52s. The U.S. lost 52 planes—17 to SAMs, 11 to AAA, three to small arms, 14 to MiGs, and seven to unknown causes. The enemy fired 777 SAMs in April, 429 in May, and 366 in June. At first, they used ripple firing tactics—one high, one low, and one in the middle—for total coverage. Early enemy successes were later offset by U.S. countermeasures, including the use of chaff, and especially by the B-52s.[25]

Linebacker I. As Nixon prepared to send Henry Kissinger, his foreign policy adviser, back to Paris for a negotiating session with North Vietnam's lead negotiator, Le Duc Tho, on 2 May 1972, he considered a three-day series of B-52 raids against Hanoi, to commence on 5 May. But Kissinger, fearing domestic reaction, and General Abrams, commander MACV, declaring his need for the B-52s in the south to curb the enemy offensive, convinced the president otherwise. Instead, Nixon opted for a plan from Kissinger's military assistant, Major General Alexander Haig, that called for sustained bombing by tactical bombers and mining of Haiphong and other North Vietnamese harbors. Similar in design to Operation ROLLING THUNDER, its main force was to be tactical aircraft from the carriers of Task Force 77 and from the 7th AF. Only a handful of B-52s were to be used, mostly in the south. The operation, called LINEBACKER I, began on 10 May 1972.[26]

On 11 May, with the eastern part of the city of An Loc under attack, one enemy prisoner of war recalled that B-52s struck about 0500 hours. They pounded the eastern approaches to the city every hour on the hour for 25 hours, bombing several targets more than once. Entire units were wiped out. Five days later, a North Vietnamese People's Army of Vietnam (PAVN) column supported by 20 tanks attacked an ARVN force just south of Kontum City on Route 14. Three cells of B-52s attacked each enemy column and obliterated them. On the 26th, the enemy made one last assault on Kontum City, which failed because of a tenacious ARVN defense and B-52 support.[27] By 29 June, President Nixon announced at the Paris Peace Talks that the South Vietnamese were now on the offensive.

Operation LINEBACKER I continued through the summer of 1972 and Nixon authorized B-52 and TACAIR attacks in North Vietnam along the DMZ which averaged about 30 sorties a day. On 8 August, he authorized the military to "hit the North harder" which resulted in an increase to about 48 sorties per day, although most were flown by

TACAIR. The attacks eventually helped bring the North Vietnamese back to Paris to negotiate an end to the war. Bombing was halted during the talks and by 21 October Kissinger announced that "Peace is at hand."

LINEBACKER I was effectively ended on 22 October 1972. However, the North Vietnamese took the bombing halt as an opportunity to build-up its forces and soon resumed offensive operations in South Vietnam. Thus, SAC B-52 bombers and TACAIR continued to hit targets south of the 19th parallel until January 1973 in response to North Vietnamese offensives. Operation BULLET SHOT also continued throughout the spring and summer and resulted in more than 150 bombers, including 98 G models, and 12,000

B-52G in a revetment on the flightline at Andersen AFB, Guam. (Ronny Young Collection)

personnel being deployed to Guam from across SAC. The bombers continued to support ARC LIGHT sorties concurrently with their LINEBACKER I commitments.

While LINEBACKER I generally had fewer restrictions than Operation ROLLING THUNDER, it was still subject to strict guidelines. Restrictions included a no-bombing-buffer area on the Sino-Vietnamese border, as well as on northern dams, dikes, civilian watercraft, civilian population centers, and non-Vietnamese seaborne shipping. All attacks had to be initially approved by the JCS.[28]

B-52D (55-0110) call sign Olive 2 became the first aircraft lost to enemy fire for more than six years. It was struck on 22 November 1972 by an SA-2 SAM near Vinh on the central coast of North Vietnam and was seriously damaged. The pilot, Captain Norbert J. Ostrozny, turned the aircraft toward Thailand and attempted to make it to U-Tapao. After crossing the Thai border, he realized the aircraft was too damaged and ordered the crew to eject. All six crew members escaped the aircraft and were rescued by an HH-52 Search and Rescue (SAR) helicopter. The aircraft crashed 15 miles southwest of Nakhon Phanom, Thailand.

B-52D (55-0110) on approach at U-Tapao RTNAF, Thailand on 30 October 1972. It was lost to enemy fire over Vietnam on 22 November becoming the first aircraft lost in more than six years. (U.S. Air Force)

Linebacker II. By November peace talks had again stalled as the North Vietnamese dragged their feet to prevent the U.S. from renewing bombing operations in the north while they continued operations in the south. In mid-December the North Vietnamese presented 17 changes to the previously agreed concessions and refused to delete them. Peace talks broke off again on 13 December. President Nixon responded to renewed enemy offensives and stalled peace negotiations by ordering Operation LINEBACKER II on 14 December 1972. Unlike Operation LINEBACKER I, which was primarily an interdiction effort directed at supply routes, Nixon ordered SAC to conduct maximum effort strategic strikes against targets in North Vietnam including Hanoi and Haiphong designed to bring the North Vietnamese back to the negotiating table.

The operation began on the night of 18 December with 129 bombers launched from Andersen and U-Tapao, including 52 B-52G models. The bombers hit targets in Hanoi and Haiphong in three waves. Targets included MiG bases, railway yards, vehicle repair and storage areas, and other strategic targets. Tactics dictated by SAC Headquarters required the bombers to fly straight and level until bombs were dropped over the targets. All aircraft in all cells followed the same ingress and egress patterns. This meant the bombers could not perform evasive maneuvers to avoid the SAMs. Two B-52G models and one B-52D were shot down, while two were damaged and diverted to U-Tapao. Day two included the launch of 93 aircraft in three waves. Over 180 SAMs were fired with no bombers hit. The next five days would not prove so lucky with six aircraft, two B-52D and four B-52G, being lost on Day three and two more B-52D aircraft were lost on 21 December before the President called a bombing halt for Christmas day.

When the bombing continued on 26 December, SAC authorized new tactics including different ingress and egress routes. All 120 aircraft, flying in cells of three each, were engaged in a single mass assault with a common bomb release point. In addition, B-52G models, with less capable ECM gear than the D models, were sent over less heav-

A B-52D awaits take-off as a B-52G (58-0244) lands at Andersen AFB, Guam after completing a bombing mission over North Vietnam during Operation LINEBACKER II. (U.S. Air Force)

ily defended targets. Two B-52D aircraft were lost, but they were aircraft in cells of two aircraft where the third aircraft had aborted enroute to target. These two-aircraft cells were no longer authorized after that night. On the 27th, another 60 bombers were launched against targets in Hanoi and the Lang Dang rail yards near the Chinese border. Two more B-52D aircraft were downed, bringing the total to 15.

By 28 December, most targets were destroyed, and the 60 bombers launched were able to fly over Hanoi and Haiphong without encountering enemy defenses. The operation ultimately ended on 29 December, and the North Vietnamese announced they were ready to resume negotiations on 30 December. The U.S. leadership had finally

Andersen AFB flightline during Operation LINEBACKER II. (Ronny Young Collection)

done what General LeMay and the JCS had recommended in 1964. The JCS had identified 94 targets in North Vietnam that they believed would destroy the enemy's ability to conduct war and their will to continue. According to LeMay, "We could have ended it in any ten-day period you wanted to, but they never would bomb the target list we had."[29]

B-52D bombers in revetments on the north ramp of the Andersen AFB flightline during LINEBACKER II. (U.S. Air Force)

B-52s from both Andersen and U-Tapao flew a combined total of 729 sorties against 34 target complexes including railyards, shipyards, air bases, radars, and missile sites. SAC bombers dropped 15,237 of the 20,370 tons of bombs dropped by U.S. aircraft during the 11-day war. Bomb damage assessment showed 1,600 military structures, 383 pieces of rolling stock, three million gallons of petroleum products, and 80 percent of electrical production damaged or destroyed, as well as 500 rail interdictions and ten airfield and runway interdictions.

The bombers encountered SAMs, MiG fighters, and AAA in one of the most heavily defended areas in the world at the time. The SAMs were the biggest threat and the NVA fired at least 884 missiles.[30] They hit 15 B-52 aircraft, most in the first few days, with ten shot down over North Vietnam while the rest made it to Laos or Thailand airspace before going down.[31] A total of 92 crew members were aboard resulting in 29 listed as Missing in Action (MIA), 33 captured and listed as Prisoners of War (POW), and 24 recovered by SAR teams. Four crew members were killed and two survived during a crash landing at U-Tapao.

The Paris Peace Talks resumed on 8 January 1973, but SAC B-52 bombers continued attacks on logistics targets in North Vietnam south of the 20th parallel as a reminder to the North Vietnamese negotiators. The bombing ended on 15 January. An agreement was signed on 27 January and B-52s flew their final mission over South Vietnam. B-52D bombers continued Operation ARC LIGHT missions against targets in Laos until 22 February under an agreement with the Laotian government. Enemy cease-fire violations brought the bombers back on 23 February, and 15-18 April, as requested by the Laotian government.

Meanwhile, rebel Cambodians with NVA and Viet Cong forces advanced on Phnom Penh, Cambodia and were repeatedly attacked by SAC B-52 bombers. The bombers also struck logistics targets, gun positions, and troop emplacements in Cambodia to slow the enemy advance towards Phnom Penh. This continued until 15 August 1973 when all SEA bombing operations were terminated. B-52 bombers had logged 126,615 sorties in SEA during the war. They dropped 2.63 million tons of ordnance and lost 18 aircraft to enemy defenses with 13 more to other operational causes.[32]

More than 150 B-52 bombers packed the Andersen AFB flightline during Operation LINEBACKER II. (U.S. Air Force)

Gulf War

Desert Shield. Soon after Iraq's leader Saddam Hussein invaded Kuwait on 2 August 1990, SAC committed B-52G aircraft to support the build-up of forces for U.S. Central Command (CENTCOM) operations under Operation DESERT SHIELD. They initially moved seven aircraft from the 69th BS/42nd BMW at Loring AFB. ME to Diego Garcia air base in the Indian Ocean. By 16 August, at least 20 bombers from the 42nd BMW and 328th BS/93rd BMW at Castle AFB, CA were standing strip alert on the island.[33] SAC planners also made an agreement with Saudi Arabia to secretly base B-52 bombers at Prince Abdullah AB[34] in Jeddah. But the Saudis were reluctant to allow the bombers to enter the country until combat operations were initiated, known as H-hour.

SAC designated the 524th BS/379th BMW at Wurtsmith AFB, MI as the lead element for the 1708th BMW (Provisional) which would operate at Jeddah. Since their aircraft could not immediately deploy to Saudi Arabia, the wing placed 10 aircraft, including three from the 93rd BMW and two from the 42nd BMW, in Employ/Deploy status meaning they would conduct bombing runs enroute to their deployment location at Jeddah. In addition, the 42nd and 93rd BMW deployed six aircraft to Andersen AFB, Guam in October 1990 to stand strip alert and await H-hour. SAC also based bombers at Moron AB, Spain beginning in January 1991 after being denied air bases in Egypt.

Before Operation DESERT STORM began on 17 January 1991, SAC had deployed at least 20 B-52G aircraft to form the 4300th BMW (Provisional) at Diego Garcia and 20 aircraft to form the 801st BMW (Provisional) at Moron, Spain, as well as the 10 aircraft prepositioned at Wurtsmith and six on Guam. Another eight aircraft were planned to arrive at RAF Fairford in February to form the 806th BMW (Provisional) giving at least 64 available aircraft. However, the numbers at each of these bases fluctuated and at least 89 aircraft were eventually committed to combat operations (not simultaneously).

Desert Storm. In the early hours of the war on 17 January, 12 B-52G aircraft from Diego Garcia conducted the first combat low-level bombing runs, in four waves of three aircraft each, entering Iraqi airspace at 300 feet and flying across the desert at less than 100 feet. They hit Iraqi air bases at Tikrit, Mudaysis, Wadi al Khirr, Ghalayson and As Salman with 1,000 pound bombs using delayed-action fuses as well as CBU-87 and CBU-89 cluster bombs rendering the bases inoperable for several days.[35] These bombers were among the first aircraft, including F-117, F-15, F-111, and other strike aircraft, to strike targets in Iraq and included the six B-52G aircraft previously deployed to Diego Garcia. These six aircraft recovered in Jeddah after completing their strikes.[36]

Long before the official start of the war, seven B-52G aircraft (57-6475, 58-0177, 58-0183, 58-0185, 58-0238, 59-2564, and 59-2582) of the 2nd BMW departed Barksdale

AFB, LA at 0636 hours on 16 January 1991 marking the beginning of the first Operation DESERT STORM operational mission, code named Operation SENIOR SURPRISE.[37] The bombers were armed with the first AGM-86C CALCMs to see operational service. They traveled over 7,000 miles in about 16 hours, arriving in western Saudi Arabia near southern Iraq at dawn local time on 17 January where they prepared their missiles for launch. The launch was timed for all missiles to hit their targets[38] during the gap between waves of aircraft after they were clear of Iraqi airspace. The bombers launched 35 missiles[39] on eight high priority targets including power generating facilities, transmission lines, and communications centers which disrupted radar and communications and "blinded" the Iraqis to other on-going operations. They then headed back to Barksdale, bucking 130 to 140-knot headwinds, after an already grueling mission landing 35.4 hours[40] later and covering 14,000 miles, marking the longest bombing mission in history at that time.

The ten B-52G aircraft[41] in Employ/Deploy status at Wurtsmith AFB, MI took off in three cells on 18 January equipped with M117 750-pound iron bombs and newly developed CBU-87 cluster bombs. After refueling off the coast of Rhode Island and over Moron, Spain one of the cells performed high-altitude strikes against Republican Guard targets in and near Kuwait before landing in Jeddah after a 17.5-hour flight. The remaining aircraft encountered bad weather and were diverted directly to Jeddah. By the second day of operations four additional aircraft had recovered in the country and Jeddah had received its full complement of 20 bombers.

B-52G (58-0182) prepares for take-off from RAF Fairford during Operation DESERT STORM. (U.S. Air Force)

Initially, bombers from Diego Garcia, Jeddah, and Moron conducted low level night strikes over Kuwait using EVS with FLIR and LLTV to evade Iraqi radar. The aircraft flew in cells of three and each cell was supported by F-4G Wild Weasels for SAM radar suppression and F-15C Eagles for MiG Combat Air Patrol (MiGCAP). Each cell was also headed by a NAVSTAR GPS equipped aircraft from the 42nd BMW at Loring AFB, ME or the 93rd BMW at Castle AFB, CA since they were the only wings then equipped. But they were met with AAA and SAMs and on the third night one aircraft was apparently

A B-52G loaded with iron bombs prepares for take-off during Operation DESERT STORM. (U.S. Air Force)

hit, although it returned to base safely. After this close call CENTCOM directed the bombers to 35,000 feet to get out of harm's way.[42]

Cells were launched every three hours, 24/7, and by mid-February 40 to 50 strikes a day kept maximum pressure on troop concentrations and the Republican Guard who could not even see or hear the bombers before bombs exploded around them. The bombers dropped dual-fuzed Mk-82 bombs to breach huge berms the Iraqis had built-up to fend off the expected amphibious attack. They attacked front line troops, artillery firebases, SAM sites, and vehicle parks. They orbited for hours above the desert on "Scud hunts" awaiting the call to conduct surprise attacks on newly discovered Scud launchers. They also flew interdiction missions against ammunition factories, supply dumps, storage areas, oil refineries, fuel depots, Scud missile storage and production facilities, industrial sites, and air bases.[43]

The B-52G also worked well as a psychological weapon. Days before the land battle began, B-52G aircrews dropped psychological warfare leaflets to warn Iraqi forces that the B-52s were coming. After the attack, aircrews dropped more leaflets telling the Iraqi troops they would be back. General Norman Schwarzkopf favored using the B-52 against the massed Republican Guard, but he rejected the term "carpet bombing," which he said, "tends to portray something totally indiscriminate, en masse with regard to the target." Despite some inaccuracy, it was estimated from POW interviews, during and after the war, that the B-52G influenced 24 percent of Iraqi soldiers to desert.[44]

A fully loaded B-52G takes-off from Jeddah during Operation DESERT STORM. (U.S. Air Force)

B-52G bombers were also put to good use during the Battle of Khafji. On 29 January 1991, Iraqi forces crossed the Kuwait border in three areas including Al Khafji. Coalition forces successfully repelled the attacks supported by a variety of CAS aviation assets. At the request of the Marine Commander, Lieutenant General Boomer, a B-52 strike and two TACAIR packages were brought in to attack approximately 100 Iraqi armored vehicles moving to reinforce the initial Iraqi penetration. According to a field report, the effectiveness of the B-52 strike was "like turning on a light in a cockroach infested apartment." The B-52 strike sent the vehicles scurrying for survival only to find their movement awaited by TACAIR.[45]

The bombers flew 1,741 sorties, including 846 from Jeddah, and dropped 72,289 weapons including Mk-82, M117, CBU-52, CBU-58, CBU-71, CBU-87, CBU-89, and UK 1,000-pound bombs. They dropped over 27,000 tons of explosive which was about 30 percent of the total coalition[46] tonnage. The bombers were credited with damaging or destroying large numbers of Iraqi tanks, armored personnel carriers, and artillery. They flew 99 offensive counterair strikes against airfields, aircraft on ground, and airfield-supporting infrastructure. B-52 bombers flew 303 strikes against strategic targets and interdiction targets, most flown at high level using unguided general purpose and cluster bombs. They also flew 1,175 strikes against Republican Guard, armor, and mechanized and infantry units in the Kuwait Theater of Operations (KTO).[47]

No aircraft were lost to combat fire but one B-52G (59-2593) was lost on 3 February 1991 when it crashed just short of landing at Diego Garcia due to electrical system failure. Three crew members were lost. A SAM hit one aircraft and AAA hit another, but both aircraft made it to Jeddah for a successful landing and were later RTS. A third B-52G (58-0248) was struck by a friendly AGM-88 HARM anti-radiation missile fired from an F-4G Wild Weasel assigned to escort the bombers. The missile inadvertently locked on to the bomber, ultimately tearing off the last seven feet of the fuselage, when it mistook the bomber's gun-laying radar emissions for an Iraqi AAA site.[48] Despite this close call, the aircraft landed successfully at Jeddah and was repaired just enough for a one-time flight to Andersen AFB, Guam via Diego Garcia. The aircraft was repaired by replacing the damaged tail section with parts of a B-52G (58-0234) which was grounded at Andersen due to a wing crack. The aircraft was RTS with the 93rd BMW at Castle AFB, CA after several months of repair work.

Iraq Operations

Desert Strike. On 31 August 1996, Saddam Hussein, still the leader of Iraq five years after the end of the Gulf War, continued to defy the United Nations (UN) and taunt American leaders. He boldly moved a Republican Guard mechanized unit consisting of 40,000 troops into northern Iraq in breach of National Security Council (NSC) Resolution 688, which prohibited repression of the Kurdish people. His troops quickly overtook the town of Irbil and began firing SAMs on Air Force aircraft patrolling the "No-Fly Zone'.

The U.S. quickly responded by deploying four B-52H (60-0014, 60-0025, 60-0054, and 60-0059) bombers from the 2nd BW at Barksdale AFB, LA to Andersen AFB, Guam on 31 August. Three aircraft (60-0014, 60-0025, and 60-0054), each loaded with a CSRL and eight CALCMs, took off from Guam on 3 September. The spare aircraft (60-0025) returned to Guam after the first refueling and the remaining two continued to the target. They were intercepted near the target area by Iraqi Mirage F1 fighters which were quickly chased off by Navy F-14 Tomcat bomber escorts. Two Navy ships stationed in the Persian Gulf as part of Naval Forces Central Command (NAVCENT) Task Force 50 fired 14 Tomahawk missiles on targets in Iraq. The two bombers then fired three CALCMs each against targets in southern Iraq. They then pulled back to reload mission tapes and reboot their systems before turning back to the targets. They ultimately fired a total of 13 missiles, hitting their targets, while three malfunctioned and could not be launched. They were followed later that day by another 17 Tomahawks fired from four Navy ships.

This 34-hour mission covered 13,600 miles, making it the second longest bombing mission in history at the time, just slightly shorter than the opening Operation DESERT STORM mission on 17 January 1991, and resulting in the award of the McKay Trophy for 1996. The bombers received three refuelings from 17 tankers at night in stormy tropical weather and less than one mile of visibility. Two of the bombers were forward deployed from Guam to Diego Garcia on 15 September to reduce the distance in case further strikes were needed. They were subsequently joined by two more aircraft and the four remained deployed until 12 October.

Desert Thunder. It was apparent by November 1997 that Saddam Hussein was continuing to defy UN weapons inspectors. He was convinced that they were merely a front for U.S. spying activities. He demanded on 13 November 1997 that all American citizens working for the UN Inspection Command (UNISCOM) leave Iraq immediately. Three days later President Clinton ordered a build-up of forces and on 19 November six B-52H aircraft from the 2nd BW deployed to Diego Garcia. They were later joined by two more aircraft and then again on 7 February 1998 six more aircraft deployed.

Diplomatic efforts eventually calmed the crisis, and the bombers were redeployed to their home station by mid-June. However, by the end of October Iraq again failed to comply with UNISCOM inspection requests and on 11 November 1998 Clinton ordered a new build-up of forces. Nine B-52H aircraft were again deployed from Barksdale to Diego Garcia. On 14 November a B-52H force was launched toward Iraq loaded with CALCMs. When the bombers were just 20 minutes from their launch box Hussein relented and the bombers were recalled back to Diego Garcia.

Desert Fox. Eight B-52H (60-0010, 60-0016, 60-0025, 60-0043, 60-0046, 61-0006, 61-0020, and 61-0023) aircraft from Barksdale remained on Diego Garcia and were joined by seven more (60-0007, 60-0015, 60-0023, 60-0034, 60-0036, 60-0044, and 61-0007) from the 5th BW at Minot AFB, ND on 11 December 1998. President Clinton approved a series of attacks on Iraq in conjunction with the UK from 16-19 December 1998 designed to force Saddam Hussein to cooperate with UN weapons inspectors and to degrade Iraq's ability to manufacture Weapons of Mass Destruction (WMD).

Night attacks on 16 December consisted solely of Navy Tomahawks and carrier-based aircraft. Air Force aircraft, including 14 B-52H aircraft from the 2nd Air Expeditionary Group (AEG), joined the attacks on 17 December. The bombers launched from Diego Garcia in two flights of six aircraft (and one spare) six hours apart. They carried eight Block 1 CALCMs each and launched a total of 74 of the missiles in the early hours of 18 December. The Block 1 CALCMs featured a new 3,000-pound warhead with increased accuracy and caused significant damage against presidential palaces, Republican Guard barracks, military, and industrial targets. Two more bombers took off on the night of 18 December and launched 16 CALCMs against targets in Iraq including the al-Taji missile manufacturing and repair facility causing significant damage.

United States and UK leaders claimed success on 19 December after striking a total of 97 targets using various weapons including the 90 CALCMs fired by B-52H bombers. Most of the bombers redeployed to their home stations on 22 December with four remaining on Diego Garcia. The bombers maintained a forward deployed presence until 21 April 1999 when the 2nd AEG was deactivated.

Balkan War

Near the end of the Cold War in October 1990 the U.S. National Intelligence Estimate predicted Yugoslavia would cease to function within one year and would dissolve into independent republics. By 1992, this prediction came to fruition. Slovenia and Croatia became independent republics after relatively brief wars, but Bosnia and Herzegovina were soon caught up in a bloody conflict for independence. The UN established a no-fly zone over Bosnia-Herzegovina on 9 October 1992, and NATO began operation DENY FLIGHT on 12 April 1993 to enforce the UN Security Council Resolution 781. The U.S. deployed several fighter aircraft and the 917th Wing of the AFRC deployed B-52H aircraft to Aviano AB, Italy in December 1993 to supplement the aircraft of the 16th AEG. They also deployed two more times, in August 1994 and May 1995, but the bombers were ultimately not required and returned to home station each time without action. Consequently, when the U.S. and NATO launched Operation DELIBERATE FORCE on 30 August 1995 the B-52H was not included in the deployment package.

By October 1998, Kosovo was demanding its independence from Yugoslavia and Yugoslavia responded with intimidation and a campaign of ethnic cleansing. NATO responded with the threat of an air campaign scheduled to begin 17 October and included seven B-52H (60-0010, 60-0032, 60-0043, 60-0059, 61-0006, 61-0010, and 61-0023) bombers deployed to RAF Fairford in the UK. Given this ultimatum by US diplomat Dick Holbrook, the Yugoslav government soon backed down and signed a ceasefire agreement. The bombers thus returned to Barksdale. Unfortunately, the ceasefire broke down in December and subsequent peace talks failed on 23 March 1999.

Allied Force. President Clinton thus ordered air attacks, which began on 24 March 1999 and included B-52H aircraft, under Operation NOBEL ANVIL (also known by the NATO code name Operation ALLIED FORCE). In preparation, eight B-52H (60-0010, 60-0016, 60-0020, 60-0022, 60-0049, 60-0059, 61-0016, and 61-0031) aircraft had already deployed to RAF Fairford on 21-22 February. The aircraft were sourced from both the 2nd and 5th BW and were assigned to the 2nd AEG along with B-1B and KC-135 aircraft. The unit launched six mission aircraft and two spares each loaded with eight CALCMs. The spares returned to base after the initial launch, and the mission aircraft struck targets in Pristina, Kosovo and nearby airfields to shut down the power grid and Serbian IADS radar sites and communication nodes.

Several more missions were launched over the next few days with several aircraft returning to Barksdale to reload their CALCMs and return to Fairford. This rolling changeover was required because the CALCMs could only be transported on the bombers and not by Air Mobility Command (AMC) transports. The changeover was also supported by five aircraft from the 5th BW at Minot and three 2nd BW aircraft that were previously deployed to Diego Garcia (since the end of Operation DESERT FOX). Ultimately, 25 aircraft ended up rotating in and out of Fairford including the original eight and 17 more (60-0009, 60-0011, 60-0014, 60-0018, 60-0033, 60-0037, 60-0043, 60-0044, 60-0046, 60-0051, 60-0052, 60-0062, 61-0002, 61-0011, 61-0020, 61-0023, and 61-0039).

The CALCM equipped B-52H bombers were assigned to target "soft" command and control facilities, air defense radars, and SAM batteries. Despite previous CALCM successes none of the aircraft were able to fire all eight missiles and one aircraft fired only two of its eight missiles. Still, 78 missiles were fired in 21 sorties which finally exhausted the Air Force CALCM inventory. Two aircraft (60-0020 and 61-0002) returned from Barksdale with GBU-10 LGBs and two (60-0049 and 60-0062) with AGM-142 HAVE NAPs on 3 May. However, no GBU-10s and only two HAVE NAPs were fired.

Instead, the B-52s were reassigned to the conventional bombing role and dropped free fall weapons on airfields, depots, and ground forces for the remainder of the operation. On 5-6 May, they struck suspected Yugoslav positions engaging ten armor concentrations. On 10 May, B-52 and B-1 bombers conducted a joint attack and on 7 June B-52 attacks destroyed two battalions of Yugoslav forces. B-52s had flown 184 combat missions and dropped 8,878 weapons including 500-, 750-, and 2,000-pound bombs, leaflet bombs, CALCMs, and two HAVE NAP missiles when the peace agreement was signed on 9 June. This accounted for 23.8 percent of all NATO, and 31.8 percent of all U.S. Air Force, weapons expended. All aircraft remaining at Fairford returned to their home stations on 23 June.

Afghan War

Enduring Freedom Phase I. On 16 September 2001, just five days after 9/11, President Bush vowed to hunt down the terrorists wherever they were hiding and hold accountable every government that harbored them. U.S. intelligence believed the attacks were carried about by al-Qaeda terrorists harbored in Afghanistan. Bush ordered the U.S. military to go on the offensive against terrorist organizations and their training camps. They quickly developed a plan for Air Force and Navy aircraft, and ship-based Tomahawk missiles, to conduct overwhelming air strikes. However, both Saudi Arabia and Pakistan refused to allow aircraft to be launched in their territories. This eliminated the ability to use land-based fighter aircraft and forced the Air Force to rely on long-range B-52H, B-1B, and B-2A bombers.

On 19 September, Bush ordered B-52H bombers from the 2nd BW and the 917th Wing (AFRC), both based at Barksdale AFB, LA, to deploy to Diego Garcia. Eight B-52H (60-0022, 60-0030, 60-0046, 60-0049, 61-0008, 61-0022, 61-0023, and 61-0039) aircraft joined eight B-1Bs as part of the 28th Air Expeditionary Wing (AEW). Unlike previous operations, no target list existed, and combat plans called for the aircraft to be retargeted in-flight based on the latest intelligence data. The B-52s were fitted with Combat Track II datalink to allow the Combined Air Operations Center (CAOC) to pass target data and reprogram the weapons inflight.

The first Operation ENDURING FREEDOM missions with six B-52H and six B-1B bombers launched from Diego Garcia on 7 October 2001. The B-52s were equipped with GBU-31 JDAMs and CBU-103 WCMDs for the first time in combat, as well as 27 Mk-82 500-pound bombs. They hit terrorist training camps, stores, safe houses, and mountain hideouts, as well as 25 drug factories in southern Afghanistan.[49]

After the first two days, the bombers fell into a sustained rate of four B-52H and four B-1B sorties per day. They typically performed Immediate Close Air Support (XCAS) missions without preplanned targets. Each aircraft orbited over a set territory and received target positions, often obtained from Forward Air Controllers (FAC) on mules or horseback, using Combat Track II intelligence data or SATCOM voice. They remained over hostile territory for hours being redirected to various targets until all stores were expended. For example, on 9 October a B-52H (61-0022) loaded with 12 GBU-31 JDAMs and 27 Mk-82 bombs attacked a terrorist camp near Kandahar with five JDAMs, flew

B-52H aircraft landing at Diego Garcia after missions over Afghanistan during Operation ENDURING FREEDOM. (U.S. Air Force)

north to drop one JDAM on a radar site, dropped six more JDAMs on a terrorist camp, and returned to Kandahar to drop its 27 Mk-82 bombs on the terrorist camp before returning to Diego Garcia 13 hours later.

The B-52H aircraft typically flew at 10,000 feet out of reach of AAA and what few SAMs remained. They struck both fixed and mobile targets. In addition to precision strikes, the bombers dropped free-fall bombs on strategic targets and carpet-bombed Taliban front lines and troop concentrations. On 31 October, a B-52H flew the most intense carpet-bombing attack on Taliban positions outside Bagram AB. They also dropped leaflet bombs warning of carpet-bombing attacks causing desertions and collapse of morale.[50]

On 2 November, the crew of a B-52H (60-0022) was on their seventh routine patrol over Afghanistan when they were alerted of a ground FAC under heavy enemy fire. The aircraft was quickly rerouted to the area west of Kabul and made three passes dropping two GBU-31 JDAMs on each run. However, the FAC had passed inaccurate coordinates, and the bombs were ineffective. Worse, the bomber was low on fuel and was forced to leave the FAC under fire for refueling. When they returned to the area the FAC had escaped but this time passed accurate coordinates to the bomber, which successfully attacked the Taliban position.

The same aircraft was then rerouted to attack a local Taliban commander in a bunker located on the high point of a ridge. GPS coordinates were suspect, and no laser was available. Instead, the bomber received video imaging from an overhead RQ-1 Predator to determine coordinates for a successful JDAM attack in one of the most technologically innovative attacks of the war.[51] As a parting shot, the aircraft then dropped a load of 27 Mk-82 bombs on a Taliban cavalry regiment before returning to Diego Garcia. The 16.6-hour flight was one of the longest during the operation.

B-52 and B-1 bombers continued to harass Taliban and al-Qaeda troops throughout November. One B-52H orbiting overhead quickly took out a Taliban tank and troop formation without warning using 16 CBUs just 19 minutes after it was requested by a key Afghan ally General Dostum. The General was "singularly impressed".[52] The bombers continually bombed trenches and artillery positions resulting in heavy losses. The trenches were dug out after each attack and were quickly bombed again. The Taliban defensive lines quickly crumbled after U.S. aircraft including the B-52H destroyed 44 bunker complexes, numerous supply dumps, 12 tanks, and 51 trucks. Taliban forces began abandoning cities across the country on 9 November and by 23 November they agreed to surrender Kunduz. Taliban and al-Qaeda forces retreated to Kandahar, but the city eventually fell after several weeks of fighting.

Enduring Freedom Phase II. The Bonn Agreement, signed on 5 December 2001, formed an interim governing committee, and named Hamid Karzai as its chairman. But the Taliban and al-Qaeda forces continued to resist. They began to reform in Kandahar and at least 2,000, including Osama Bin Laden, occupied the Tora Bora cave complex in the mountains near Jalalabad which was previously financed by the CIA to support the Afghan fight against the Soviets. B-52 aircraft had already begun bombing this complex on 16 November in an attempt to drive them out.

B-52 aircraft also continued to pound Taliban positions around Kandahar logging 375 sorties by 1 December, accounting for 10 percent of the strike missions over Afghanistan and dropping 65 percent of the munitions. On the evening of 1 December, a B-52H supporting XCAS missions southwest of Kandahar was called in by an airborne FAC to suppress enemy shoulder fired missiles who were firing on F-14 Tomcats. The bomber made five passes dropping a total of 27 Mk-82 bombs from 38,000 feet logging "shacks" as the bombs hit their targets.[53]

The United States began an all-out 72-hour offensive against Tora Bora on 3 December using B-52H and B-1B bombers, F/A-18C and F-14 carrier-based aircraft, AC-130 gunships, and Predators. A "Jawbreaker" group of 20 CIA and special forces soldiers were

B-52H dropping a load of free-fall bombs similar to those dropped in Operation ENDURING FREEDOM during a training run over Nevada. (U.S. Air Force)

inserted by helicopter to mark targets with laser designators. The bombers dropped JDAMs and free-fall bombs causing massive landslides and a 15,000-pound BLU-82/B bomb was used to seal a portion of the complex. A B-52H also launched an AGM-142 HAVE NAP to seal another cave entrance. Carrier-based aircraft hit pinpointed targets with LGBs and JDAMs, while the Predators remained overhead, and gunships racked ridgelines with deadly accurate cannon fire causing some al-Qaeda fighters to beg for mercy over their radios. Some 2,000 local tribal militias, supported by CIA, U.S., British, and German special forces progressively cleared enemy positions, over-running the last cave complex by 17 December. Meanwhile, Bin Laden had escaped into Pakistan.[54]

B-52 XCAS strikes also made a significant contribution to once again expelling Taliban forces from Kandahar. U.S. Marines and special forces entered the city and fully secured it on 9 December after two weeks of intensive ARC LIGHT style B-52H and B-1B bombing raids. The bombers, and carrier-based aircraft, continued to loiter overhead at Kandahar and Tora Bora but no munitions were dropped. This continued well into February 2002, and the bombers routinely returned to Diego Garcia without being called to drop any of their weapons.

Despite this relative calm in the air war, on 3 January 2002 the bombing effort shifted to a terrorist training camp in a cave complex at Zhawar Kili. B-52H and B-1B bombers, along with F/A-18 and AC-130 aircraft, dropped 250 bombs on the caves, AAA defenses, tanks, and artillery over four days. The bombing continued for almost two weeks before the complex was fully destroyed.

B-52H bombers and other U.S. airpower were called out once again in March to support Operation ANACONDA, the first large scale use of ground troops in Afghanistan. Between 2 and 16 March, 1,700 US troops and 1,000 Afghan militia fought between 300 and 1,000 Taliban fighters in the Shahi-Kot Valley. Resistance was heavy and one CH-47 Chinook helicopter was shot down. The extremely rough terrain meant the U.S. and Afghan forces had no artillery. Initially, the ground forces planned to depend on rotary wing aircraft to provide the needed fire support, but these low-flying aircraft proved susceptible to enemy fire.

Fixed wing aircraft support was soon requested and resulted in highly congested airspace above the valley. Various aircraft were stacked eight miles high with B-52s at the highest altitude of 39,000 feet dropping JDAMs through stacks of B-1s and fighters orbiting at 22,000 to 25,000 feet, EP-3Es and AC-130s still lower, and the Predators, A-10s, and attack helicopters down in the weeds. To further complicate things, three civil air routes ran through the airspace. At least once, a B-52 was directed not to drop its bombs because an AC-130 was orbiting directly below.[55]

In one exciting incident during the Battle of Takur Ghar in early March, a beleaguered Enlisted Terminal Attack Controller (ETAC) told a circling B-52, "I want you to put every fucking bomb you have on that fucking ridgeline, right fucking now!" His commander warned him someone was probably recording the radio exchange, and he replied, "Sir, if I survive this, they can court martial me for poor radio discipline." The commander responded, "Roger that. B-52, you heard the man…". The B-52 crew quickly checked the coordinates and cleared the airspace below before releasing their bombs on target. The ETAC calmly replied, "Thanks, that did it."[56]

By 15 April 2002, the 50th anniversary of the first B-52 flight, the B-52 had dropped nearly 35 percent of all ordnance expended in Operation ENDURING FREEDOM, with 100 percent launch rate and safety, while flying less than 3 percent of all bombing sorties. On 1 May, President Bush made his infamous "Mission Accomplished" speech on the aircraft carrier *USS Abraham Lincoln* declaring an end to the fighting in Iraq. On the same day, SECDEF Donald Rumsfeld said the U.S. had moved from major combat operations in Afghanistan to a period of stability and stabilization. Both statements proved premature. B-52 operations over Afghanistan continued until May 2006, when they were relieved by B-1Bs.[57]

Despite the declared end to war, fighting continued. B-52 operations also continued on a limited basis, although most missions returned without dropping their weapons. By 2006 the B-52H was withdrawn and all missions were supported by B-1B aircraft deployed to Al Udeid AB in Qatar. This continued for the next several years with B-52s deployed to Diego Garcia supporting missions in Afghanistan as needed. The B-52 fleet underwent several upgrades during its absence from the theater that significantly increased its flexibility and lethality including the CRL and CONECT, which added new datalinks and the ability to retarget weapons in flight. The aircraft were also upgraded to carry the laser guided GBU-54 LJDAM along with the GPS guided GBU-31 JDAM allowing it to hit moving and stationary targets.[58]

Over the next several years, the U.S. became bogged down and began to lose interest in the war. President Obama vowed to end the war and announced that combat operations would end at the end of 2014, and all troops would withdraw by the end of 2016. Of course, these dates came and went, and the U.S. remained engaged.

Jagged Knife. The new B-52H CRL capability proved very successful and was used in Afghanistan during Operation JAGGED KNIFE in February 2018. The CRL equipped bombers dropped unparalleled numbers of PGMs to destroy nearly 70 Taliban drug labs and other narcotics related targets in the Helmand province. One CRL equipped B-52 dropped 24 guided munitions, including 19 JDAMs, on Taliban positions and vehicles stolen from the Afghan army setting a record for the most precision weapons dropped on a single B-52 sortie.[59]

The B-1B Lancers began returning to Al Udeid in 2018 having completed their required upgrades, with the first two aircraft arriving on 31 March. All B-52H bombers departed the base by 11 April. During their service at Al Udeid the B-52H had flown 3,279 sorties and dropped more than 16,000 weapons. However, they were not done yet and soon returned to operations on 8-9 May 2019 when four aircraft from Barksdale deployed to Al Udeid. They were part of a build-up of forces including F-15 fighters and carrier air in response to saber rattling from Iran. A similar operation occurred again on 7 January 2020 although this time the aircraft were deployed to Diego Garcia.

Withdrawal. When President Biden took office in 2021, he vowed to withdraw all troops by 11 September 2021, 20 years after the initial 9/11 attacks. Taliban forces soon took over most of Afghanistan as U.S. forces withdrew and B-52s were called in for air cover. Six aircraft from the 5th BW at Minot AFB, ND were deployed to Qatar from April to September. They, along with AC-130 gunships, initially supported the Afghan military as they tried to halt the Taliban advances.

As the Taliban overtook the Afghan forces, the bombers transitioned to an armed overwatch role to protect coalition forces as they moved from dispersed locations to Kabul. Their Sniper targeting pods provided situational awareness to ground troops and their secure messaging systems became data-sharing hubs. The bombers also flew show of force missions to ensure the Taliban continued to cooperate. On 7 August, they attacked a gathering as the Taliban overtook Sheberghan. B-52H bombers flew 240 sorties over more than 3,100 hours before all troops were withdrawn on 30 August 2021.

Iraq War

Iraqi Freedom. In January 2003, the U.S. accused Saddam Hussein of continuing to hide its WMD from UN inspectors. President Bush threatened in his second State of the Union Address "...let there be no misunderstanding. If Saddam Hussein does not fully disarm... we will lead a coalition to disarm him." B-52H aircraft soon deployed to Diego Garcia and RAF Fairford, UK in anticipation. Fourteen aircraft deployed from Barksdale to Diego Garcia to supplement the 40th AEW. Twelve B-52 aircraft were also deployed from Barksdale to Guam to form the 7th AEW along with 12 B-1B aircraft. Another 14 aircraft were deployed from Minot to RAF Fairford to supplement the 16th Air Expeditionary Task Force. The Fairford deployment brought significant protests against the B-52 and the coming war action. Over 20,000 protesters climbed fences and attempted to barge through the main gate at Fairford and as many as 750,000 to two million organized in London.

When Operation IRAQI FREEDOM was launched on 19 March 2003 the battlefield had already been "shaped" by aircraft supporting the no-fly zone under Operation SOUTHERN FOCUS (previously SOUTHERN WATCH). Several radars, air defense weapons, and fiber optic links were already destroyed significantly degrading the Iraqi air defense and command and control systems. This essentially gave the U.S. forces control of the air and the initial strikes by Tomahawk missiles, F-117A stealth fighters, and other aircraft were virtually unchallenged.

B-52 aircraft entered the fight on 21 March with the largest CALCM strike in history with the bombers firing 76 of over 140 missiles against strategic command, control, and communications targets, as well as military and air defense targets in and around Baghdad. The 40th AEW launched ten aircraft from Diego Garcia including the first aircraft to carry both JDAMs and CALCMs on the same mission, and the first combat firing of AGM-86C CLACM Block 1A and AGM-86D CALCM Block 2. Another seven aircraft from RAF Fairford also joined the first wave of attacks. After these initial strategic attacks, the missions soon turned to tactical efforts to prepare the battlefield for ground troops. The focus of this "Shock and Awe" campaign was directed against deployed Iraqi Army and Republican Guard positions, as well as interdiction and CAS sorties coordinated with ground forces designed to decapitate the regime.[60]

On 30 March 2003, a B-52H (60-0015) loaded with 12 GBU-31 JDAMs and 27 Mk-82 bombs took off from Diego Garcia and ran into immediate trouble. Its No. 7 engine failed after the first aerial refueling and it encountered heavy, but ineffective, AAA about 100 kilometers south of Baghdad. On top of that, the planned air suppression package consisting of F-15C and EA-6B aircraft was cancelled. The crew continued to the target inside the heavily defended Baghdad Super Missile Engagement Zone (MEZ), missed

B-52H (60-0043) loaded with JDAMs prepares for take-off during Operation IRAQI FREEDOM. (U.S. Air Force)

by several unguided SAMs fired by Iraqi defenses, and dropped 12 JDAMs on Republican Guard positions. They dodged two more SAMs as they departed the target area with one exploding near the bomber. But the bomber continued undeterred to the next target, an ammo storage area inside the Super MEZ and dropped its 27 Mk-82 bombs. As their luck would have it that day, the No. 8 engine also failed before the first refueling on the return leg and the aircraft returned though darkness and bad weather back on Diego 16 hours later.[61]

The following day, on 1 April, a B-52H took off from RAF Fairford to hit heavily defended targets in Northern Iraq. It attacked and nearly destroyed a Baath Party building and a fuel storage area using four JDAMs. It then conducted a second bomb run and destroyed these targets with four more JDAMs before attacking a third nearby target with M117 750-pound bombs. The bomber was then directed to fly south and hit artillery sites and a chemical weapons facility north of Kirkuk using its four remaining JDAMs. When all seemed done, the aircraft was then redirected to return north and drop its remaining M117 bombs on the previously bombed target.

On 2 April, a B-52H (60-0007) flying out of Diego Garcia was credited with the first use of the CBU-105 Sensor Fused Weapon (SFW) in combat. The aircraft was loaded with 16 SFWs and 27 Mk-82 bombs. It dropped the Mk-82 bombs on an ammunition dump and then dropped two SFWs on a column of Iraqi armored vehicles. The weapons worked perfectly and destroyed half the column. The surviving Iraqis immediately surrendered to the Marines.[62] The aircraft then rerouted to attack a mass of armored vehicles near Baghdad with four more CBU-105 SFWs.

On 10 April, another B-52H faced Iraqi SAMs without suppression support despite flying with a hung landing gear. The aircraft successfully evaded SAM radars and AAA batteries before dropping its weapons on five separate armored emplacements strung out along a ridgeline. They destroyed all five targets after two bomb runs and were then directed back in to destroy the SAM sites they had just evaded. They successfully evaded all defenses again and destroyed all targets in a single run.

The first B-52H (61-0021) flying with an operational Northrop Grumman AN/AAQ-28 Litening II Targeting Pod took from RAF Fairford on 11 April 2003. It was loaded with 16 CBU-103 cluster bombs and the targeting pod on the wings and three GBU-12 LGBs in the bomb bay. The pod allowed the B-52 to laser designate its own targets directly instead of relying on other aircraft or ground personnel. The aircraft flew north over Iraq towards Syria before targeting a command center at the Al Sahra airfield near Tikrit and destroying it with a single GBU-12 LGB. The crew then hit a radar complex with the two remaining GBU-12 LGBs and cluster bombs. The aircraft and crew were credited with the first combat drop of an LGB from a B-52.

Meanwhile, Baghdad had fallen to coalition troops on 9 April and four days later U.S. Marines captured Saddam Hussein's hometown of Tikrit. On 1 May, President Bush declared mission accomplished aboard the *USS Abraham Lincoln*. U.S. and coalition aircraft had flown 41,404 sorties and dropped 29,199 weapons during the operation. The B-52H bombers flew 298 sorties and dropped 3.2 million tons of ordnance including GBU-32 JDAMs, GBU-31 JDAMs, CBU-103 Combined Effect Munitions (CEM), CBU-105 SFWs, Mk-82 and M117 bombs, M129 leaflet bombs, and 153 CALCMs.[63]

Inherent Resolve. Despite the declared end to war, fighting continued. B-52 operations also continued on a limited basis, although most missions returned without dropping their weapons. By 2006 the B-52H was withdrawn for modifications and all missions in Iraq were supported by B-1B aircraft deployed to Al Udeid AB in Qatar. This continued for the next several years with B-52s deployed to Diego Garcia supporting missions as needed.

On 15 October 2014, combat operations in Iraq refocused against the Islamic State (ISIS) under Operation INHERENT RESOLVE. By 2016, the B-1B fleet was withdrawn and returned home for the Integrated Battle Station Upgrade and the newly modified B-52H aircraft were deployed to pick up the mission. The first B-52 to deploy to a CENTCOM base since 1991 arrived at Al Udeid AB on 9 April 2016. Combat strikes began on 20 April against a weapons storage facility in northwestern Iraq. This initial mission was followed over the next several months by additional missions consisting mostly of CAS strikes against multiple targets. On 19 November 2017, a B-52H from the 69th Expeditionary Bomb Squadron (EBS) at Al Udeid AB became the first aircraft to employ the CRL on an operational mission.

Competitions and Exercises

Through most of its history SAC used the B-52 as a nuclear bomber. The command conducted competitions annually to hone the accuracy and discipline of its aircrews and make them the best airmen in the world. Beginning in the late 1970s and into the 1980s, based on its experiences supporting the Vietnam War, SAC realized conventional warfare would become an important mission. This was reinforced by a series on CINCSACs with fighter backgrounds. They believed SAC needed to train the way it would fight. To that end, B-52 units started participating in various exercises like those used by tactical forces to test capability in the actual theaters of potential conflict. The following competitions and exercises, although not all-encompassing, represent the evolution of this philosophy.

SAC Bomb Comp. In 1948, General George Kenney, SAC Commanding General, decided to hold the first SAC Bomb Comp from 20-27 June at Castle AFB, CA hoping to stimulate interest in improving bombing accuracy. The competition was so successful at increasing competitive spirit among the crews and bombing accuracy, that General LeMay made it an annual event when he took command of SAC in 1949. It later evolved to include navigation and was renamed the SAC Bomb-Nav Competition in 1951, although it was still typically called the SAC Bomb Comp.

B-52 aircraft from two bombardment wings competed for the first time, along with B-47, RB-47, and B/RB-36 aircraft during the eighth annual Bomb Comp from 24-30 August 1956. The 11th BMW (Carswell AFB, TX), a B/RB-36 unit, beat out the B-47 and B-52 units to win the Fairchild Trophy which was named for General Muir S. Fairchild, former Air Force Vice Chief of Staff, and awarded to the best bomber unit in the combined fields of navigation and bombing.

Except for crew and wing navigation awards, B-47 units beat out the B/RB-36 units in all events of the SAC Bomb Comp held from 30 October to 6 November 1957. B-52 aircraft from five wings participated. The 321st BMW (McCoy AFB, FL), a B-47 unit, won the Fairchild Trophy. The final appearance of the B/RB-36 in a SAC Bomb Comp occurred from 13-18 October 1958. B-52 aircraft from ten wings participated. The RAF also entered their new Vickers Valiant four-jet bombers and ten crews. B-47 units dominated the competition and the 306th BMW won the Fairchild Trophy.

The eleventh, and largest ever, bombing competition was held from 25-30 October 1959. Two crews each from 27 B-47 and 20 B-52 wings participated along with 47 Aerial Refueling Squadrons. The B-52 crews failed to make a mark, and the B-47 once again dominated. The 307th BMW, a B-47 unit, won the Fairchild Trophy.

The twelfth Bomb Comp was held from 12-16 September 1960. While the eleventh competition was the largest ever held, this one featured a reduced number of participants.

The winners of the 1959 SAC Bomb Comp proudly display the Fairchild Trophy. (U.S. Air Force)

It was officially called the SAC Combat Competition, a name that was relatively short-lived and became the SAC Aircraft Combat Competition (as opposed to missiles) before it slipped away in 1971. Each NAF held a preliminary event to select their two best B-52, B-47, KC-135, and KC-97 wings to enter the competition. The B-52 finally proved itself when the 11th BMW, a B-52E unit from Altus AFB, OK, won the Fairchild Trophy with the best combined score for alert activities, bombing, navigation, ECM, and aerial refueling.

The thirteenth competition was held from 16-22 September 1961 using the same ground rules as the 1960 competition. The NAFs entered their two best B-52, B-47, KC-135, and KC-97 wings. Once again, a B-52 unit, the 4137th SW a B-52D unit from Robins AFB, GA won the Fairchild Trophy with the best combined score for alert exercise, bombing, navigation, ECM, aerial refueling, pilot techniques, and munitions loading.

No Bomb Comp was held in 1962 due to increased operational commitments for the CUBAN MISSILE CRISIS. The competition was also cancelled in 1963 due to increased

The Fairchild Trophy was named for General Muir S. Fairchild, former USAF Vice Chief of Staff, and was awarded at every bomb competition beginning 1951. The trophy, which was donated by Hughes Aircraft Corporation, was awarded to the best bomber unit in the combined fields of navigation and bombing. (U.S. Air Force)

operational commitments and cost reductions. However, General Thomas Power, CINCSAC, decided to award the Fairchild Trophy to the wing with the best overall combat capability record for Fiscal Year 1963. A special board of officers reviewed the records of all wings and selected the 2nd BMW, a B-52F unit from Barksdale AFB, LA, as the winner. This was a significant achievement since the 2nd BMW was extremely busy flying B-47 aircraft at Hunter AFB, GA until 1 April 1963, and then moved to Barksdale AFB, LA to absorb the B-52F resources of the deactivated 4238th SW.

The competition was again cancelled in 1964 due to cost reductions. This time the Fairchild Trophy was awarded to the 70th BMW, a B-52E unit from Clinton-Sherman AFB, OK, using the same process as in 1963. The Bomb Comp continued in 1965 and was held from 12-18 September. One crew participated from each of the 44 bomb wings (two B-58, five B-47, and 37 B-52). The 454th BMW, a B-52F unit from Columbus AFB, MS won the Fairchild Trophy. This was the last competition for the B-47.

By the time the fifteenth competition was held from 2-8 October 1966, the event was becoming a predominantly B-52 affair. Participants included one crew each from 35 B-52 wings and two B-58 wings. Three RAF Vulcan crews also participated. With the addition of international competition, the event was thereafter often referred to as the World Series of Bombing. The 8th AF was the big winner with the first four places in the overall competition and its 19th BMW, a B-52H unit from Homestead AFB, FL, won the Fairchild Trophy.

The Bomb Comp was again cancelled in 1967 due to operational commitments, and no Fairchild Trophy was awarded. No competition was held in 1968 due to SEA combat operations. It continued the following year from 5-15 October 1969 and included 22 B-52, two B-58 wings, and three RAF crews flying Vulcan bombers. Several B-52D wings did not participate due to SEA operations. This was also the last competition for the B-58. The 319th BMW, a B-52H unit from Grand Forks, ND, won the Fairchild Trophy.

The seventeenth SAC Bomb Comp was held from 15-20 November 1970, and participants included 23 B-52 wings and three RAF crews flying the Vulcan. It also included two FB-111 aircraft from the 340th BG, an FB-111 training unit at Whiteman AFB, MO. Once again, several B-52D wings did not participate due to SEA combat operations. The Fairchild Trophy, which for the first time was awarded based on the combined score of the B-52 wing and its assigned KC-135 Aerial Refueling Squadron, went to the 93rd BMW, a B-52F unit from Castle AFB, CA. One of the FB-111 crews gave a strong initial showing by winning the Bombing Trophy.

The event became known as the SAC Aircraft Combat Competition for a short while before it was renamed GIANT VOICE in 1971.

The eighteenth SAC Bomb Comp was held under the new exercise code name of GIANT VOICE '71 from 12-17 December 1971. Although the GIANT VOICE moniker would officially persist until 1987, most SAC crews and maintainers continued to simply call it the SAC Bomb Comp. Participants included 22 B-52 units, two FB-111, and three RAF crews flying Vulcan bombers. The 449th BMW, a B-52H unit from Kincheloe AFB, MI, won the Fairchild Trophy. No Bomb Comp was held in 1972 due to SEA operational commitments. The planned Bomb Comp was cancelled again in 1973 due to SEA operational commitments.

The annual SAC Bomb Comp grew from a SAC competition into an international event involving dozens of aircraft and crews. (U.S. Air Force)

The SAC Bomb Comp/GIANT VOICE '74 was back on again and held from 10-16 November. SAC participants included 20 B-52 wings and two FB-111 wings. The RAF entered four Vulcan bombers, and TAC entered two F-111 aircraft for the first time. The Fairchild Trophy and the Bombing Trophy went to the 380th BMW, an FB-111 unit from Plattsburg AFB, NY. The RAF won the Mathis Trophy, for best bombing and celestial navigation, and the Navigation Trophy. The Bomb Comp was again cancelled in 1975, and the command conducted an exercise called Operation HIGH NOON instead. This exercise was intended to test SAC's ability to conduct contingency operations with minimal preparations.

The SAC Bomb Comp/GIANT VOICE '76 was conducted in three phases. In Phase I, from July to August, each unit selected its best three crews and one wild card to enter the Phase II Semi-Finals held from 27-30 September. Phase III was the Finals held from

4-5 October, and participants were the top crews from each unit as determined by the Phase II competition. The 380th BMW won the Fairchild Trophy for the second year. It also had the highest score for the Mathis Trophy, awarded for best high- and low-level bombing, but was ineligible since it won the Fairchild. The Mathis Trophy, along with the Bombing Trophy, went to the 7th BMW, a B-52D unit from Carswell AFB, TX instead.

SAC held the twenty-first SAC Bomb Comp/GIANT VOICE '77 in 1977 using the same rules as the previous year. Two crews each from 18 bomb wings entered the finals from 21-23 September, along with two RAF Vulcan and two TAC F-111 crews. Once again, the 380th BMW won the Fairchild Trophy. The 509th BMW, also an FB-111 wing from Pease AFB, NH, won the Mathis Trophy.

SAC held the twenty-second SAC Bomb Comp/GIANT VOICE '78, now 30 years since the first event, with the final phase held 18-19 October. Participants included one aircraft and crew from 16 B-52 and two FB-111, one RAF Vulcan, and two TAC F-111 units. Once again, the FB-111 units dominated. The Fairchild Trophy went to the 380th BMW. The 380th BMW also won the Meyer Trophy and the Bombing Trophy for the top F/FB-111 unit in low-level bombing and ECM activity. The 509th BMW won the Mathis Trophy and the Bombing Trophy for best combined high- and low-level bombing score. However, the 28th BMW, a B-52H wing from Ellsworth AFB, SD, won the Crumm Trophy for most points in high altitude bombing. The 319th BMW won the General Russell E. Dougherty SRAM Trophy, which was awarded for the first time for the best score in simulated SRAM launching.

SAC held the twenty-third SAC Bomb Comp/GIANT VOICE '79 in November with one aircraft and crew from 16 B-52 and two FB-111, one RAF Vulcan, and two TAC F-111 units participating. Once again, the FB-111 units dominated and for the first time an F-111 unit won two events. The Fairchild Trophy went to the 509th BMW and the 380th BMW won the Bombing Trophy. The 27th Tactical Fighter Wing (TFW), a TAC F-111 unit, won the Meyer Trophy and the Bombing Trophy for the top F/FB-111 unit in low-level bombing and ECM activity. The 27th TFW also won the Mathis Trophy for the best combined high- and low-level bombing

The original competition evolved to include navigation and was renamed the SAC Bomb-Nav Competition, although it was still typically called the SAC Bomb Comp.

The FB-111A with its state-of-the-art bombing system dominated the SAC Bomb Comp during the 1970s. (U.S. Air Force)

score. However, B-52 aircraft performed well as the 28th BMW won the Crumm Trophy and the 379th BMW won the General Russell E. Dougherty SRAM Trophy.

The final phase of the twenty-fourth SAC Bomb Comp/GIANT VOICE '80 was held in November. Participants included one aircraft and crew from 16 B-52 and two FB-111 units, four RAF Vulcan aircraft and crews, and two TAC F-111 aircraft and crews. The Royal Australian Air Force (RAAF) also entered two F-111C aircraft and crews for the first time. The 320th BMW, a B-52G unit from Mather AFB, CA, took home the Fairchild Trophy marking the first time a B-52 unit had won it in the last five competitions. In fact, B-52 units made a clean sweep, also winning the Mathis Trophy, Dougherty Trophy, and Crumm Trophy. However, the 380th BMW, an FB-111 unit, won the Bombing Trophy, which was renamed the General Curtis E. LeMay Bombing Trophy and awarded for the first time.

The twenty-fifth SAC Bomb Comp/GIANT VOICE '81 marked the first time in several years that only SAC units participated. Participants included one aircraft and crew from 16 B-52 and two FB-111 units. There were two final phases this time, from 5-7 November and 7-10 November. Once again, the FB-111 units dominated. The 509th BMW took home the Fairchild Trophy, the Mathis Trophy, and the LeMay Trophy. The 320th BMW won the Dougherty Trophy with the most points in simulated SRAM launches and the 97th BMW, a B-52G unit from Blytheville AFB, AR, won the Crumm Trophy for high altitude bombing.

The FB-111 was again the big winner of the twenty-sixth SAC Bomb Comp/GIANT VOICE '82. Participants in the final phase included the top two bomber and tanker crews from Phase I, which evaluated planning and execution of simulated Emergency War Order missions. Two crews and aircraft from 15 B-52 and one FB-111 units, and four RAAF F-111 aircraft, competed. FB-111 units continued their winning streak when the 509th BMW set a competition record by winning five awards including the Fairchild Trophy, Mathis Trophy, and Meyer Trophy. The 380th BMW won the LeMay Bombing Trophy. The 416th BMW, a B-52G unit from Griffiss AFB, NY won the General John D. Ryan Trophy, which was awarded, for the first time, to the B-52 unit with the most high- and low-level bombing points.

TAC F-111 crews returned to the SAC Bomb Comp/GIANT VOICE '83. The 509th BMW an FB-111 wing took home the Fairchild Trophy. The 28th BMW was the big B-52 winner. It won the Mathis Trophy, Ryan Trophy, Crumm Trophy, and the LeMay trophy.

The RAF returned to the SAC Bomb Comp/GIANT VOICE '84 after a three-year absence. This time, they were equipped with their new Tornado aircraft which had replaced their Vulcan bombers. TAC and RAAF F-111 aircraft also competed. The 380th BMW won the Fairchild Trophy continuing FB-111 domination. B-52 units fared better than in the last few years. The 416th BMW won the Mathis Trophy, the Ryan Trophy, and the Dougherty Trophy. The 379th BMW, a B-52G wing from Wurtsmith AFB, MI won the Crumm Trophy.

An OAS modified B-52G from the 97th BMW won the Fairchild Trophy and the Mathis Trophy during the SAC Bomb Comp/GIANT VOICE '85. This marked only the second time a B-52 unit had won the trophy, which had been dominated by the more modern and accurate FB-111, since 1974. This proved the wisdom of upgrading to the more accurate OAS, which was also required to launch the new ALCM stand-off missiles. The 7th BMW, now a B-52H unit from Dyess AFB, TX won the Ryan Trophy. The 319th BMW, now a B-52G wing from Grand Forks AFB, ND won the Dougherty Trophy while the 416th BMW won the Crumm Trophy. The RAF performed strong with their new Tornado aircraft winning both the LeMay Trophy and the Meyer Trophy.

The SAC Bomb Comp/GIANT VOICE '86 competition caused some controversy when the 92nd BMW at Fairchild AFB, WA dominated all others and won nine trophies. This caused immediate rule changes that prohibited aircraft being specially modified for the event and going forward the SAC operations staff would select the competing crews, not the wings. Participants included 16 SAC bomb wings and F-111 aircraft from TAC. There were no international competitors for the first time since 1983.

The SAC Bomb Comp/GIANT VOICE '87 competition was the last one using the GIANT VOICE exercise name. The next competition took the code name SAC Bomb Comp/PROUD SHIELD '88. It would be the first time for EC-135 aircraft to compete, and the first time TAC F-111 aircraft could compete for the Fairchild Trophy. A new Billy Mitchell Trophy was added to recognize the top bomber unit in conventional bombing accuracy, survival in an ECM environment, and avoiding fighter intercepts. B-1B units were allowed to compete for the first time, although they could not compete for the Billy Mitchell Trophy because they had a strictly nuclear mission.

A B-52H lands at Minot AFB, ND after the SAC Bomb Comp/PROUD SHIELD '88 competition. (National Archives)

The SAC Bomb Comp/PROUD SHIELD '89 marked the first year a B-1B unit took the Fairchild Trophy. The competition was cancelled for the next two years due to Operation DESERT STORM. The final SAC Bomb Comp/PROUD SHIELD '92 in SAC was held from 13-15 April. Participants included 21 SAC units flying B-52, B-1, KC-135, EC-135, and KC-10 aircraft, nine ANG KC-135 units, and a TAC KC-10 squadron. After refueling, the bombers performed two single weapon releases over the Utah Test and Training Range at 20,000 feet. They then descended to 300 to 500 feet to perform terrain avoidance and low-level drops at two different complexes while avoiding SAM and AAA threats. The 92nd BMW at Fairchild AFB, WA was the last to receive the Fairchild Trophy.[64]

RAF Bombing Competition. SAC B-52 crews participated in the RAF Bombing Competition for the first time from 14-20 May 1957 when the 92nd BMW at Fairchild AFB, WA entered six B-52 aircraft and crews. The wing made a strong showing and won five of six eligible awards. SAC also entered six crews from the 6th BMW from Walker AFB, NM in the RAF Bombing Competition held from 1-3 May 1960. The 6th was selected based on its being the top B-52 unit in the 1959 Bomb Comp. It won the best unit award for bombing and navigation, and several individual crews won awards for top scores in bombing and top combined scores in bombing and navigation.

SAC B-52 crews again participated in the RAF Bombing Competition held from 13-15 March 1967 after a seven-year absence. Each NAF sent one B-52 aircraft and crew. Participants were the 449th BMW a B-52H unit from Kincheloe AFB, MI (2nd AF), 19th BMW a B-52H unit from Homestead AFB, FL (8th AF), and 93rd BMW a B-52F unit from Castle AFB, CA (15th AF). None of the SAC participants brought home a trophy.

SAC also entered the RAF Bombing Competition held in May 1970 with four B-52 aircraft and crews from the 2nd BMW a B-52G unit at Barksdale AFB, LA, the 319th BMW a B-52H unit at Grand Forks AFB, ND, the 320th BMW a B-52G unit at Mather AFB, CA, and the 379th BMW a B-52H unit at Wurtsmith AFB, MI. These wings represented the top two B-52G and B-52H wings based on 1969 SAC Bomb Comp standings. The SAC team won the Blue Steel Trophy, an inter-air force award for best combined bombing and navigation score.

SAC also entered the RAF Bombing Competition held from 17-24 April 1971 with four B-52 aircraft and crews from the 2nd BMW, the 320th BMW, the 379th BMW, and the 410th BMW a B-52H unit from K.I. Sawyer AFB, MI. The two FB-111 units participated in a demonstration capacity only. The 410th BMW won the Blue Steel Trophy.

Despite cancelling the SAC Bomb Comp in 1972 due to SEA commitments, SAC entered the RAF Bombing and Navigation Competition held from 14-20 May with four B-52 aircraft and crews from the 2nd BMW, the 17th BMW a B-52H unit at Wright-Patterson AFB, OH, the 28th BMW a B-52G unit at Ellsworth AFB, SD, and the 449th BMW a B-52H unit at Kincheloe AFB, MI. The 28th BMW won the Blue Steel Trophy.

The planned SAC Bomb Comp was also cancelled in 1973, but SAC again entered the RAF Bombing and Navigation Competition held from 29 April-5 May with four B-52 aircraft and crews from the 5th BMW a B-52H unit at Minot AFB, ND, the 17th BMW, the 319th BMW a B-52H unit at Grand Forks AFB, ND, and the 410th BMW. None of the SAC units won the Blue Steel Trophy. They placed 2nd, 7th, 8th, and 9th in the competition.

SAC entered the 1976 RAF Bombing Competition with four B-52 aircraft and crews from the 2nd BMW, the 92nd BMW a B-52G unit at Fairchild AFB, WA, the 97th BMW a B-52G unit at Blytheville AFB, AR, and the 320th BMW. The 320th BMW won the Blue Steel Trophy. It also won the Camrose Trophy for the best score in bombing, marking the first time for a SAC unit. SAC also entered four B-52H wings in the 1977 RAF Bombing Competition. Participants were the 5th BMW, the 319th BMW, the 410th BMW, and the 449th BMW. The units failed to bring home any awards.

SAC entered four B-52 units in the RAF Bombing Competition held from 26 June to 3 July 1978, but they failed to bring home any trophies. SAC again entered three B-52H units in 1979, and they failed to bring home trophies. SAC sent four B-52G units to the RAF Bombing Competition in 1980, determined to turn around their recent losses. However, they failed once again to bring home any trophies. Unfortunately, 1981 was a repeat performance and marked SAC's last participation in the event.

High Noon. SAC conducted a new exercise called Operation HIGH NOON from 30 September to 2 October 1975 in place of the SAC Bomb Comp. The exercise tested SAC's ability to conduct contingency operations with minimum preparation. Participants included two aircraft and crews from 21 SAC bomb wings (limited to CONUS only), four RAF Vulcan bombers, and six TAC F-111 aircraft. The 92nd BMW, a B-52G unit from Fairchild AFB, WA, won the Outstanding SAC Unit Award for overall performance in bombing and aerial refueling.

Red Flag. SAC participated for the first time in the TAC RED FLAG exercise in April 1976. The exercise was conducted out of Nellis AFB, NV on a complex in the Nevada desert designed to provide realistic combat training. SAC crews flew numerous B-52 and FB-111 sorties to practice defensive tactics under simulated combat conditions. They avoided threats from TAC's "aggressor" aircraft by initiating appropriate defensive actions. They also defended against simulated ground threats from the range's SAM and AAA radars. The exercise enabled crews to practice navigation and weapon-delivery tactics at low altitudes over rugged terrain to penetrate the simulated threats. The payoff was the improved coordination within the crew that resulted from reacting to the stresses of this demanding environment.[65]

SAC B-52 and FB-111 wings continued to participate in the RED FLAG exercises conducted by TAC in 1978. This exercise was conducted in the Nevada desert and was so successful it was expanded to include ranges near Cold Lake, Canada. SAC participated in this new exercise, called MAPLE FLAG, for the first time on 28 April 1978. B-52 crews flew over vast, unpopulated areas covered with thousands of lakes and geographic features that do not exist in the U.S.. Tactical forces also flew air intercepts against the penetrating B-52s.[66]

SAC completed a new training range called the Strategic Training Range Complex (STRC) on 1 July 1981. The new range, with mobile radar bombing sites in Wyoming and Idaho, was designed to capture the benefits of the RED FLAG range while providing better simulation of strategic penetration of the Soviet Union.

Global Shield. SAC exercised every function in its SIOP, short of nuclear war, for the first time from 8-16 July 1979 during GLOBAL SHIELD '79. This no-notice exercise was the most comprehensive in SAC history and included hundreds of bombers, tankers, and missiles generated to alert, deployed to pre-assigned dispersal bases, and flying over bombing ranges. For the first time in many years, large numbers of B-52 crews executed MITO procedures. Prior to GLOBAL SHIELD, only two or three aircraft normally practiced these procedures. During GLOBAL SHIELD, crews flew most of the aircraft in their units, more than twenty in some cases, to accomplish the MITO. The exercise helped identify and correct problems associated with MITO procedures and improved the confidence of the crews in its execution.[67]

General Richard Ellis, CINCSAC, said "Initial evaluation of the exercise indicates that all of our objectives were achieved. Everyone had an opportunity to gain valuable training in the performance of our Emergency War Order mission and, at the same time, to help identify ways to improve our plans and procedures." SAC repeated the no-notice exercise from 20-29 June 1980 during GLOBAL SHIELD '80. It again exercised the entire SIOP and involved nearly 100,000 personnel at 44 bases with 437 aircraft flying 5,506 hours over 1,035 sorties.

SAC conducted GLOBAL SHIELD '81 from 26 January to 16 February 1981. The no-notice exercise included most aspects of the first two and included bombers and tankers on Guam. More than 120 bombers and tankers were dispersed to 30 pre-determined locations, and the bombers flew airborne alert missions on low-level routes in Colorado, New Mexico, Kansas, and Texas. About 400 aircraft performed MITOs from 70 locations demonstrating SAC's ability to get the entire force airborne within minutes.

GLOBAL SHIELD '82, held from 15-23 July 1982, was once again SAC's largest annual exercise. This time, in addition to exercising SAC's ability to implement the SIOP, the exercise included forces outside the command and was designed to test SAC's ability to operate in a prolonged nuclear war. The North American Air Defense Command also conducted a concurrent exercise pitting U.S. and Canadian fighters against SAC bombers. GLOBAL SHEILD exercises proved extremely successful in the early 1980s, and SAC continued to conduct them until near then of the decade. They were often conducted by the JCS and combined with other on-going exercises of the Air Force and other services.

B-52 weapons download during GLOBAL SHIELD '84 at Ellsworth AFB, SD on 15 April 1984. (National Archives)

Busy Prairie. SAC tested the SPF for the first time from 22-25 September 1980 during an exercise called BUSY PRAIRIE. Fourteen 5th BMW B-52H aircraft deployed from Minot AFB, ND to a simulated FOB at Whiteman AFB, MO. The bombers were joined by SAC refueling and reconnaissance aircraft, as well as MAC C-5 airlifters. The bombers flew simulated combat operations and attacked three airfields on a RED FLAG training range near Nellis AFB, NV. The crews used low-altitude penetration to bomb targets while under simulated attack by various ground threats and aggressor aircraft.[68] The 5th BMW B-52H aircraft were joined by 319th BMW B-52H bombers flying from their home base at Grand Forks AFB, ND.

Bright Star. On 23 November 1981, eight B-52H bombers of SAC's SPF made the longest non-stop bombing mission in the history of SAC at that time. Two flights of four aircraft each from the 319th BMW at Grand Forks AFB, ND and 5th BMW at Minot AFB, ND flew nonstop to Egypt and dropped conventional munitions on a simulated runway in Egypt during the exercise designated BRIGHT STAR '82. The aircraft completed the 15,000-mile flight with five refueling hook-ups in 32 hours and demonstrated the SPF's ability to rapidly strike targets in any part of the world. BRIGHT STAR began in 1981 as an annual joint military exercise led by the U.S. and Egypt. The exercise soon grew to include forces from surrounding countries and beginning in 1983 was changed to every two years due to the significant planning required. B-52 units continued to support the exercise which continued to be conducted as of 2023.

A B-52H lands at Cairo, Egypt during Operation BRIGHT STAR '83. (National Archives)

Busy Brewer. SAC began participating in a new exercise in Europe called BUSY BREWER in 1980 designed provide BMW staffs and aircrews the opportunity to plan, brief, and execute B-52 conventional missions from FOBs in the UK. Two-to-five-week deployments of three to eight aircraft were conducted several times per year.[69] The first exercise was conducted from 10-16 June when three B-52D (55-0073, 56-0600, and 56-0658) aircraft and personnel from the 7th BMW at Carswell AFB, TX deployed to RAF Marham, UK.[70] Two aircraft (56-0600 and 55-0073) completed the only flight of the exercise on 12 June. The aircraft flew a high-altitude training mission in support of the NATO Allied Air Forces Central Europe exercise called CLOUDY CHORUS. The exercise tested the effectiveness of NATO fighters and the bombers, along with other aircraft, acted as adversaries. This deployment was followed by three B-52D (55-0677, 56-0676, and 56-0659) aircraft from the 96th BMW at Dyess AFB, TX to RAF Brize Norton in September 1980.

BUSY BREWER quickly became a regular exercise, and various B-52 units supported multiple deployments each fiscal year. B-52D (55-0069, 55-0070, and 56-0690) aircraft from Carswell continued to support BUSY BREWER '82A when they deployed to RAF Fairford from 20 October to 13 November 1981. They supported various NATO exercises during the three-week deployment including CLOUDY CHORUS. Some aircraft landed at RAF Brize Norton for Rest Over Night (RON) after missions before returning to RAF Fairford.

The G model units soon entered the line-up as B-52D aircraft were retired from service. B-52G (58-0172, 58-0192, and 58-0197) aircraft deployed to RAF Fairford from the 19th BMW at Robins AFB, GA from 7 September to 11 October 1982 for BUSY BREWER '82D. They were joined by two more B-52G (58-0207 and 58-0236) aircraft from the 42nd BMW at Loring AFB, ME from 9-23 September. The aircraft participated in several NATO exercises including COLD FIRE, DISPLAY DETERMINATION, NORTHERN WEDDING, BOLD GUARD, NATINADS and DAMSEL FAIR. The Robins aircraft were deployed to RAF Brize Norton on 8 October before flying home on 11 October.

B-52G model units continued to support BUSY BREWER deployments to several UK and European bases over the next several years. For example, three B-52G aircraft from the 320th BMW at Mather AFB, CA deployed to RAF Upper Heyford for BUSY BREWER '84A from 24 April to 18 May 1984, while the 42nd BMW deployed to Moron AB, Spain for BUSY BREWER '84B, and three B-52G (57-6520, 58-0253, and 59-2599) aircraft from the 2nd BMW at Barksdale AFB, LA returned the deployment to RAF Fairford for BUSY BREWER '84C from 13 September to 12 October 1984.

The 2nd BMW deployed seven more B-52G (57-6489, 57-6505, 57-6512, 57-6519, 58-0210, 59-2586, and 59-2599) aircraft to RAF Fairford for BUSY BREWER '85A from 24 May to 20 June 1985. This was the largest number of bombers in the UK at one time. The bombers participated in NATO exercises CENTRAL ENTERPRISE, NATINADS, EWEX and LOCKED GATE. Another seven aircraft deployed to RAF Fairford for BUSY BREWER '85B from 6 September to 11 October 1985 to take part in NATO exercises OCEAN SAFARI, COLD FIRE, DAMSEL FAIR, OUTIGUE and DISPLAY DETERMINATION. This time five B-52G (57-6468, 57-6476, 58-0258, 59-2565, and 59-2598) aircraft came from the 42nd BMW and two (57-6469 and 58-0226) came from the 320th BMW. Three of the aircraft (57-6468, 57-6469, and 58-0226) deployed to Sidi Slimane AB, Morocco from 19-23 September marking the first time SAC bombers had visited the base since SAC withdrew in 1963.

By the mid-1980s RAF Fairford became the main FOB for B-52 aircraft operating in Europe and became the main staging base for BUSY BREWER deployments. The main units deployed became the 2nd and 42nd BMW, which SAC had by then designated as its conventional wings. This proved highly beneficial when the base was pressed into service as a B-52 FOL during the Gulf War. BUSY BREWER '86B was conducted from 24 August to 6 October 1986 when three 2nd BMW B-52G (58-0216, 58-0240, and 59-2569) aircraft deployed to Fairford for NATO exercises NORTHERN WEDDING, COLD FIRE and DISPLAY DETERMINATION.

In 1987, six 42nd BMW B-52G (58-0172, 58-0241, 59-2578, 58-0202, 58-0240, and 57-6510) aircraft deployed to RAF Fairford for BUSY BREWER '87A from 1 May to 8 June. Three of the aircraft (58-0172, 58-0241, and 59-2578) deployed from 1-20 May and the next three replaced them for the rest of the deployment instead of the planned BUSY BREWER '87B. The aircraft participated in NATO exercises DRAGON HAMMER,

A 42nd BMW B-52G (59-2565) lands at RAF Fairford, UK during BUSY BREWER '85B. (American Aviation Historical Society)

OPEN GATE, CENTRAL ENTERPRISE and GIANT SQUID. One B-52G (58-0202) aircraft deployed to RAF Upper Heyford on 5 June for static display before returning home. B-52G aircraft from the 42nd BMW (58-0172, 58-0224, and 59-2596) and the 2nd BMW (57-6506, 58-0216, and 59-2588) deployed to RAF Fairford for BUSY BREWER '87C from 1 September to 5 October 1987. They supported NATO exercises OCEAN SAFARI, COLD FIRE and DISPLAY DETERMINATION.

BUSY BREWER '88A was conducted from 29 April to 16 May 1988. It consisted of seven 42nd BMW B-52G (58-0224, 58-0226, 58-0232, 58-0240, 58-0241, 58-0251, and

59-2596) aircraft deployed to RAF Fairford. BUSY BREWER '88B from 3-13 June 1988 included five 2nd BMW B-52G (58-0212, 59- 2569, 59-2588, 59-2595, and 57-6476) aircraft. The first four aircraft deployed straight to RAF Fairford, while 57-6576 arrived on 6 June after supporting the RAF Mildenhall and RAF Upper Heyford airshows.

In August 1988, the exercise became known as MIGHTY WARRIOR. The NATO support taskings were still called BUSY BREWER and the deployments were called BUSY WARRIOR. Seven 42nd BMW B-52G (58-0186, 58-0224, 58-0225, 58-0226, 58-0241, 58-0251, and 59-2573) aircraft deployed for the first MIGHTY WARRIOR exercise from 1-10 August 1988. The crews were very busy and typically flew two missions each day over an island off the Dutch/Belgian coast and dropped two live bombs on each mission. One B-52G (58-0225) aircraft suffered a cracked windscreen on the flight to RAF Fairford and was unable to fly any of the missions.

BUSY BREWER '88C, 30 August to 19 September 1988, was the final busy brewer coded exercise and included three 42nd BMW B-52G (58-0195, 58-0226, and 58-0235) aircraft and two 320th BMW B-52G (58-0213 and 58-0255) aircraft. The aircraft supported NATO exercises including TEAMWORK and COLD FIRE.

Two 42nd BMW B-52G (58-0241 and 59-2595) aircraft deployed under BUSY WARRIOR from 18-21 April 1989 but no missions were flown due to a Spanish Air Traffic Control (ATC) strike. Five 2nd BMW B-52G (58-0189, 58-0216, 58-0233, 59-2565, and 59-2570) aircraft and two 42nd BMW B-52G (58-0232 and 58-0240) aircraft deployed to RAF Fairford for the final MIGHTY WARRIOR exercise from 7-25 September 1989. The aircraft supported NATO exercises including COLD FIRE and SHARP SPEAR. The two 42nd BMW aircraft deployed to RAF St. Mawgan marking the first time the base was used to launch B-52 aircraft. Aircraft 59-2570 deployed to RAF Leuchars on 21 September for static display before returning to home station on 25 September.

Gallant Eagle. SAC bombers and tankers participated in the largest military exercise in the United States since 1962, called GALLANT EAGLE '84, from 4-12 September 1984. It was a conventional forces exercise conducted by U.S. CENTCOM and included SAC, TAC, MAC, and Army elements operating from various bases in the CONUS. B-52 aircraft from various units flew a total of 182 sorties and B-52H aircraft from the 28th BMW at Ellsworth AFB, SD deployed to Biggs Army Air Field (AAF), TX.

Other Exercises. SAC participated in several large-scale exercises during the 1970s and throughout the 1980s including TEAM SPIRIT, GIANT WARRIOR, DISTANCE MARINER, and BULL RIDER. The first operational OCONUS deployment of a B-52 unit since the Vietnam War occurred from 9-23 May 1979 when three B-52D aircraft deployed to RAF Upper Heyford, UK to participate in the FLINTLOCK and DAWN PATROL exercises.

TEAM SPIRIT was a multi-service exercise conducted by the JCS in collaboration with South Korea. It was first conducted in 1978 and continued until 1993 as a show of force and to demonstrate to North Korea the U.S. commitment to the Republic of Korea (ROK) in the south. The exercise typically included over 150,000 personnel from all U.S. services and the ROK and provided realistic training.

A B-52G aerial refueling during TEAM SPIRIT '82 on 1 March 1982. (National Archives)

The 410th BMW from K.I. Sawyer AFB, MI deployed B-52 bombers to RAAF Darwin, Australia for the first time. The deployment, called BUSY BOOMERANG, followed several years of negotiation with the Australian government to arrive at an agreement in March 1981. The B-52H arrived on 5 May for two days of static display. An operational B-52 sortie from Darwin occurred on 22 June for a sea-search mission. BUSY BOOMERANG exercises continued with regular deployments for the next several years.

Three B-52G aircraft from the 42nd BMW at Loring AFB, ME deployed to Moron AB, Spain for the first time in 1983. The bombers performed high- and low-level bombing, sea search, and fighter-intercept missions during exercise ETERNAL TRIANGLE from 27-28 October and CRISEX from 31 October to 8 November.

SAC B-52 bombers conducted aerial mining operations and participated in a concurrent Naval exercise called MINEX. GIANT WARRIOR was a large-scale PACOM air operation including forces from SAC and other Air Force commands. One highlight during

B-52G aircraft parked on the flightline at Hickam AFB, HI on 1 August 1990 during GIANT WARRIOR '90. (National Archives)

GIANT WARRIOR '89 was a massive 16 aircraft MITO from Andersen AFB, Guam including B-52 and B-1B bombers, KC-135 tankers, and E-3 AWACS.

Nuclear Tests

Between 1952 and 1956, various Air Force aircraft supported nuclear tests including the B-36, B-47, and B-50. B-36 aircraft dropped atomic weapons for nuclear tests under several operational code names including Operations TEXAN, IVY, UPSHOT-KNOTHOLE, CASTLE, and TEAPOT. Sometimes they were used for "effects aircraft" to determine the effect of high-yield nuclear bombs (heat, blast, and overpressure) on aircraft in flight. Other times they were used as "sampler aircraft" to measure radiation released from nuclear bombs. The B-36 was used primarily because it was still SAC's front-line bomber, and it could carry large amounts of equipment over long distances.

The first air drop of a deliverable H-bomb occurred during Operation REDWING on 21 May 1956 at Eniwetok during Shot Cherokee. A B-36 was planned as the drop aircraft, but analysis indicated it would be too slow and an NB-52B (52-0013) called *Barbara Grace* assigned to the 4925th Test Group (Atomic) at Kirtland AFB, NM was used instead. This was the first drop of a live thermonuclear bomb from a B-52. The target was a point on Namu Island, Bikini Atoll. However, the aircrew misidentified an observation facility on a different island for their targeting beacon. The bomb missed Namu Island by four miles and detonated at 4,350 feet above the ocean causing significant damage to the aircraft. Most test data was lost due to the error in targeting. Another Operation REDWING test was an airdrop of a 1.7 kiloton weapon during Shot Osage on 16 June 1956. A JRB-52B (52-0004) participated throughout the testing period for monitoring the nuclear detonations. A B-52B (53-0383) was also on temporary duty for Shot Cherokee from 16 April to 21 May 1956.

Operation REDWING, Shot Cherokee fireball, 0551 hours, 21 May 1956. (National Nuclear Security Administration)

One B-52D (56-0591) named "Tommy's Tigator" was chosen to support nuclear test runs. The aircraft was one of several pulled off the assembly line for "special purposes." Tommy's Tigator was assigned temporary duty to Eniwetok from 1 April to 18 July 1958 to participate in Operation HARDTACK I, during which nuclear tests were conducted at the Pacific Proving Ground between 28 April and 18 August 1958. AMC and the AEC detonated 35 nuclear devices at Bikini Atoll, Eniwetok Atoll in the Marshall Islands, and Johnston Island in the Pacific Ocean. They included balloon, surface, underwater, and rocket-borne high-altitude tests. Tommy's Tigator, under the command of Captain Thomas M. "Tommy" Summer, was loaded with monitoring equipment and flew sorties through the radioactive atmosphere to collect radiation effects information during nine of the nuclear explosions.[71]

The U.S. and Soviet Union entered a de facto moratorium on atmospheric nuclear testing on 1 November 1958 in response to these tests. However, the moratorium was shaky at best and both sides introduced issues that threatened to end the talks in Geneva. The Soviet Union insisted on veto power over any proposed inspection. American efforts to reopen the technical issues of the detection system did not help matters. On 29 December 1959, President Eisenhower announced that the "voluntary moratorium" on testing would expire on 31 December, and that although the U.S. was now free to resume testing it would not do so without announcing any resumption in advance.[72]

The Soviets finally broke the agreement when they conducted an atmospheric test of a weapon with a yield of 150 kilotons on 1 September 1961. This was quickly followed by fifty more tests in the next 60 days including a 50-megaton weapon. The total yield from these combined tests exceeded all previous tests, by all nations, combined.[73] It was essentially the excuse the U.S. needed to renew its own testing program. Planning began on 10 October 1961 after approval by President Kennedy and final approval for the tests was given on 2 March 1962. The NB-52B (52-0013) airdropped 29 of the 36 test shots of nuclear weapons during Operation DOMINIC from 25 April to 4 November 1962. The 29 airdrops were weapons development tests, intended to evaluate advanced designs that the labs had been working on during the moratorium. Five rocket-launched tests were conducted to gather further weapons effects data on high-altitude phenomena. Two tests of operational weapon systems were conducted – the Polaris SLBM and the Anti-Submarine Rocket (ASROC).

The first drop occurred on 25 April during Shot Adobe when the aircraft dropped a 190-kiloton bomb from 2,900 feet. Shot Androscoggin was conducted on 2 October when the NB-52B (52-0013) dropped a device 210 miles southwest of Johnston Island. This marked the first time the test was monitored with all instrumentation carried aboard aircraft including RC-121, C-130, and KC-135.

The last airdrop of a U.S. nuclear weapon occurred during Shot Housatonic on 30 October 1962 when the NB-52B (52-0013) dropped an 8.3 megaton weapon on the Johnston Island test site from 12,130 feet. The bombing error was only 100 feet, and the mushroom cloud rose to 54,000 feet.

Following these tests, the U.S. and the Soviet Union agreed to a Limited Test Ban treaty on 25 July 1963. This treaty eliminated all atmospheric testing. In response, the U.S. created a new program called FLASHBACK to rapidly build and demonstrate a high yield hydrogen bomb in the event the Soviets violated the treaty, as they were likely expected to do. FLASHBACK was a test of such a bomb… but without the actual atomic explosive, testing the associated hardware, ballistics, and electronics. The bomb casing and other hardware would be tested and ready to go, with the fission/fusion explosive expected to be plugged in and ready to detonate within 90 days of the Soviet test.

The FLASHBACK Test Vehicle was tested at Kirtland Air Force Base in New Mexico in January 1965. These tests did not involve the dropping of a nuclear bomb which then detonated. Instead, the susceptibility of the FLASHBACK device to electromagnetic emissions from the ARC-58 transmitter in the B-52C (54-2669) carrier aircraft and the device's own telemetry transmitters was the point of the test. During the flight tests all high explosive and nuclear components were deleted, and a simulator replaced the warhead. The FLASHBACK device itself was large: large enough that to fit into the B-52 bomb bay the bomb bay doors had to be removed, and even then, the device protruded from the belly of the aircraft. It was about 96 inches in diameter and 297 inches long, not counting the protruding parachute pack or antenna.[74]

FLASHBACK Test Vehicle mounted in the bomb bay of the B-52C (54-2669) carrier aircraft. (U.S. Air Force)

Broken Arrows

The term Broken Arrow was used within the Defense Department, and consequently SAC, to describe an incident involving a nuclear warhead. Typically, it involved aircraft crashes or accidents with a nuclear weapon onboard. SAC reported several broken arrows over the years including six involving B-52 bombers.

B-52F (57-0036). A B-52F with two nuclear weapons on board crashed on 15 October 1959 after colliding with a KC-135 (57-1513) during refueling over Hardinsburg, KY during an Operation CHROME DOME airborne alert mission. Both aircraft were from Columbus AFB, MS. The refueling was accomplished at night, but the weather was clear and there was no turbulence. All four tanker crew members were killed. The copilot, navigator, instructor navigator, and gunner failed to leave the aircraft, but the remaining four crew members survived. Both weapons were unarmed and recovered intact. One was partially burned but no radiation or contamination was released. The bombs were subsequently moved for inspection and dismantlement.

B-52G (58-0187). A B-52G with two Mk-39 nuclear weapons onboard crashed on 24 January 1961 during landing at Seymour-Johnson AFB, NC. The aircraft experienced structural failure and fuel leak of the right wing during an Operation CHROME DOME airborne alert mission. The tanker crew noticed the leak during aerial refueling and notified the bomber pilot. The leak was severe, and the aircraft lost more than 5,400 gallons of fuel in less than three minutes.

Most of the Mk-39 bomb was uncovered from an excavation in the waterlogged farmland near Goldsboro, NC. (U.S. Air Force)

The pilot headed back to Seymour-Johnson for an emergency landing, but the wing failed when the flaps were engaged, and the aircraft went out of control. Five of the eight crew members ejected safely and survived. The aircraft broke up at 2,000 to 10,000 feet altitude and the wreckage was strewn over two square miles. The two nuclear weapons broke loose. One parachuted essentially undamaged to the ground and one sank to at least 50 feet in a marsh.

The recovery team worked for eight days to uncover most of the bomb. But a portion of one weapon, containing uranium, could not be recovered despite evacuation in the waterlogged farmland. The Air Force subsequently purchased an easement requiring permission for anyone to dig. There was no detectable radiation and no hazard in the area.

B-52F (57-0166). A B-52F with two Mk-39 nuclear bombs onboard crashed approximately 15 miles west of Yuba City, CA on 14 March 1961. The aircraft took-off from Mather AFB, CA on 13 March for a 24-hour Operation CHROME DOME airborne alert mission. The mission took the aircraft north over water and over Alaska and required two aerial refueling hook-ups. The aircraft began experiencing problems within 20 minutes after take-off when the crew was unable to control excessive heating blowing from the vents below the instrument panel. The heat was almost unbearable and forced the crew to depressurize the cabin, don oxygen masks, and bring in ram air to cool down the cockpit.

But the troubles continued. The rate of climb indicator failed, and the No. 3 engine lost 30 percent of its power. Later the forward fuel tank gauge failed and about 14 hours after take-off the pilots outside window panel shattered. The extreme heat cracked the glass in the turn-and-slip indicators and dehydrated the crew who had run out of water 15 hours after take-off. The pilot descended to 12,000 feet altitude to allow the crew to better tolerate the depressurized cabin but this severely increased fuel consumption.

The pilot requested another tanker 21 hours into the flight when he realized they would have only 14,000 pounds when arriving at Mather. The controller at Mather said a tanker was enroute and would rendezvous over Red Bluff, but the tanker was late and all eight engines on the bomber flamed out before it could be refueled. All crew members ejected safely but the aircraft crashed in a rice field southwest of Yuba City and the two nuclear weapons were ejected from the aircraft. Although the bombs were severely damaged the high explosives did not detonate and no radioactive materials were released.

B-52D (55-0060). A B-52D from Turner AFB, GA crashed into Savage Mountain near Barton, MD on 13 January 1964 during a ferry flight enroute to its home base from Westover AFB, MA, where it had landed after an Operation CHROME DOME patrol over Europe. Two nuclear weapons were onboard, and both were in the "Tactical Ferry Configuration" with no electrical connections and the safing switches were in the SAFE position. The aircraft encountered severe turbulence during a blizzard which caused the tail fin and rudder assembly to separate from the aircraft. The pilot and copilot ejected safely and survived. The navigator and gunner ejected safely but died of exposure in the sub-zero temperatures on the ground. The radar navigator failed to eject from the aircraft. The two nuclear weapons remained in the aircraft until it crashed and were recovered relatively intact.

B-52G (58-0256). On 17 January 1966, a B-52G from Seymour-Johnson AFB, NC carrying four Mk-28 nuclear weapons crashed during an Operation CHROME DOME airborne alert mission. The bomber collided with a KC-135 (61-0273) during refueling near Palomares, Spain. The bomber had approached the tanker too fast and collided with the refueling boom, which penetrated the bomber's fuselage, and the wing broke off. The bomber exploded and the tanker, on fire, went into a steep dive and exploded at 1,600 feet. Three of the seven bomber crew and all four tanker crew members were killed. The four nuclear weapons fell free from the bomb bay. Three were found on land and two of these released radioactive material as the result of non-nuclear explosions. Clean-up crews recovered about 1,400 tons of slightly contaminated soil and vegetation and sent it to the U.S. for disposal. One of the bombs fell into the Mediterranean Sea and was found by a U.S. Navy submarine on 15 March after an extensive search. The bomb was located about five miles offshore in an underwater canyon under 2,500 feet of water. It was recovered intact on 7 April after several unsuccessful attempts caused the bomb to slip deeper and deeper, finally settling at 2,850 feet.

B-52G (58-0188). A B-52G from Plattsburg AFB, NY crashed on 22 January 1968 at North Star Bay, Greenland during an Operation CHROME DOME airborne alert mission. A faulty heater caused an electrical fire and knocked out power. The pilot attempted an emergency landing at Thule AB, but the crew ultimately ejected, and the aircraft crashed on the sea ice. Five of the six crew members were rescued and one perished. Four nuclear weapons detonated (non-nuclear) and caused widespread contamination. Operation CRESTED ICE was launched in cooperation with the Government of Denmark to clean up contaminated ice and snow. This operation was completed on 13 September.

Displays and Disposition

Displays

SAC phased out its B-52 force over the years in various increments affecting different models. Most were eventually scrapped but a few were placed on display in several locations. Some of the most notable are shown below.

B-52D (55-0100). The original ARC LIGHT Memorial was a static display of aircraft 55-0100, named "Old 100", and was dedicated on 14 February 1974, one year after the first prisoners of war were released from North Vietnam prison camps. The memorial was dedicated to the 75 B-52 crew members killed during Operation ARC LIGHT from June 1965 to August 1973. Old 100 flew more than 5,000 hours in SEA and was one of the final three aircraft to bomb North Vietnam in Operation LINEBACKER II during its final mission on 29 December 1972. Ten years after the memorial dedication Old 100 was determined unsafe due to severe corrosion from the salty air and damp climate of Guam. The decision was made to replace the aircraft with another one and B-52D (56-0586), still on Guam despite the ongoing conversion of the 43rd SW to G models, and

Original B-52D (55-0100) "ARC LIGHT Memorial" on display at Andersen AFB, Guam. (Tip Klamberg Collection)

the last B-52D to be retired, was chosen. Its tail number was redesignated 55-0100 by the National Museum of the U.S. Air Force. The memorial was redesignated on 9 June 1983 with the new aircraft in place.

Old 100 was moved to the west side of base and placed in the "Andersen Boneyard" along with other aircraft that were unable to leave the island. It was earmarked for destruction under the Strategic Arms Limitation Talks (SALT) II agreement and subsequently dismantled on 16 July 1986. However, Typhoon Roy scattered the sections of the aircraft into the jungle where they were concealed until Typhoon Paka uncovered the tail and assorted debris on 17 December 1997.

Unfortunately, the second aircraft was finally overcome by corrosion too and was removed from the memorial in 2014. The rededicated memorial no longer has an aircraft as its centerpiece. Instead, it has a full-sized silhouette of a B-52D on the ground and the tail of the second aircraft on a raised platform. The original plaque remains and a B-52H tail mounted on a raised platform was added to honor the continuous bomber presence at Andersen. Sections of Old 100 are also on display at the Andersen Heritage Hall in the passenger terminal.

NB-52B (52-0013). This aircraft was delivered directly to Kirtland AFB, NM on 4 May

NB-52B (52-0013) at the National Museum of Nuclear Science & History being restored for display in a new outdoor visitor exhibit. (U.S. Air Force)

1955 where it remained assigned throughout its service life. This individual airplane has dropped more than a dozen live nuclear bombs during weapons testing. In addition to Operation REDWING, the aircraft participated in Operation DOMINIC in 1962, which involved 29 B-52 air drops. The airplane carried the name Deterrent I painted on its nose but flew missions with the call sign "Cow Slip Two." It is the only B-52 still in existence to have dropped a nuclear bomb. It was withdrawn from nuclear testing service in 1963 after the signing of the Limited Nuclear Test Ban treaty. It was formally retired on 8 December 1970 and transferred to the National Atomic Museum (now, the National Museum of Nuclear Science & History) at Kirtland in 1971, where it was placed on display. It began to show its wear after several years and a restoration project began in 2016 to bring the aircraft back to its former condition.

B-52D (56-0687). This B-52D is on display at the B-52 Memorial Park, Orlando International Airport, previously McCoy AFB, Orlando, FL. It made the last flight of a B-52D (non-operational) from Carswell AFB, TX to McCoy for display on 20 February 1984. It was delivered to the 28th BMW at Ellsworth AFB, SD on 10 October 1957. It underwent the AGM-28 Hound Dog missile modification at Kelly AFB, TX before being transferred to the 92nd BMW at Fairchild AFB, WA on 31 March 1963.

B-52D (56-0687) on display at B-52 Memorial Park, Orlando International Airport, Orlando, FL. (Unknown)

The aircraft returned to Ellsworth in January 1966 and received the "BIG BELLY" modification in March 1966 at Kelly AFB, TX before being returned to Fairchild in February 1967. It served its first tour in SEA and performed Operation ARC LIGHT missions from 23 March to 23 September 1968 when it was transferred to the 93rd BMW at Castle AFB, CA. The aircraft transferred again on 9 February 1971, this time to the 22nd BMW at March AFB, CA, before serving its second ARC LIGHT tour from 6 April to 29 June 1972. It underwent PDM at Kelly AFB, TX and did a short stint back at Castle before its third SEA tour from 17 November 1972 to 10 July 1973 where it flew Operation ARC LIGHT and LINEBACKER II missions. The aircraft then spent the next several years assigned to the 7th BMW at Carswell AFB, TX with short assignments to U-Tapao and Andersen AFB, Guam before making its final flight to McCoy for display in 1984.

B-52 STRATOFORTRESS: THE IRON FIST OF STRATEGIC AIR COMMAND

A total of 56 B-52 aircraft were placed on display over the years, either as complete aircraft or just nose and cockpit sections, as shown in the following table.

Tail #	Type	Museum/Site	Location	Comments
49-231	YB-52	National Museum of the U.S. Air Force	Dayton, OH	Donated 27-Jan-58. Removed in mid-1960s.
52-0003	NB-52A	Pima Air & Space Museum	Tucson, AZ	
52-0005	B-52B	Wings Over the Rockies Air & Space Museum	Lowry AFB, CO	
52-0008	NB-52B	Air Force Test Center (AFTC) Museum Airpark	Edwards AFB, CA	Displayed at the north gate. Named "The Challenger"
52-0013	NB-52B	National Museum of Nuclear Science & History	Albuquerque, NM	
52-8711	B-52B	Strategic Air Command & Aerospace Museum	Ashland, NE	
53-0394	B-52B	National Museum of the U.S. Air Force	Dayton, OH	Donated 6-Jan-66. Removed in mid-1984.
55-0057	B-52D	B-52D Stratofortress Historical Marker	Maxwell AFB, AL	
55-0062	B-52D	K.I. Sawyer Heritage Air Museum	K.I. Sawyer AFB, MI	Former base
55-0063	B-52D	Southwest Aerospace Museum	Fort Worth, TX	Scrapped on site when the museum closed in the early 1990s.
55-0067	B-52D	Pima Air & Space Museum	Tucson, AZ	
55-0068	B-52D	USAF History & Traditions Museum	Lackland AFB, TX	Located at Lackland Airpark.
55-0071	B-52D	USS Alabama Battleship Memorial Park	Mobile, AL	
55-0083	B-52D	U.S. Air Force Academy	Colorado Springs, CO	Second B-52 credited as MiG Killer when A1C Albert Moore shot down a MiG-21 on 24-Dec-72.
55-0085	B-52D	Museum of Aviation	Robins AFB, GA	
55-0094	B-52D	Kansas Aviation Museum	Wichita, KS	
55-0095	B-52D	Valiant Air Command Warbird Museum	Titusville, FL	Nose and cockpit section only.
55-0100	B-52D	Arc Light Memorial	Andersen AFB, Guam	Original aircraft. Removed due to corrosion. Replaced with 56-0586.
55-0105	B-52D	War Service Memorial	Seoul, South Korea	
55-0677	B-52D	Michigan Flight Museum	Ypsilanti, MI	Formerly Yankee Air Museum.
55-0679	B-52D	March Field Air Museum	Riverside, CA	
56-0585	B-52D	AFTC Museum Airpark	Edwards AFB, CA	Removed in 2016.
56-0586	B-52D	Arc Light Memorial	Andersen AFB, Guam	Second aircraft. Removed due to corrosion. Not replaced.
56-0589	B-52D	Sheppard AFB Airpark	Sheppard AFB, TX	Removed in 2012. Replaced by B-52G (58-0200).
56-0612	B-52D	Castle Air Museum	Atwater, CA	
56-0629	B-52D	Barksdale Global Power Museum	Barksdale AFB, LA	
56-0657	B-52D	South Dakota Air & Space Museum	Ellsworth AFB, SD	
56-0659	B-52D	Heritage Park	Davis-Monthan AFB, AZ	Removed in 2011.
56-0665	B-52D	National Museum of the U.S. Air Force	Dayton, OH	

Tail #	Type	Museum/Site	Location	Comments
56-0666	B-52D	National Museum of the Mighty Eighth Air Force	Savanah, GA	Vertical fin only
56-0676	B-52D	Armed Forces & Aerospace Museum	Fairchild AFB, WA	First B-52 credited as MiG Killer when SSgt Samuel O. Turner shot down a MiG-21 on 18-Dec-72.
56-0682	B-52D	National Museum of the U.S. Air Force	Dayton, OH	Nose and cockpit section only.
56-0683	B-52D	Captain Robert Morris Memorial – Arnold Gate	Whiteman AFB, MO	Previously displayed at Pease AFB, NH.
56-0685	B-52D	Dyess Linear Airpark	Dyess AFB, TX	
56-0687	B-52D	Orlando International Airport B-52 Memorial Park	Orlando, FL	Former McCoy AFB. The last B-52D to fly when it arrived in 1984.
56-0689	B-52D	Imperial War Museums	Duxford, England	Housed in the American Air Museum building with a B-29A.
56-0692	B-52D	Kelly Field Heritage Airpark	San Antonio, TX	Previously displayed at Tinker AFB, OK (South side).
56-0695	B-52D	Charles B. Hall Airpark	Tinker AFB, OK	Near Air Depot Blvd gate.
56-0696	B-52D	Heritage Center	Travis AFB, CA	
57-0038	B-52F	Joe Davies Heritage Airpark	Palmdale, CA	Previously displayed at Oklahoma City, OK Fairgrounds (1974-2006).
57-0042	B-52F	Yanks Air Museum	Chino, CA	Nose and cockpit section only (under restoration for display). Previously displayed at Museum of Flying, Santa Monica, CA.
57-0101	B-52E	Pearl Harbor Aviation Museum	Honolulu, HI	Nose section only.
57-6468	B-52G	Zorinsky Memorial Airpark	Offutt AFB, NE	Displayed at the STRATCOM gate.
57-6509	B-52G	Barksdale Global Power Museum	Barksdale AFB, LA	
58-0158	B-52G	Heritage Park Museum	Fairchild AFB, WA	Removed in 1995.
58-0183	B-52G	Pima Air & Space Museum	Tucson, AZ	
58-0185	B-52G	Air Force Armament Museum	Eglin AFB, FL	
58-0191	B-52G	Hill Aerospace Museum	Hill AFB, UT	
58-0200	B-52G	Sheppard AFB Airpark	Sheppard AFB, TX	Replaced B-52D (56-0589)
58-0225	B-52G	Mohawk Valley B-52 Memorial	Rome, NY	
58-0232	B-52G	Hangar 25 Museum	Big Spring, TX	Nose and cockpit section only.
59-2577	B-52G	Heritage Center	Grand Forks AFB, ND	
59-2579	B-52G	Tillamook Air Museum	Tillamook, OR	Nose and cockpit section only. Previously displayed at Southern Utah Air Museum, Washington, UT.
59-2584	B-52G	Museum of Flight	Tukwila, WA	Displayed at Boeing Field, WA
59-2596	B-52G	Darwin Aviation Museum	Darwin, Australia	
59-2601	B-52G	Tactical Air Command Memorial Park	Langley AFB, VA	

Disposition

In 1991, the U.S., Russia, Belarus, Kazakhstan, and Ukraine signed the START to reduce the number of strategic nuclear delivery vehicles. This treaty only allowed the Air

Force to retain 71 B-52 bombers equipped to carry 20 cruise missiles. About 365 B-52 bombers of various models sat on the desert floor at AMARC. These bombers were preserved in long-term storage and able to return to service in the event of a contingency or as fleet replacements. Therefore, these aircraft were counted in the total bombers available and were earmarked for destruction.

Crews at AMARC set about the awesome task of eliminating the aircraft under the treaty timelines beginning in August 1993. Originally, the crews used a 13,500 pound "guillotine" blade to slice through each bomber. Each aircraft was stripped of engines, reusa-

B-52 aircraft awaiting destruction at AMARC. (Ronny Young Collection)

ble parts, and hazardous material and fluids were removed. The blade was then hoisted to 80 feet using a crane and dropped on each predefined cut point leaving the aircraft cut into four sections including the tail section, each wing, and the forward fuselage section. The sections were then arranged for easy identification by Russian satellites for 90 days before being sold to local scrap companies.

A total of 217 B-52C, D, E, and F models were destroyed by 15 December 1994. The crew typically processed three aircraft per week, at one point resulting in about 100 aircraft lined up on the desert floor. About 150,000 pounds of aluminum and other metals valued at about $20,000 were available for recycling from each aircraft.

B-52D aircraft being cut up at AMARC. (Ronny Young Collection)

Unfortunately, the guillotine method was a rather "brute-force" approach that made it difficult to preserve the interior of the aircraft. AMARC crews eventually switched to a more surgical approach using power saws. This approach allowed retrieval of spare parts from the B-52G aircraft that could be used on operational B-52H aircraft. The power saws allowed precision cuts, and the sections were then sealed until the parts were retrieved.

On 8 April 2010, a new treaty commonly called "New START" was signed which agreed to further reductions. A B-52G (58-0197) became the first B-52 destroyed under the New START agreement in 2011. The aircraft was dissected into two large pieces by making a surgical cut through the rear empennage, severing the tail and horizontal stabilizer. New START called for the elimination of 38 more of the stored B-52G bombers.[1] The final B-52G (58-0224) was destroyed on 19 December 2013.

Appendix A: Specifications

B-52A Stratofortress

The B-52A was the first production aircraft and featured side-by-side cockpit seating. (U.S. Air Force)

Aircraft Specifications					
Type:	B-52	Series:	B-52A	Name:	Stratofortress
Wingspan:	185 feet	Length:	156.5 feet	Height:	48.3 feet
Empty Weight:	167,424 pounds	Combat Weight:	264,610 pounds	Max TO Weight:	390,000 pounds
Combat Radius:	3,565 miles	Combat Ceiling:	47,050 feet	Service Ceiling:	48,600 feet
Cruise Speed:	526 mph	Max Speed:	624 mph	Max Payload:	43,000 pounds
OEM:	Boeing	Produced:	3	SAC Inventory:	0
First Flight:	5-Aug-54	First Delivery:	Jun-54	Phase Out:	Apr-61
Missions:	Intercontinental Strategic Bombardment				
Tail Number(s):	52-0001 to 52-0003				
Propulsion:	Eight P&W J57-P-1W jet engines with 8,250 pounds normal thrust and 11,100 pounds take-off thrust each with water injection. Two engines each mounted on four wing pylons.				
Accommodations:	Total 6. Pilot, copilot, navigator, radar navigator, EWO, tail gunner.				
Payload:	Up to 43,000 pounds of bombs; plus four .50-caliber guns.				
Comments:	Specifications estimated based on RB-52B Standard Aircraft Characteristics with -1W engines, 9-Oct-53. Max speed was 624 mph at 19,500 feet. Combat Radius, Combat Ceiling, Service Ceiling, Cruise Speed and Max Speed were calculated with 3,700-pound bombload (flash bombs) at combat weight without refueling (Basic Mission). Included 1,000-gallon droppable fuel tanks near each wing tip giving a total fuel capacity of 37,385 gallons. Used side by side seating vs. tandem cockpit. First B-52A rolled out of the factory in Seattle on 18-Mar-54. Accepted in Jun-54 and it, and the next two B-52A aircraft, remained at Boeing for flight testing. All three aircraft were redesignated NB-52A on 8-Oct-57. NB-52A (52-003) was used by NASA at Edwards AFB, CA as the original X-15 mothership.				

B-52B Stratofortress

The B-52B was the first operational model and the first to see service with SAC. (U.S. Air Force)

		Aircraft Specifications			
Type:	B-52	Series:	B-52B	Name:	Stratofortress
Wingspan:	185 feet	Length:	156.6 feet	Height:	48.3 feet
Empty Weight:	164,081 pounds	Combat Weight:	272,000 pounds	Max TO Weight:	420,000 pounds
Combat Radius:	3,537 miles	Combat Ceiling:	47,100 feet	Service Ceiling:	47,700 feet
Cruise Speed:	522 mph	Max Speed:	635 mph	Max Payload:	43,000 pounds
OEM:	Boeing	Produced:	23	SAC Inventory:	23
First Flight:	7-Jul-55	First Delivery:	9-Nov-55	Phase Out:	29-Jun-66
Missions:	Intercontinental Strategic Bombardment				
Tail Number(s):	53-0373 to 53-0376; 53-0380 to 53-0398				
Propulsion:	Eight P&W J57-P-19W, -29W, or -29WA with 9,000 pounds normal thrust and 12,100 pounds take-off thrust (-29W 11,500 pounds) each with water injection. Two engines each mounted on four wing pylons.				
Accommodations:	Total 6. Pilot, copilot, navigator, radar navigator, electronic warfare officer (EWO), tail gunner.				
Payload:	Up to 43,000 pounds total of bombs and/or AGM-28 Hound Dogs (53-0380 to 53-0398); plus two 20-mm cannons (53-0373 to 53-0376; 53-0380 to 53-0391) or four .50-caliber guns (53-0392 to 53-0398).				
Comments:	Specifications based on B-52B (Bomber Version) Standard Aircraft Characteristics, 1-Oct-58. Max speed was 635 mph at 20,300 feet. Combat Radius, Combat Ceiling, Service Ceiling, Cruise Speed and Max Speed were calculated with 10,000-pound bombload at combat weight without refueling (Basic Mission). First operational B-52 model. First aircraft produced as a B-52B (53-0373) delivered to 93rd BMW, Castle AFB, CA on 9-Nov-55. Aircraft 53-0373 to 53-0376 equipped with J57-P-1W engines (see RB-52B for specs). Most remaining aircraft were equipped with the -29W or -29WA engines but the last five (53-0394 to 53-0398) were equipped with -19W engines.				

B-52 STRATOFORTRESS: THE IRON FIST OF STRATEGIC AIR COMMAND

RB-52B Stratofortress

The RB-52B was essentially a B-52B with -1 engines and capability to carry an internal recon pod. (U.S. Air Force)

Aircraft Specifications

Type:	B-52	Series:	RB-52B	Name:	Stratofortress
Wingspan:	185 feet	Length:	156.6 feet	Height:	48.3 feet
Empty Weight:	168,332 pounds	Combat Weight:	278,700 pounds	Max TO Weight:	405,000 pounds
Combat Radius:	3,565 miles	Combat Ceiling:	46,050 feet	Service Ceiling:	46,750 feet
Cruise Speed:	522 mph	Max Speed:	629 mph	Max Payload:	43,000 pounds
OEM:	Boeing	Produced:	27	SAC Inventory:	25
First Flight:	Dec-54	First Delivery:	29-Jun-55	Phase Out:	29-Jun-66
Missions:	Intercontinental Strategic Bombardment, Photo Reconnaissance, ELINT, weather, and observation				
Tail Number(s):	52-0004 to 52-0013; 52-8710 to 52-8716; 53-0366 to 53-0372; 53-0377 to 53-0379				
Propulsion:	Eight P&W J57-P-1W, -1WA, or -1WB jet engines with 8,250 pounds normal thrust and 11,400 pounds take-off thrust each with water injection. Two engines each mounted on four wing pylons.				
Accommodations:	Total 8. Pilot, copilot, navigator, radar navigator, electronic warfare officer (EWO), tail gunner, plus two more specialist crewmembers for reconnaissance/data gathering missions.				
Payload:	Up to 43,000 pounds of bombs or recon pod and 24 M-120 photoflash bombs; plus two 20-mm cannons (52-0009; 52-8710 to 52-8716; 53-0366 to 53-0372; 53-0377 to 53-0379) or four .50-caliber guns (52-0004 to 52-0008; 52-0010 to 52-0013).				
Comments:	Specifications based on B-52B (Reconnaissance Version) Standard Aircraft Characteristics, 22-Jan-58. Max speed was 629 mph at 19,800 feet. Combat Radius, Combat Ceiling, Service Ceiling, Cruise Speed and Max Speed were calculated with 3,700-pound bombload (Flash Bombs) at combat weight without refueling (Basic Mission). First RB-52B (52-0004) flew in Dec-54 and delivered in place to Boeing on 3-Mar-55. First RB-52B (52-8711) assigned to SAC flew on 25-Jun-55 and delivered to 93rd BMW on 29-Jun-55. Aircraft 53-0377 to 53-0379 equipped with J57-P-29W engines (see B-52B for specs). Most RB-52B aircraft were redesignated B-52B upon assignment to SAC units.				

B-52 Stratofortress: The Iron Fist of Strategic Air Command

B-52C Stratofortress

The B-52C featured 3,000-gallon droppable wing tanks which increased combat radius by 270 miles. (U.S. Air Force)

Aircraft Specifications					
Type:	B-52	Series:	B-52C	Name:	Stratofortress
Wingspan:	185 feet	Length:	156.6 feet	Height:	48.3 feet
Empty Weight:	164,486 pounds	Combat Weight:	283,100 pounds	Max TO Weight:	450,000 pounds
Combat Radius:	3,807 miles	Combat Ceiling:	46,350 feet	Service Ceiling:	46,950 feet
Cruise Speed:	522 mph	Max Speed:	635 mph	Max Payload:	43,000 pounds
OEM:	Boeing	Produced:	35	SAC Inventory:	33
First Flight:	9-Mar-56	First Delivery:	16-Jun-56	Phase Out:	29-Sep-71
Missions:	Intercontinental Strategic Bombardment				
Tail Number(s):	53-0399 to 53-0408; 54-2664 to 54-2688				
Propulsion:	Eight P&W J57-P-19W or -29WA turbojet engines with 9,000 pounds normal thrust and 12,100 pounds take-off thrust each with water injection. Two engines each mounted on four wing pylons.				
Accommodations:	Total 6. Pilot, copilot, navigator, radar navigator, electronic warfare officer (EWO), tail gunner.				
Payload:	Up to 43,000 pounds total of bombs and/or AGM-28 Hound Dogs; plus four .50-caliber guns.				
Comments:	Specifications based on B-52C & D Standard Aircraft Characteristics, 24-Mar-58. Max speed was 635 mph at 20,200 feet. Combat Radius, Combat Ceiling, Service Ceiling, Cruise Speed and Max Speed were calculated with 10,000-pound bombload at combat weight without refueling (Basic Mission). Increased droppable fuel tanks near wing tip to 3,000 gallons which increased total fuel capacity to 41,550 gallons. Capable of carrying the recon pod designed for the RB-52B. Later upgrades made it equivalent to the B-52D. First B-52C (53-0399) delivered in place to Boeing on 28-Feb-56. First B-52C (53-0400) assigned to SAC delivered to 42nd BMW at Loring AFB, ME on 16-Jun-56. Fleet phased out by 29-Sep-71. NB-52C (53-0399) used for testing and remained in service at Edwards AFB, CA until 28-Jul-75.				

B-52 Stratofortress: The Iron Fist of Strategic Air Command

B-52D Stratofortress

The B-52D was the second most produced model with 170 aircraft, second only to the B-52G. (U.S. Air Force)

Aircraft Specifications					
Type:	B-52	Series:	B-52D	Name:	Stratofortress
Wingspan:	185 feet	Length:	156.6 feet	Height:	48.3 feet
Empty Weight:	164,486 pounds	Combat Weight:	283,100 pounds	Max TO Weight:	450,000 pounds
Combat Radius:	3,807 miles	Combat Ceiling:	46,350 feet	Service Ceiling:	46,950 feet
Cruise Speed:	522 mph	Max Speed:	635 mph	Max Payload:	43,000 pounds
OEM:	Boeing	Produced:	170	SAC Inventory:	169
First Flight:	14-May-56	First Delivery:	26-Jun-56	Phase Out:	11-Jul-84
Missions:	Intercontinental Strategic Bombardment				
Tail Number(s):	55-0049 to 55-0117; 55-0673 to 55-0680; 56-0580 to 56-0630; 56-0657 to 56-0698				
Propulsion:	Eight P&W J57-P-19W or -29WA turbojet engines with 9,000 pounds normal thrust and 12,100 pounds take-off thrust each with water injection. Two engines each mounted on four wing pylons.				
Accommodations:	Total 6. Pilot, copilot, navigator, radar navigator, electronic warfare officer (EWO), tail gunner.				
Payload:	Up to 43,000 pounds total of bombs and/or AGM-28 Hound Dogs; plus four .50-caliber guns. "BIG BELLY" bomb bay and wing pylons increased total conventional bombload to 60,000 pounds.				
Comments:	Specifications based on B-52C & D Standard Aircraft Characteristics, 24-Mar-58. Max speed was 635 mph at 20,200 feet. Combat Radius, Combat Ceiling, Service Ceiling, Cruise Speed and Max Speed were calculated with 10,000-pound bombload at combat weight without refueling (Basic Mission). The B-52D was built in both Seattle and Wichita. The first Wichita-built B-52D (55-0049) rolled off the assembly line on 7-Dec-55. First flight of a Wichita built aircraft occurred on 14-May-56 and the first Seattle built 28-Sep-56. First B-52D (55-0049) delivered the 93rd BMW on 26-Jun-56. Last B-52D (55-0094) retired was placed on display at the Kansas Aviation Museum on 11 July 1984.				

B-52E Stratofortress

The B-52E was the first model to use the new low-level equipment including the BIG FOUR modification. (U.S. Air Force)

Aircraft Specifications					
Type:	B-52	Series:	B-52E	Name:	Stratofortress
Wingspan:	185 feet	Length:	156.6 feet	Height:	48.3 feet
Empty Weight:	163,752 pounds	Combat Weight:	282,600 pounds	Max TO Weight:	450,000 pounds
Combat Radius:	3,825 miles	Combat Ceiling:	46,350 feet	Service Ceiling:	46,950 feet
Cruise Speed:	522 mph	Max Speed:	635 mph	Max Payload:	43,000 pounds
OEM:	Boeing	Produced:	100	SAC Inventory:	99
First Flight:	3-Oct-57	First Delivery:	3-Dec-57	Phase Out:	13-Mar-70
Missions:	Intercontinental Strategic Bombardment				
Tail Number(s):	56-0631 to 56-0656; 56-0699 to 56-0712; 57-0014 to 57-0029; 57-0095 to 57-0138				
Propulsion:	Eight P&W J57-P-19W or -29WA turbojet engines with 9,000 pounds normal thrust and 12,100 pounds take-off thrust each with water injection. Two engines each mounted on four wing pylons.				
Accommodations:	Total 6. Pilot, copilot, navigator, radar navigator, electronic warfare officer (EWO), tail gunner.				
Payload:	Up to 43,000 pounds total of bombs, ADM-20 Quails, and/or AGM-28 Hound Dogs; plus four .50-caliber guns.				
Comments:	Specifications based on B-52E Standard Aircraft Characteristics, 1-Oct-58. Max speed was 635 mph at 20,200 feet. Combat Radius, Combat Ceiling, Service Ceiling, Combat Speed and Max Speed were calculated with 10,000-pound bombload at combat weight without refueling (Basic Mission). First model to use the new low-level equipment including the BIG FOUR modification, A/A42G-11 AFCS, and AN/APN-159 Radar Altimeter. First B-52E flight was a Seattle-built aircraft on 3-Oct-57 followed by the first Wichita-built aircraft on 17-Oct-57. First B-52E (56-0631) delivered in place to Boeing on 7-Oct-57. First B-52E (56-0700) assigned to SAC delivered to the 93rd BMW on 3-Dec-57.				

B-52F Stratofortress

The B-52F major improvement was J57-P-43 engines with increased thrust. (U.S. Air Force)

Aircraft Specifications					
Type:	B-52	Series:	B-52F	Name:	Stratofortress
Wingspan:	185 feet	Length:	156.6 feet	Height:	48.3 feet
Empty Weight:	164,936 pounds	Combat Weight:	283,600 pounds	Max TO Weight:	450,000 pounds
Combat Radius:	3,853 miles	Combat Ceiling:	46,600 feet	Service Ceiling:	47,400 feet
Cruise Speed:	523 mph	Max Speed:	637 mph	Max Payload:	43,000 pounds
OEM:	Boeing	Produced:	89	SAC Inventory:	88
First Flight:	6-May-58	First Delivery:	14-Jun-58	Phase Out:	7-Dec-78
Missions:	Intercontinental Strategic Bombardment				
Tail Number(s):	57-0030 to 57-0073; 57-0139 to 57-0183				
Propulsion:	Eight P&W J57-P-43W, -43WA, or -43WB with 9,500 pounds normal thrust and 13,750 pounds take-off thrust each with water injection. Two engines each mounted on four wing pylons.				
Accommodations:	Total 6. Pilot, copilot, navigator, radar navigator, electronic warfare officer (EWO), tail gunner.				
Payload:	Up to 43,000 pounds total of bombs, ADM-20 Quails, and/or AGM-28 Hound Dogs; plus four .50-caliber guns.				
Comments:	Specifications based on B-52F Standard Aircraft Characteristics, Jul-64. Max speed was 637 mph at 20,500 feet. Combat Radius, Combat Ceiling, Service Ceiling, Cruise Speed and Max Speed were calculated with 10,000-pound bombload at combat weight without refueling (Basic Mission). Major improvement was J57-P-43 engines with increased thrust. Upgraded with new low-level equipment and AN/APQ-38. First Seattle-built B-52F flight was on 6-May-58 followed by the first Wichita-built on 14-May-58. First B-52F (57-0030) was delivered in place to Boeing on 12-May-58. First two B-52F (57-0139 and 57-0140) assigned to SAC reached the 93rd BMW on 14-Jun-58.				

B-52 Stratofortress: The Iron Fist of Strategic Air Command

B-52G Stratofortress

The B-52G was a huge improvement and drastic departure from previous B-52 designs. (U.S. Air Force)

Aircraft Specifications					
Type:	B-52	Series:	B-52G	Name:	Stratofortress
Wingspan:	185 feet	Length:	157.6 feet	Height:	40.7 feet
Empty Weight:	166,555 pounds	Combat Weight:	286,366 pounds	Max TO Weight:	488,000 pounds
Combat Radius:	3,819 miles	Combat Ceiling:	46,600 feet	Service Ceiling:	47,250 feet
Cruise Speed:	523 mph	Max Speed:	635 mph	Max Payload:	43,000 pounds
OEM:	Boeing	Produced:	193	SAC Inventory:	193
First Flight:	31-Aug-58	First Delivery:	13-Feb-59	Phase Out:	3-May-94
Missions:	Intercontinental Strategic Bombardment				
Tail Number(s):	57-6468 to 57-6520; 58-0158 to 58-0258; 59-2564 to 59-2602				
Propulsion:	Eight P&W J57-P-43WB with 9,500 pounds normal thrust and 13,750 pounds take-off thrust each with water injection. Two engines each mounted on four wing pylons.				
Accommodations:	Total 6. Pilot, copilot, navigator, radar navigator, electronic warfare officer (EWO), tail gunner. Gunner eliminated 1-Oct-91.				
Payload:	Up to 43,00 pounds total of bombs, ADM-20 Quails, and/or AGM-28 Hound Dogs (later modified for up to 50,000 pounds multiple weapons); plus four .50-caliber guns.				
Comments:	Specifications based on B-52G Standard Aircraft Characteristics, Jul-64. Max speed was 635 mph at 20,800 feet. Combat Radius, Combat Ceiling, Service Ceiling, Cruise Speed and Max Speed were calculated with 10,000-pound bombload at combat weight without refueling (Basic Mission). Major improvement reduced weight to improve performance. Tail height was reduced by ~8 feet and the chord (width) increased. Later ECM modifications increased fuselage length to 160.9 feet. Gunner moved to forward crew compartment. First B-52 (57-6468) delivered to ARDC at Eglin AFB, FL on 1-Nov-58. First B-52G (57-6478) assigned to SAC delivered to 5th BMW at Travis AFB, CA on 13-Feb-59.				

B-52 Stratofortress: The Iron Fist of Strategic Air Command

B-52H Stratofortress

B-52H major improvement was TF-33 engines which drastically increased thrust without water injection. (U.S. Air Force)

Aircraft Specifications					
Type:	B-52	Series:	B-52H	Name:	Stratofortress
Wingspan:	185 feet	Length:	156.0 feet	Height:	40.7 feet
Empty Weight:	169,822 pounds	Combat Weight:	281,905 pounds	Gross Weight:	488,000 pounds
Combat Radius:	4,481 miles	Combat Ceiling:	47,200 feet	Service Ceiling:	47,800 feet
Cruise Speed:	525 mph	Max Speed:	639 mph	Max Payload:	43,000 pounds
OEM:	Boeing	Produced:	102	SAC Inventory:	102
First Flight:	6-Mar-61	First Delivery:	9-May-61	Phase Out:	N/A
Missions:	Intercontinental Strategic Bombardment				
Tail Number(s):	60-0001 to 60-0062; 61-0001 to 61-0040				
Propulsion:	Eight P&W TF33-P-3 engines with 14,500 pounds normal thrust and 17,000 pounds take-off thrust each. Two engines each mounted on four wing pylons.				
Accommodations:	Total 6. Pilot, copilot, navigator, radar navigator, electronic warfare officer (EWO), tail gunner. Gunner eliminated 1-Oct-91.				
Payload:	Up to 43,000 pounds total of bombs, ADM-20 Quails, and/or AGM-28 Hound Dogs (later modified for up to 70,000 pounds multiple weapons); plus one 20-mm M-61 cannon.				
Comments:	Specifications based on B-52G Standard Aircraft Characteristics, Feb-63. Max speed was 639 mph at 20,700 feet. Combat Radius, Combat Ceiling, Service Ceiling, Cruise Speed and Max Speed were calculated with 10,000-pound bombload at combat weight without refueling (Basic Mission). Tail height was reduced like the B-52G. Later ECM modifications increased fuselage length to 159.3 feet. Major improvement was TF-33 engines which increased thrust and range. Installed new 120 kVA alternators, transformer-rectifier, and 20-mm cannon. Last 18 aircraft produced with forward fired rocket launchers. First B-52H (60-0001) assigned to SAC delivered to 379th BMW, Wurtsmith AFB, MI on 9-May-61.				

Appendix B: Colors and Markings

The B-52 fleet sported various paint schemes throughout its service life. It originally had natural metal surfaces all around. The natural metal top was accompanied by a white-gloss anti-flash underside and nose in the late 1950s. This underside was painted in production for some B-52B and all C through H models and retrofitted on most B models. The aircraft also received a SAC shield and blue "Milky Way" banner on one side with the unit emblem and banner on the other side.

The white underside on the B-52F was quickly replaced with a gloss black underside during its participation in the Vietnam War in 1965 through March 1966. The natural metal top was retained. The B-52D received a gloss black underside and tail with dark green (FS 34079), green (FS 34159), and tan green (FS 34201), sometimes called "SAC Bomber Tan", camo top during the "BIG BELLY" modification and this was retained for the rest of its lifetime. Some aircraft were apparently painted with tan (30219), instead of tan green, like the SEA "tactical" camo used on fighter aircraft. The D models also had red (sometimes orange, white or yellow) tail number markings. The banner was removed but the SAC shield and unit emblems were retained. The tail number was moved forward on the nose.

During the early 1970s, the top surfaces of the G and H models, along with some C, E, and F models, were painted with SIOP camouflage, consisting of dark green (FS 34079), green (FS 34159), and tan green (FS 34201), and the white anti-flash underside. Some aircraft were apparently painted with tan (30219), instead of tan green, although this appears to be very rare. The top of the nose was originally painted dark green, but the entire nose was eventually painted white during EVS installations. However, the white nose was easily seen from above during low level attack and was repainted with dark gray (FS 36081). Dark green areas (FS 34079) were also repainted dark gray (FS 36081) at the same time. Eventually, this paint was applied to the entire forward fuselage.

Beginning in 1984, SAC adopted a new "Strategic Camouflage" scheme which featured dark green (FS 34086) and dark gray (FS 36081) top surfaces with a dark gray (FS 36081) and gunship gray (FS 35118) underside.[1] This paint was so dark at night that some bases were forced to hang reflective streamers from the wing tips to prevent collisions by maintenance vans.

In 1987, SAC elected to paint the entire aircraft with dark gray (FS 36081) paint to achieve a more uniform paint scheme. This paint was also extremely dark at night like the Strategic Camo. However, by 1991 SAC agreed to use gunship gray (FS 36118) on the entire aircraft like the AC-130 and F-15E. This soon became standard for most Air Force aircraft and remains the standard B-52H paint scheme.

B-52C in original SAC paint scheme. The natural metal top was accompanied by a white-gloss anti-flash underside and nose in the late 1950s. This underside was painted in production for some B-52B and all C through H models and retrofitted on most B models. (U.S. Air Force)

B-52G with natural metal top and white underside. All aircraft models also received a SAC shield and blue "Milky Way" banner on one side with the wing emblem and banner on the other side. The tail number was added on the forward fuselage. Some aircraft had a white top over the cabin area, especially at hot weather bases, to reduce the heat inside. (U.S. Air Force)

The white underside on the B-52F was quickly replaced with a gloss black underside during its participation in the Vietnam War in 1965 through March 1966. The natural metal top and other markings were retained. (U.S. Air Force)

The B-52D received a gloss black underside and tail with dark green (FS 34079), green (FS 34159), and tan green (FS 34201) camo top during the "BIG BELLY" modification and this was retained for the rest of its lifetime. (National Archives)

Some B-52D aircraft were apparently painted with tan (30219) paint, instead of tan green, like the SEA tactical camo used on fighter aircraft. (U.S. Air Force)

During the early 1970s, the top surfaces of the G and H models, along with some C, E, and F models, were painted with SIOP camouflage, consisting of dark green (FS 34079), green (FS 34159), and tan green (FS 34201) and the white anti-flash underside. The entire nose was painted white after EVS installations. (U.S. Air Force)

Some B-52G aircraft were apparently painted with tan (30219), instead of tan green, although this appears to be very rare. The top of the nose was also painted dark green before EVS installation. (U.S. Air Force)

The white nose was easily seen from above during low-level attack and was repainted with dark gray (FS 36081). Dark green areas (FS 34079) were also repainted dark gray (FS 36081) at the same time. Eventually, this paint was applied to the entire forward fuselage. (U.S. Air Force)

Beginning in 1984, SAC adopted a new "Strategic Camouflage" scheme which featured dark green (FS 34086) and dark gray (FS 36081) top surfaces with a dark gray (FS 36081) and gunship gray (FS 35118) underside. (U.S. Air Force)

In 1987, SAC elected to paint the entire aircraft with dark gray (FS 36081) paint to achieve a more uniform paint scheme. (American Aviation Historical Society)

By 1991 SAC agreed to use gunship gray (FS 36118) on the entire aircraft like the AC-130 and F-15E. This soon became standard for most Air Force aircraft and remains the standard B-52H paint scheme. (U.S. Air Force)

Appendix C: Unit Assignments

B-52 Stratofortress: The Iron Fist of Strategic Air Command

Unit	Components	Base	Aircraft	Notes
2nd BW (1993-) 2nd Wing (1991-93) 2nd BMW (1963-91) 4238th SW (1958-63)	11th BS (1994-) 96th BS (1993-) 20th BS (1992-) 596th BS (1968-93) 62nd BS (1965-93)	Barksdale AFB, LA	B-52H (1992-) B-52G (1965-92) B-52F (1958-65)	The 4238th SW was activated at Barksdale AFB, LA in 1958 and received its first B-52 in 1958. The 2nd BMW converted from B-47 and moved to Barksdale from Hunter AFB, GA in 1963, It absorbed the resources of the 4238th SW which was deactivated. Redesignated 2nd Wing in 1991 and 2nd BW in 1993.
5th BW (1992-) 5th Wing (1991-92) 5th BMW (1959-91)	69th BS (2010-) 23rd BS (1959-) 72nd BS (1994-96)	Minot AFB, ND (1968-) Travis AFB, CA (1959-68)	B-52H (1968-) B-52G (1959-68)	Converted from B-36 to B-52 in 1959. Moved to Minot from Travis AFB, CA in 1968 and absorbed resources of the 450th BMW which was inactivated. Redesignated 5th Wing in 1991 and 5th BW in 1992.
6th SAW (1962-67) 6th BMW (1957-62)	24th BS (1957-67) 40th BS (1957-67) 39th BS (1957-63) 4129th CCTS (1959-63)	Walker AFB, NM	B-52E (1957-67)	Converted from B-36 to B-52 in 1957. Received Atlas missiles in 1962 and redesignated 6th SAW. Walker closed in 1967; moved to Eielson AFB, AK as 6th SW and converted to RC-135.
7th Wing (1991-92) 7th BMW (1958-91)	9th BS (1971-93) 20th BS (1965-92) 9th BS (1958-68) 492nd BS (1958-59) 4018th CCTS (1974-85)	Carswell AFB, TX	B-52H (1982-92) B-52D (1969-82) B-52C (1969-71) B-52F (1958-69)	Converted from B-36 to B-52 in 1958. Moved to Dyess AFB, TX in 1993 and converted to B-1.
11th SAW (1962-68) 11th BMW (1957-62)	26th BS (1957-68) 42nd BS (1957-60) 98th BS (1957)	Altus AFB, OK (1957-68) Carswell AFB, TX (1958-57)	B-52E (1958-68)	Moved from Carswell AFB, TX to Altus AFB, OK in 1957 and converted from B-36 to B-52 in 1958. Received Atlas missiles in 1961-62 and redesignated as 11th SAW. Redesignated 11th ARW in 1968 and converted to KC-135.
17th BMW (1962-76) 4043rd SW (1959-63)	34th BS (1962-76) 42nd BS (1959-63)	Beale AFB, CA (1975-76) Wright-Patterson AFB, OH (1959-75)	B-52G (1975-76) B-52H (1968-75) B-52E (1960-68)	The 4043rd SW was activated in 1959 at Wright-Patterson AFB, OH and received its first B-52 in 1960. The 17th BMW was reactivated in 1962 at Wright-Patterson AFB, OH and absorbed resources of the 4043rd SW which was deactivated in 1963. Moved to Beale AFB, CA in 1975 and absorbed resources 456th BMW. Inactivated in 1976.

Unit	Components	Base	Aircraft	Notes
19th BMW (1961-83)	28th BS (1961-83)	Robins AFB, GA (1968-83) Homestead AFB, FL (1961-68)	B-52G (1968-83) B-52H (1962-68)	Converted from B-47 to B-52 at Homestead AFB, FL in 1961-62. Moved to Robins AFB, GA in 1968 and absorbed resources of 465th BMW which was deactivated. Redesignated 19th ARW in 1983 and converted to KC-135.
22nd BMW (1963-82)	486th BS (1966-71) 2nd BS (1963-82)	March AFB, CA	B-52E (1968-70) B-52C (1967-71) B-52D (1966-82) B-52B (1963-66)	Converted from B-47 to B-52 in 1963. Redesignated 22nd ARW in 1982 and converted to KC-135 and KC-10.
28th BMW (1957-87)	37th BS (1977-82) 77th BS (1957-87) 717th BS (1957-60) 718th BS (1957-60)	Ellsworth AFB, SD	B-52H (1977-86) B-52G (1971-77) B-52D (1957-71)	Converted from B-36 to B-52 in 1957 and B-1 in 1987.
39th BMW (1963-65) 4135th SW (1958-63)	62nd BS (1963-65) 301st BS (1958-63)	Eglin AFB, FL	B-52G (1959-65)	The 4135th SW was activated at Eglin AFB, FL in 1958 and received its first B-52 in 1959. The 39th BMW was activated at Eglin in 1962 and in 1963 absorbed resources of the 4135th SW which was inactivated. The 39th BMW was inactivated in 1965: Eglin transferred to TAC.
42nd BW (1992-94) 42nd Wing (1991-92) 42nd BMW (1956-91)	69th BS (1956-94) 70th BS (1956-66) 75th BS (1956-59)	Loring AFB, ME	B-52G (1959-94) B-52D (1956-59) B-52C (1956-57)	Converted to from B-36 to B-52 in 1956. Inactivated in 1994; Loring closed.
43rd BMW (1986-90) 43rd SW (1970-86) 4133rd BMW (P) (1966-70) 3960th SW (1965-70)	63rd BS (P) (1972-73) 60th BS (1971-90) 63rd BS (1965-70) 64th BS (1965-70) 65th BS (1965-70)	Andersen AFB, Guam	B-52G (1983-90) B-52D (1966-83) B-52F (1965-66)	Deactivated at Little Rock AFB, AR in 1970 and assigned B-58 aircraft retired. Moved to Andersen AFB, Guam in 1970 and reactivated as 43rd SW. Took over the deployed B-52 aircraft of the 3960th SW and assumed the tasks of the 4133rd BMW Provisional (P) at Andersen AFB, Guam. Employed attached aircraft and aircrews of other SAC units to participate in SAC's "ARC LIGHT" combat missions in SEA. Also supported LINE-BACKER I and II in 1972. Redesignated as 43rd BMW in 1986 and deactivated in 1990.
68th BMW (1963-82) 4241st SW (1958-63)	51st BS (1958-82) 73rd BS (1958-63)	Seymour-Johnson AFB, NC	B-52G (1959-82)	The 4241st SW was activated at Seymour-Johnson AFB, NC in 1958 and received first B-52

Unit	Components	Base	Aircraft	Notes
				in 1959. The 68th BMW converted from B-47 and moved to Seymour-Johnson from Chennault AFB, LA in 1963. It absorbed the resources of the 4241st SW which was deactivated. Redesignated 68th ARG in 1982 and converted to KC-135 and KC-10.
70th BMW (1963-69) 4123rd SW (1957-63)	6th BS (1963-69) 98th BS (1957-63)	Clinton-Sherman AFB, OK	B-52C (1968-69) B-52D (1968-69) B-52E (1959-68)	The 4123rd SW was activated in 1957 at Clinton-Sherman AFB, OK and received first its B-52 in 1959. The 70th SRW converted from RB-47 and moved to Clinton-Sherman from Lockbourne AFB, OH as the 70th BMW in 1963. It absorbed resources of the 4123rd SW which was inactivated. The 70th BMW was inactivated in 1969; Clinton-Sherman closed.
72nd BMW (P) (1972-73) 72nd BMW (1959-71)	64th BS (P) (1972-73) 65th BS (P) (1972-73) 329th BS (P) (1972-73) 486th BS (P) (1972-73) 60th BS (1959-71)	Andersen AFB, Guam (1972-73) Ramey AFB, Puerto Rico (1959-71)	B-52G (1972-73) B-52G (1959-71)	Converted from B-36 to B-52 at Ramey AFB, Puerto Rico in 1959. Inactivated in 1971: Ramey closed, and portion transferred to U.S. Coast Guard. Moved to Andersen AFB, Guam in 1972 as provisional wing supporting LINEBACKER I and II.
91st BMW (1963-68) 4141st SW (1958-63)	322nd BS (1963-68) 326th BS (1958-63)	Glasgow AFB, MT	B-52C (1967-68) B-52D (1961-68)	The 4141st SW was activated in 1958 at Glasgow AFB, MT and received its first B-52 in 1961. The 91st BMW was reactivated in 1963 at Glasgow and absorbed the resources of the 4141st SW which was deactivated. Moved to Minot AFB, ND in 1968 as 91st SMW; Glasgow closed.
92nd BW (1992-94) 92nd Wing (1991-92) 92nd BMW (1972-91) 92nd SAW (1962-72) 92nd BMW (1957-62)	325th BS (1957-94) 327th BS (1957-60) 326th BS (1957-58)	Fairchild AFB, WA	B-52H (1986-94) B-52G (1970-86) B-52C (1967-71) B-52D (1957-71)	Converted from B-36 to B-52 in 1957. Received Atlas missiles in 1962 and redesignated 92nd SAW. Redesignated 92nd BMW in 1972. Redesignated 92nd ARW in 1994 flying KC-135.

Unit	Components	Base	Aircraft	Notes
93rd BW (1992-95) 93rd Wing (1991-92) 93rd BMW (1955-91)	330th BS (1988-91) 329th BS (1986-94) 328th BS (1947-94) 329th BS (1947-71) 330th BS (1947-63) 4017th CCTS (1955-56)	Castle AFB, CA	B-52G (1974-94) B-52H (1974-93) B-52E (1967-70) B-52G (1966) B-52F (1958-74) B-52E (1957-58) B-52D (1956-58) B-52B (1955-65)	Converted from B-47 to B-52 in 1955. Primary training base for B-52 aircrews. Inactivated in 1995; Castle closed.
95th BMW (1959-66)	334th BS (1959-66)	Biggs AFB, TX	B-52B (1959-66)	Converted from B-36 to B-52 in 1959. Inactivated in 1966; Biggs closed and transferred to US Army as Biggs AAF.
96th BMW (1972-85) 96th SAW (1963-72)	337th BS (1963-85)	Dyess AFB, TX	B-52H (1982-85) B-52D (1969-82) B-52C (1969-71) B-52E (1963-70)	Received Atlas missiles in 1962 and redesignated from 96th BMW to 96th SAW. Converted from B-47 to B-52 in 1963. Redesignated 96th BMW in 1972. Converted to B-1 in 1985.
97th Wing (1991-92) 97th BMW (1959-91)	340th BS (1959-92)	Eaker AFB, AR	B-52G (1960-92)	Reactivated in 1959 at Blytheville AFB, AR (later renamed Eaker AFB) and received first B-52 in 1960. Inactivated in 1992; Eaker closed.
99th BMW (1956-74)	346th BS (1956-74) 348th BS (1956-73) 347th BS (1956-61)	Westover AFB, MA	B-52D (1966-73) B-52B (1958-59) B-52D (1957-61) B-52C (1956-71)	Moved from Fairchild AFB, WA to Westover AFB, MA in 1956 and converted from B-36 to B-52. Inactivated in 1974; Westover transferred to AFRES as Westover ARB.
306th BMW (1963-74) 4047th SW (1961-63)	367th BS (1963-74) 347th BS (1961-63)	McCoy AFB, FL	B-52C (1967-71) B-52D (1961-73)	The 4047th SW was activated in 1961 at McCoy AFB, FL and received its first B-52 in 1961. The 306th BMW converted from B-47 and moved to McCoy in 1963. It absorbed the resources of the 4047th SW which was inactivated. The 306th BMW was inactivated in 1974; McCoy closed.
307th BW (2011-) 307th SW (1970-75) 4258th SW (1966-70)	93rd BS (2011-) 343rd BS (2011-) 364th BS (P) (1972-75) 365th BS (P) (1972-74) 486th BS (P) (1970-71)	Barksdale AFB, LA (2011-) U-Tapao RTNAF, Thailand (1966-75)	B-52H (2011-) B-52D (1967-75)	The 4258th SW was activated in 1966 at U-Tapao RTNAF, Thailand to support deployed KC-135. It began supporting deployed B-52D in 1967. The 307th BMW reactivated as the 307th SW in 1970 at U-Tapao and absorbed resources of the 4258th SW which was deactivated. The 307th SW was inactivated in 1975. Reactivated

Unit	Components	Base	Aircraft	Notes
				as the 307th BW (AFRC) in 2011 at Barksdale AFB, LA and absorbed B-52H resources of the 917th Composite Wing (AFRC) which was deactivated.
319th BMW (1963-87) 4133rd SW (1958-63)	46th BS (1963-87) 30th BS (1958-63)	Grand Forks AFB, ND	B-52G (1982-86) B-52H (1962-82)	The 4133rd SW was activated in 1958 at Grand Forks AFB, ND and received first B-52 in 1962. The 319th BMW was reactivated at Grand Forks in 1963 and absorbed the resources of the 4133rd SW which was inactivated. Converted to B-1 in 1987.
320th BMW (1963-89) 4134th SW (1958-63)	441st BS (1963-89) 72nd BS (1958-63)	Mather AFB, CA	B-52G (1968-89) B-52F (1958-68)	The 4134th SW was activated in 1958 at Mather AFB, CA and received its first B-52 in 1958. The 320th BMW discontinued B-47 operations at March AFB, CA in 1960 and moved to Mather in 1963. It absorbed the resources of the 4134th SW which was inactivated. Inactivated in 1989; Mather closed 1991.
340th BMW (1963-66) 4130th SW (1958-63)	486th BS (1963-66) 335th BS (1958-63)	Bergstrom AFB, TX	B-52D (1959-66)	The 4130th SW was activated in 1958 at Bergstrom AFB, TX and received its first B-52 in 1959. The 340th BMW converted from B-47 and moved to Bergstrom in 1963. It absorbed the resources of the 4130th SW which was inactivated. Inactivated in 1966.
366th Wing (1992-94)	34th BS (1992-94)	Mountain Home AFB, ID	B-52G (1992-94)	Composite wing (aircraft located at Castle AFB, CA). The wing was part of the ACC Rapid Intervention Force experiment designed to develop strike packages that would train in peacetime to support combat commands as they would in wartime. First integration of bombers in a predominately fighter-equipped wing. Converted to B-1B in 1994.
376th SW (1970) 4252nd SW (1965-70)	4180th BS (P)	Kadena AB, Okinawa	B-52D (1968-70)	The 376th BMW was reactivated as the 307th SW in 1970 at Kadena AB, Okinawa and absorbed resources of the

Unit	Components	Base	Aircraft	Notes
				4252nd SW which was inactivated. Briefly flew missions over Khe Sanh in 1968 and ARC LIGHT missions 1968-70 until B-52D withdrew from Kadena.
379th BMW (1961-93) 4026th SW (1958-61)	524th BS (1961-93)	Wurtsmith AFB, MI	B-52G (1977-92) B-52H (1961-77)	The 4026th SW was activated in 1958 at Wurtsmith AFB, MI but never received B-52 aircraft. The 379th BMW converted from B-47 and moved to Wurtsmith in 1961. It the resources of the 4026th SW which was inactivated. Received first B-52 in 1961. Inactivated in 1993; Wurtsmith closed.
380th SAW (1966-71)	528th BS (1966-71)	Plattsburg AFB, NY	B-52G (1966-71)	The 380th BMW received Atlas missiles in 1963 and was redesignated 380th SAW in 1964. Converted from B-47 to B-52 in 1966. Converted to FB-111 in 1971.
397th BMW (1963-68) 4038th SW (1960-63)	596th BS (1963-68) 341st BS (1960-63)	Dow AFB, ME	B-52G (1960-68)	The 4038th SW was activated in 1960 at Dow AFB, ME and received its first B-52 in 1960. The 397th BMW was activated in 1962 and organized at Dow in 1963. It absorbed the resources of the 4038th SW which was inactivated. Inactivated in 1968; Dow closed.
410th BMW (1963-95) 4042nd SW (1958-63)	644th BS (1963-94) 526th BS (1958-63)	K.I. Sawyer AFB, MI	B-52H (1961-94)	The 4042nd SW was activated in 1958 at K.I. Sawyer AFB, MI and received its first B-52 in 1961. The 410th BMW was activated in 1962 and organized at K.I. Sawyer in 1963. It absorbed the resources of the 4042nd SW which was inactivated. Inactivated in 1995; K.I. Sawyer closed.
416th BMW (1963-95) 4039th SW (1958-63)	668th BS (1963-95) 75th BS (1958-63)	Griffiss AFB, NY	B-52H (1992-95) B-52G (1960-92)	The 4039th SW was activated in 1958 at Griffiss AFB, NY and received its first B-52 in 1960. The 416th BMW was activated at Griffiss in 1963 and absorbed the resources of the 4039th SW which was inactivated. Inactivated in 1995; Griffiss closed.

Unit	Components	Base	Aircraft	Notes
449th BMW (1963-77) 4239th SW (1959-63)	716th BS (1963-77) 93rd BS (1959-63)	Kincheloe AFB, MI	B-52H (1961-77)	The 4239th SW was activated in 1959 at Kincheloe AFB, MI and received its first B-52 in 1961. The 449th BMW was activated in 1962 and organized at Kincheloe in 1963. It absorbed the resources of the 4239th SW which was inactivated. Inactivated 1977; Kincheloe closed.
450th BMW (1963-68) 4136th SW (1958-63)	720th BS (1963-68) 525th BS (1958-63)	Minot AFB, ND	B-52H (1961-68)	The 4136th SW was activated in 1958 at Minot AFB, ND and received its first B-52 in 1961. The 450th BMW was activated in 1962 and organized at Minot in 1963. It absorbed the resources of the 4136th SW which was inactivated. Inactivated in 1968 and replaced at Minot by the 5th BMW.
454th BMW (1963-69) 4228th SW (1958-63)	736th BS (1963-69) 492nd BS (1958-63)	Columbus AFB, MS	B-52D (1966-69) B-52F (1959-66)	The 4228th SW was activated in 1958 at Columbus AFB, MS and received its first B-52 in 1959. The 454th BMW was activated in 1962 and organized at Columbus in 1963. It absorbed the resources of the 4228th SW which was inactivated. Inactivated in 1969; Columbus transferred to ATC.
456th BMW (1963-75) 4126th SW (1959-63)	744th BS (1963-75) 31st BS (1959-63)	Beale AFB, CA	B-52G (1960-75)	The 4126th SW was activated in 1959 at Beale AFB, CA and received its first B-52 in 1960. The 456th BMW was activated in 1962 and organized at Beale in 1963. It absorbed the resources of the 4126th SW which was inactivated. Inactivated in 1975; replaced at Beale by the 17th BMW.
461st BMW (1963-68) 4128th SW (1958-63)	764th BS (1963-68) 718th BS (1958-63)	Amarillo AFB, TX	B-52C (1967-68) B-52D (1960-68)	The 4128th SW was activated in 1958 at Amarillo AFB, TX and received its first B-52 in 1960. The 461st BMW was reactivated in 1962 and organized at Amarillo in 1963. It absorbed the resources of the 4128th SW which was inactivated. Inactivated in 1968; Amarillo closed.
462nd SAW (1963-66)	768th BS (1963-66)	Larson AFB, WA	B-52D (1960-66)	The 4170th SW was activated

Unit	Components	Base	Aircraft	Notes
4170th SW (1958-63)	327th BS (1958-63)			in 1958 at Larson AFB, WA and received its first B-52 in 1960. The 462nd SAW was activated in 1962 and organized at Larson in 1963. It absorbed the resources of the 4170th SW which was inactivated. It also provided training for Titan II missiles until 1965. Inactivated in 1966; Larson closed.
465th BMW (1963-68) 4137th SW (1958-63)	781st BS (1963-68) 342nd BS (1958-63)	Robins AFB, GA	B-52G (1960-68)	The 4137th SW was activated in 1958 at Robins AFB, GA and received its first B-52 in 1960. The 465th BMW was activated in 1962 and organized at Robins in 1963. It absorbed the resources of the 4137th SW which was inactivated. Inactivated in 1968. Replaced at Robins by the 19th BMW.
484th BMW (1963-67) 4138th SW (1958-63)	824th BS (1963-67) 336th BS (1958-63)	Turner AFB, GA	B-52D (1959-67)	The 4138th SW was activated in 1958 at Turner AFB, GA and received its first B-52 in 1959. The 484th BMW was activated in 1962 and organized at Turner in 1963. It absorbed the resources of the 4138th SW which was inactivated. Inactivated in 1967; Turner closed. Transferred to U.S. Navy in 1968 as NAS Albany.
494th BMW (1963-66) 4245th SW (1959-63)	864th BS (1963-66) 717th BS (1959-63)	Sheppard AFB, TX	B-52D (1960-66)	The 4245th SW was activated in 1959 at Sheppard AFB, TX and received its first B-52 in 1960. The 494th BMW was activated in 1963 at Sheppard and absorbed the resources of the 4245th SW which was inactivated. Inactivated in 1966.
509th BMW (1966-70)	393rd BS (1966-70)	Pease AFB, NH	B-52C (1966-69) B-52D (1966-69)	Converted from B-47 to B-52 in 1966. Converted to FB-111 in 1970.
917th Wing (1993-2011)	93rd BS (1993-2011) 343rd BS (1993-2011)	Barksdale AFB, LA	B-52H (1993-2011)	Composite wing including A-10 and B-52. First AFRC wing equipped with B-52 aircraft. (Aircraft located at Barksdale, AFB, LA). Deactivated in 2011 and resources of 93rd BS assigned to reactivated 307th BW (AFRC) at Barksdale.

Unit	Components	Base	Aircraft	Notes
801st BMW (P) (1991)	69th BS (1991) 340th BS (1991) 524th BS (1991) 596th BS (1991) 668th BS (1991)	Moron AB, Spain	B-52G (1991)	Activated for Operation DESERT STORM. Composed of aircraft and personnel from deployed units formed around the nucleus provided by the 596th BS/2nd BMW. Deployed units were organized into the 801st BS (P) and 802nd BS (P).
806th BMW (P) (1991)	62nd BS (1991) 328th BS (1991) 340th BS (1991) 524th BS (1991) 668th BS (1991)	RAF Fairford, UK	B-52G (1991)	Activated for Operation DESERT STORM. Composed of aircraft and personnel from deployed units formed around the nucleus provided by the 340th BS/97th BMW. Deployed units were organized into the 806th BS (P).
1500th SW (P) (1990-91)	69th BS (1990-91) 328th BS (1990-91)	Andersen AFB, Guam	B-52G (1990-91)	Activated for Operation DESERT STORM. Composed of six B-52 aircraft and personnel from the 42nd and 93rd BMW. The aircraft replaced aircraft from Diego Garcia that landed in Jeddah on 17-Jan-91.
1708th BMW (P) (1990-91)	69th BS (1991) 328th BS (1991) 524th BS (1991) 596th BS (1991) 668th BS (1991)	Prince Abdulla AB, Saudi Arabia	B-52G (1991)	Activated for Operation DESERT STORM. Composed of aircraft and personnel from deployed units formed around the nucleus provided by the 524th BS/379th BMW. Deployed units were organized into the 1708th BS (P).
4300th BMW (P) (1990-91)	69th BS (1990-91) 328th BS (1990-91)	Diego Garcia AB, Indian Ocean	B-52G (1990-91)	Activated for Operation DESERT STORM. Composed of aircraft and personnel from deployed units formed around the nucleus provided by the 69th BS/42nd BMW. Deployed units were organized into the 4300th BS (P).

Appendix D: Tail Numbers

Tail #	CN	Lineage	Status	Last Unit	Date	Comments
49-230	16248	XB-52	Retired	ADC	Unknown	Assigned to the Wright ADC at Wright-Patterson AFB, OH as a test bed beginning in Mar-57. Used for several development tests including shorter tail fin. Retirement date unknown.
49-231	16249	YB-52	Display	AF Museum	27-Jan-58	Donated to Air Force Museum for display. Later removed and scrapped in mid-1960s. Replaced with B-52B (53-0394).
52-0001	16491	B-52A JB-52A NB-52A GB-52A	Retired	OC-AMA	Unknown	Delivered in place to Boeing and used for Phase IV testing. Redesignated JB-52A on 30-Nov-55 and NB-52A on 8-Oct-57. This aircraft was used for initial testing of the shorter vertical fin eventually used on the B-52G and H models. Also tested the J57-P-43W engines used on the B-52F. Retired 2-Jul-59 and used for ground training at Chanute AFB, IL through early 1965. Used for firefighting film and training in 1965-66. Ultimately, destroyed by fire.
52-0002	16492	B-52A JB-52A NB-52A	Retired	Chanute	Unknown	Delivered in place to Boeing and used for Phase IV testing. Redesignated JB-52A on 30-Nov-55 and NB-52A on 8-Oct-57. Retired in 1960 after flight testing with gross weights up to 415,000 pounds. Scrapped at OC-AMA, Tinker AFB, OK 26-Apr-61.
52-0003	16493	B-52A JB-52A NB-52A	Display	AFFTC	16-Sep-81	Delivered in place to Boeing and used for Phase IV testing in Sep-54. Redesignated JB-52A on 30-Nov-55. Redesignated NB-52A on 8-Oct-57 and assigned to NASA at Edwards AFB, CA. Modified to launch the X-15 rocket plane beginning in Jan-58. Arrived at MASDC from AFFTC on 15-Oct-69. Departed to Pima Air & Space Museum, Tucson, AZ 16-Sep-81; Static Display (The High and Mighty One). Oldest surviving B-52 aircraft.
52-0004	16494	RB-52B JRB-52B B-52B	Retired	22nd BMW	11-Jan-66	Delivered in place to Boeing on 3-Mar-55 for various tests. Redesignated JRB-52B on 30-Nov-55. Assigned to the AFSWC, Kirtland AFB, NM and used for monitoring detonations during Operation REDWING nuclear testing from April to July 1956. Assigned to 93rd BMW at Castle AFB, CA on 30-Sep-57 and redesignated B-52B. Declared excess on 15-Jan-69 and scrapped.
52-0005	16495	RB-52B B-52B GB-52B	Display	93rd BMW	10-Feb-66	Delivered to AFFTC, Edwards AFB, CA on 13 March 1955 to support ARDC test activities. Assigned to 93rd BMW at Castle AFB, CA on 4-Oct-55 and redesignated B-52B. Arrived at MASDC on 10-Feb-66 from 93rd BMW. To Lowry AFB, CO on 28-Apr-66; RTS and designated GB-52B ground trainer. Used by 3415th Maintenance and Supply Group (MSG) at Lowry Technical Training Center until Oct-75. Displayed at Lowry from 1-Apr-84 (estimated) until the base closed in 1994. Transferred to Wings Over the Rockies Air and Space Museum.
52-0006	16496	RB-52B B-52B	Retired	22nd BMW	4-Feb-66	Delivered to AFFTC, Edwards AFB, CA on 21-Mar-55 to support ARDC test. Assigned to 93rd BMW at

Tail #	CN	Lineage	Status	Last Unit	Date	Comments
						Castle AFB, CA on 27-Jan-56 and redesignated B-52B. Declared excess on 15-Jan-69 and scrapped.
52-0007	16497	RB-52B JRB-52B B-52B	Destroyed	MDC	23-Jun-66	Delivered in place to Boeing for various tests. Redesignated JRB-52B on 30-Nov-55. Assigned to 93rd BMW at Castle AFB, CA on 27-Nov-57 and redesignated B-52B. Arrived at MASDC from 93rd BMW on 11-Feb-66. Departed to MDC, Holloman AFB, NM on 5-May-66; Tested to destruction on 23-Jun-66 during BIG MAMA explosive test.
52-0008	16498	RB-52B JRB-52B NB-52B	Display	NASA	17-Dec-04	Delivered in place to Boeing on 3-May-55 for various tests. Redesignated JRB-52B on 30-Nov-55. Redesignated as NB-52B on 12-Dec-58. Assigned to AFFTC at Edwards AFB, CA on 13-Dec-58. and used as NASA mothership for the X-15, X-38, and X-43A. Used for HL-10 lifting body program, and several critical NASA program. Removed from service on 17-Dec-04. Served with AFFTC and NASA for nearly 50 years. Oldest NASA aircraft and oldest B-52 on flying status at time of retirement. Lowest airframe hours of any operational B-52. Stored pending restoration for years until displayed at Edwards north entrance. Named "The Challenger".
52-0009	16499	RB-52B JRB-52B B-52B	Destroyed	93rd BMW	7-Feb-64	Delivered in place to Boeing on 3-May-55 for various tests. Redesignated JRB-52B on 30-Nov-55. Equipped for testing with a tail turret containing two M24A-1 20mm cannons controlled by a new MD-5 DFCS. Assigned to 93rd BMW at Castle AFB, CA on 20-Dec-57 and redesignated B-52B. Crashed on flight from Castle AFB, CA near Tranquility, CA (west of San Joaquin) due to hydraulic system fire. Pilot set heading to crash over the ocean before the crew ejected but it crashed short in a farmer's field. Total 0 fatalities/7 survivors (all crew ejected safely).
52-0010	16500	RB-52B JRB-52B B-52B	Destroyed	MDC	23-Jun-66	Delivered in place to Boeing on 3-May-55 to support various tests including initial testing of the multi-purpose reconnaissance pod. Redesignated JRB-52B on 30-Nov-55. Assigned to 93rd BMW at Castle AFB, CA on 18-May-56 and redesignated B-52B. Arrived at MASDC from 93rd BMW on 14-Feb-66; Departed to MDC, Holloman AFB, NM on 5-May-66; Tested to destruction on 23-Jun-66.
52-0011	16501	RB-52B JRB-52B B-52B	Retired	509th BMW	6-Jun-66	Delivered to Eglin AFB, FL on 23-May-55 to support ARDC test activities. Assigned to Wright-Patterson AFB, OH on 24-Aug-55. Redesignated JRB-52B on 1-Dec-55. Assigned to 93rd BMW at Castle AFB, CA on 14-Nov-57 and redesignated B-52B. Declared excess on 15-Jan-69 and scrapped.
52-0012	16502	RB-52B JRB-52B B-52B	Retired	380th BMW	1-Jul-66	Delivered to Eglin AFB, FL on 10-May-55 to support ARDC test activities. Redesignated JRB-52B on 1-Jan-56. Assigned to 93rd BMW at Castle AFB, CA on 6-May-57 and redesignated B-52B. Arrived at

Tail #	CN	Lineage	Status	Last Unit	Date	Comments
						MASDC on 15-Feb-66. Briefly RTS and then returned to MASDC on 1-Jul-66. Declared excess on 15-Jan-69 and scrapped.
52-0013	16503	RB-52B JRB-52B JB-52B NB-52B	Display	AFSWC	8-Dec-70	Assigned to the AFSWC, Kirtland AFB, NM on 4-May-55 and used for nuclear testing. Redesignated JRB-52B on 12-Dec-55. One of the few B-52 aircraft to drop a nuclear weapon when it dropped the first H-bomb during Operation REDWING on 21-May-56. Redesignated JB-52B on 16-May-58 and NB-52B on 4-Jan-62. Participated in Operation DOMINIC in 1962. Removed from service on 8-Dec-70. Displayed at the National Museum of Nuclear Science and History near Kirtland AFB, NM.
52-8710	16838	RB-52B B-52B	Retired	22nd BMW	29-Sep-65	Delivered to 93rd BMW at Castle AFB, CA on 6-Jul-55 and redesignated B-52B. Declared excess on 15-Jan-69 and scrapped.
52-8711	16839	RB-52B B-52B	Display	22nd BMW	29-Sep-65	First B-52 aircraft assigned to SAC arriving at 93rd BMW, Castle AFB, CA on 29-Jun-55. Redesignated B-52B. Removed from service on 29-Sep-65 at March AFB, CA. Displayed at Strategic Air Command and Aerospace Museum, Ashland, NE.
52-8712	16840	RB-52B B-52B	Retired	22nd BMW	31-Jan-66	Delivered to 93rd BMW at Castle AFB, CA on 26-Jul-55 and redesignated B-52B. Declared excess on 15-Jan-69 and scrapped.
52-8713	16841	RB-52B B-52B	Retired	22nd BMW	2-Feb-66	Delivered to 93rd BMW at Castle AFB, CA on 20-Aug-55 and redesignated B-52B. Declared excess on 15-Jan-69 and scrapped.
52-8714	16842	RB-52B B-52B GB-52B	Retired	22nd BMW	8-Mar-65	Delivered to 93rd BMW at Castle AFB, CA on 30-Jun-55 and redesignated B-52B. First B-52 retired. Used for ground training at Chanute AFB, IL.
52-8715	16843	RB-52B B-52B	Retired	22nd BMW	27-Jan-66	Delivered to 93rd BMW at Castle AFB, CA on 31-Aug-55 and redesignated B-52B. Declared excess on 15-Jan-69 and scrapped.
52-8716	16844	RB-52B B-52B	Destroyed	93rd BMW	30-Nov-56	Delivered to 93rd BMW at Castle AFB, CA on 23-Jul-55 and redesignated B-52B. Crashed less than three minutes after take-off and four miles north of Castle AFB, CA on a night training mission instructing former B-36 pilots. Aircraft experienced two explosions before it crashed into a farmer's field. It disintegrated as it skidded at least one mile. There were 25 photo bombs (used for illumination of night photos) on board that were deactivated before the magnesium inside exploded. Total 10 fatalities/0 survivors (Third B-52 crash at Castle in eight months).
53-0366	16845	RB-52B B-52B	Retired	22nd BMW	19-Jan-66	Delivered to 93rd BMW at Castle AFB, CA on 5-Aug-55 and redesignated B-52B. Declared excess on 15-Jan-69 and scrapped.
53-0367	16846	RB-52B B-52B	Retired	22nd BMW	5-Jan-66	Delivered to 93rd BMW at Castle AFB, CA on 16-Sep-55 and redesignated B-52B. Declared excess on 15-Jan-69 and scrapped.

Tail #	CN	Lineage	Status	Last Unit	Date	Comments
53-0368	16847	RB-52B B-52B	Retired	22nd BMW	17-Jan-66	Delivered to 93rd BMW at Castle AFB, CA on 7-Sep-55 and redesignated B-52B. Declared excess on 15-Jan-69 and scrapped.
53-0369	16848	RB-52B B-52B	Retired	22nd BMW	8-Feb-66	Delivered to 93rd BMW at Castle AFB, CA on 8-Oct-55 and redesignated B-52B. Declared excess on 15-Jan-69 and scrapped.
53-0370	16849	RB-52B B-52B	Retired	22nd BMW	25-Jan-66	Delivered to 93rd BMW at Castle AFB, CA on 7-Oct-55 and redesignated B-52B. Declared excess on 15-Jan-69 and scrapped.
53-0371	16850	RB-52B B-52B	Destroyed	93rd BMW	29-Jan-59	Delivered to 93rd BMW at Castle AFB, CA on 26-Oct-55 and redesignated B-52B. Crashed after aborted take-off when aircraft ran off the end of the runway. Damaged beyond repair. Total 0 fatalities.
53-0372	16851	RB-52B B-52B GB-52B	Retired	93rd BMW	18-Feb-67	Delivered to 93rd BMW at Castle AFB, CA on 26-Oct-55 and redesignated B-52B. Arrived at MASDC on 27-Sep-65 from 22nd BMW. Departed to Castle AFB, CA on 11-Oct-65 and used for ground training. Returned to MASDC on 18-Feb-67 from 93rd BMW. Declared excess on 15-Jan-69 and scrapped.
53-0373	16852	B-52B	Retired	22nd BMW	21-Jan-66	Declared excess on 15-Jan-69 and scrapped.
53-0374	16853	B-52B	Retired	22nd BMW	7-Jan-66	Declared excess on 15-Jan-69 and scrapped.
53-0375	16854	B-52B	Retired	22nd BMW	3-Jan-66	Declared excess on 15-Jan-69 and scrapped.
53-0376	16855	B-52B	Retired	22nd BMW	13-Jan-66	Declared excess on 15-Jan-69 and scrapped.
53-0377	16856	RB-52B B-52B GB-52B	Retired	43rd SW	29-Jun-66	Delivered to 93rd BMW at Castle AFB, CA on 21-Jun-56 and redesignated B-52B. Used for ground training at Andersen AFB, Guam.
53-0378	16857	RB-52B B-52B	Retired	306th BMW	29-Jun-66	Delivered to 93rd BMW at Castle AFB, CA on 25-Oct-55 and redesignated B-52B. Declared excess on 15-Jan-69 and scrapped.
53-0379	16858	RB-52B B-52B JB-52B B-52B	Retired	AFFTC	3-Oct-68	Delivered to 93rd BMW at Castle AFB, CA on 3-Nov-55 and redesignated B-52B. Supported nuclear tests at the AFSWC, Kirtland AFB, NM from 30-Jan-62 to 10-Aug-62. Redesignated JB-52B on 2-Mar-62. Assigned to 95th BMW at Biggs AFB, TX on 10-Aug-62 and redesignated B-52B. Removed from service on 3-Oct-68 at AFFTC. Used for barrier testing until 1970. Currently at photo range near Rogers Dry Lake south of Edwards AFB, CA.
53-0380	16859	B-52B	Destroyed	95th BMW	7-Apr-61	Crashed during a training sortie from Biggs AFB, TX during inception training with two F-100A Super Sabre from the New Mexico ANG. These aircraft were equipped with AIM-9B Sidewinder missiles, and one was accidently fired at the bomber. It hit the left inboard engine pod and exploded severing the wing. The crew ejected through the ensuing confusion. Total 3 fatalities/5 survivors.
53-0381	16860	B-52B	Retired	95th BMW	12-May-65	Declared excess on 15-Jan-69 and scrapped.
53-0382	16861	B-52B	Destroyed	93rd BMW	6-Nov-57	Crashed on the runway during touch and goes at Castle AFB, CA. Landing gear lever latch failed resulting in gear retraction on the runway. Damaged beyond repair. Total 0 fatalities.

Tail #	CN	Lineage	Status	Last Unit	Date	Comments
53-0383	16862	B-52B JB-52B B-52B	Retired	91st BMW	27-Jun-66	Supported nuclear tests at the AFSWC, Kirtland AFB, NM from 28-Dec-55 to 31-Dec-57. Redesignated JB-52B on 4-Jan-56. TDY to Eniwetok from 16-Apr to 21-May-56. Assigned to 99th BMW at Westover AFB, MA on 27-Jul-58 and redesignated B-52B. Declared excess 15-Jan-69 and scrapped.
53-0384	16863	B-52B	Destroyed	93rd BMW	16-Feb-56	Crashed when an electrical power panel exploded blowing off the cover and causing a fire during flight near Tracy, CA. The cover jammed the left-hand forward alternator regulator valve causing an overspeed failure. Total 4 fatalities/4 survivors.
53-0385	16864	B-52B	Retired	95th BMW	3-Feb-66	Declared excess on 15-Jan-69 and scrapped.
53-0386	16865	B-52B	Retired	95th BMW	30-Sep-65	Declared excess on 15-Jan-69 and scrapped.
53-0387	16866	B-52B	Retired	509th BMW	28-Jun-66	Declared excess on 15-Jan-69 and scrapped.
53-0388	16867	B-52B	Retired	92nd SAW	29-Jun-66	Declared excess on 15-Jan-69 and scrapped.
53-0389	16868	B-52B	Retired	95th BMW	7-Feb-66	Declared excess on 15-Jan-69 and scrapped.
53-0390	16869	B-52B	Destroyed	95th BMW	19-Jan-61	Aircraft crashed after fire started while cruising at 36,000 feet on a clear evening 41 minutes into flight. Aircraft began to nose down. Possibility aircraft would break up after explosion and the copilot ordered abandon aircraft. Copilot and navigator safely ejected from the aircraft. Aircraft broke into several pieces as it descended and dove straight into the ground in a field near Monticello, UT. Total 5 fatalities/2 survivors.
53-0391	16870	B-52B	Retired	95th BMW	9-Feb-66	Declared excess on 15-Jan-69 and scrapped.
53-0392	16871	B-52B	Retired	509th BMW	27-Jun-66	Declared excess on 15-Jan-69 and scrapped.
53-0393	16872	B-52B	Destroyed	93rd BMW	17-Sep-56	Crashed after fire in flight nine miles SE of Madera, CA. Total 5 fatalities/2 survivors.
53-0394	16873	B-52B	Display	95th BMW	6-Jan-66	Part of three-ship around the world flight Operation POWER FLITE in Jan-57 as "Lucky Lady III." Displayed at Air Force Museum. Later removed and scrapped in 1984. Replaced with B-52D (56-0665). The nose section was preserved on Walter Soplata's farm in Newbury, OH.
53-0395	16874	B-52B	Retired	92nd SAW	24-Jun-66	Declared excess on 15-Jan-69 and scrapped.
53-0396	16875	B-52B	Retired	340th BMW	28-Jun-66	Declared excess on 15-Jan-69 and scrapped.
53-0397	16876	B-52B	Retired	340th BMW	24-Jun-66	Part of three-ship around the world flight Operation POWER FLITE in Jan-57 as "Lucky Lady I." First and only B-52B to deploy to the UK on 17-Jan-61. Declared excess on 15-Jan-69 and scrapped.
53-0398	16877	B-52B	Retired	95th BMW	19-Nov-65	Part of three-ship around the world flight Operation POWER FLITE in Jan-57 as "Lucky Lady II." Declared excess on 15-Jan-69 and scrapped.
53-0399	16878	B-52C JB-52C NB-52C JB-52C	Retired	AFFTC	28-Jul-75	To Boeing on 28-Feb-56 and used for initial testing. Redesignated JB-52C on 11-Apr-56 and used for various tests including J57-P-43W engine testing with nacelle bulge to cover the newly installed alternator. Redesignated NB-52C on 7-May-61 and assigned to Wright-Patterson AFB, OH to support ASD

Tail #	CN	Lineage	Status	Last Unit	Date	Comments
						testing. Redesignated JB-52C on 31-Jul-62. Designated for X-20 testing before the program was cancelled in 1963. To AFFTC at Edwards AFB, CA on 27-Aug-68 and supported various tests including B-1A crew escape capsule. Declared excess on 10-Jan-94 and scrapped.
53-0400	16879	B-52C	Retired	22nd BMW	28-Sep-71	First B-52C to see service with SAC delivered to 42nd BMW at Loring AFB, ME on 16-Jun-56. To MASDC on 28-Sep-71. Declared excess on 15-Sep-93 and scrapped.
53-0401	16880	B-52C	Retired	99th BMW	23-Mar-71	Declared excess in Jan-94 and scrapped.
53-0402	16881	B-52C	Retired	22nd BMW	29-Sep-71	Last B-52C retired. Declared excess in Oct-93 and scrapped.
53-0403	16882	B-52C	Retired	96th SAW	31-Aug-71	Painted in SIOP CAMO paint scheme 22-Jun-67 to 1-Sep-67 during PDM. Declared excess in Dec-93 and scrapped.
53-0404	16883	B-52C	Retired	22nd BMW	27-Jul-71	Declared excess in Sep-93 and scrapped.
53-0405	16884	B-52C	Retired	22nd BMW	13-Aug-71	Declared excess in Oct-93 and scrapped.
53-0406	16885	B-52C	Destroyed	99th BMW	24-Jan-63	Crashed into Elephant Mountain near Greenville, ME during a low-level exercise flight from Westover AFB, MA. The aircraft was flying below 500 feet at 322 mph using terrain following radar when the aircrew encountered turbulence with wind gusts up to 40 mph. The pilot began a climb to avoid the turbulence and the vertical fin attachments suddenly failed. The aircraft began rolling right and pitching down out of control. Total 7 fatalities/2 survivors (Navigator killed hitting a tree on descent. Pilot and Navigator survived ejection with injuries).
53-0407	16886	B-52C	Retired	7th BMW	5-Aug-71	Declared excess in Jan-94 and scrapped.
53-0408	16887	B-52C	Retired	96th SAW	10-Aug-71	Declared excess in Oct-93 and scrapped.
54-2664	17159	B-52C	Retired	96th SAW	26-Aug-71	Painted in SIOP CAMO paint scheme 2-Mar-67 to 3-May-67 during PDM. Declared excess on 6-Jan-94 and scrapped.
54-2665	17160	B-52C	Retired	28th BMW	7-Jul-71	Declared excess on 2-Dec-93 and scrapped.
54-2666	17161	B-52C	Destroyed	99th BMW	7-Jan-71	Crashed into Lake Michigan during a low-level mission originating from Westover AFB, MA. Aircraft disappeared suddenly from the radar just before the "pop-up" point. Cause of crash unknown (possible wing structural failure or explosion). Radar "blossomed" before disappearing, indicating explosion on impact. Aircraft never recovered. Total 9 fatalities/ 0 survivors.
54-2667	17162	B-52C	Destroyed	306th BMW	29-Aug-68	Crashed during an evening flight from McCoy AFB, FL. Flap malfunction, total electrical failure, and fuel starvation. Crashed and exploded in a field near Cape Kennedy, FL. Total 0 fatalities/7 survivors (all crew ejected/bailed out safely).
54-2668	17163	B-52C	Retired	306th BMW	2-Sep-71	Declared excess on 2-Dec-93 and scrapped.
54-2669	17164	B-52C JB-52C	Retired	28th BMW	6-Jul-71	Initially delivered to Boeing on 9-Aug-56 and used for testing. To ARDC at Wright-Patterson on 14-Jan-

Tail #	CN	Lineage	Status	Last Unit	Date	Comments
		B-52C				57 for testing and redesignated JB-52C on 15-Jan-57. To ARDC Eglin AFB, FL on 12-Jun-57 for additional testing. Redesignated B-52C on 2-Dec-57. Assigned to Westover AFB, MA on 11-Dec-57. TDY to the AFSWC from 30-Jul to 16-Dec-63. Used as the FLACHBACK test vehicle in Jan-65. Declared excess on 19-Oct-93 and scrapped.
54-2670	17165	B-52C	Retired	99th BMW	23-Sep-71	Declared excess on 15-Sep-93 and scrapped.
54-2671	17166	B-52C	Retired	7th BMW	9-Sep-71	Declared excess on 19-Oct-93 and scrapped.
54-2672	17167	B-52C	Retired	7th BMW	24-Aug-71	Declared excess on 19-Oct-93 and scrapped.
54-2673	17168	B-52C	Retired	7th BMW	28-Sep-71	Declared excess on 2-Dec-93 and scrapped.
54-2674	17169	B-52C	Retired	28th BMW	8-Jul-71	Declared excess on 19-Oct-93 and scrapped.
54-2675	17170	B-52C	Retired	99th BMW	22-Jul-71	Declared excess on 19-Oct-93 and scrapped.
54-2676	17171	B-52C JB-52C	Destroyed	Boeing	29-Mar-57	Initially delivered on 26-Jun-56 to Boeing for testing and redesignated JB-52C. Crashed 15 miles north of Tulsa, OK during flight test from Wichita, KS. Complete AC power loss due to defective constant speed drive during negative G condition. Total 3 fatalities/1 survivor.
54-2677	17172	B-52C	Retired	99th BMW	3-Aug-71	Declared excess on 6-Jan-94 and scrapped.
54-2678	17173	B-52C	Retired	99th BMW	19-Aug-71	Painted in SIOP CAMO paint scheme 26-May-66 to 24-Aug-66 during PDM. Declared excess on 13-Oct-93 and scrapped.
54-2679	17174	B-52C	Retired	22nd BMW	14-Jul-71	Declared excess on 19-Oct-93 and scrapped.
54-2680	17175	B-52C	Retired	28th BMW	8-Jul-71	Declared excess on 13-Oct-93 and scrapped.
54-2681	17176	B-52C	Retired	306th BMW	21-Sep-71	Painted in SIOP CAMO paint scheme 21-Dec-66 to 28-Feb-67 during PDM. Declared excess on 13-Oct-93 and scrapped.
54-2682	17177	B-52C	Destroyed	99th BMW	10-Aug-59	Crashed near Goose Bay, Labrador into Spruce Swamp. Aircraft was trying to make an emergency landing at Goose Bay due to nose radome failure in flight. Total 0 fatalities/8 survivors (all crew ejected/bailed out safely).
54-2683	17178	B-52C	Retired	92nd SAW	14-Sep-71	Declared excess on 19-Oct-93 and scrapped.
54-2684	17179	B-52C	Retired	22nd BMW	16-Sep-71	Declared excess on 19-Oct-93 and scrapped.
54-2685	17180	B-52C	Retired	306th BMW	29-Jul-71	Declared excess on 15-Sep-93 and scrapped.
54-2686	17181	B-52C	Retired	99th BMW	7-Sep-71	Declared excess on 2-Dec-93 and scrapped.
54-2687	17182	B-52C	Retired	22nd BMW	20-Jul-71	Painted in SIOP CAMO paint scheme 5-Dec-66 to 17-Feb-67 during PDM. Declared excess on 19-Oct-93 and scrapped.
54-2688	17183	B-52C	Retired	306th BMW	17-Aug-71	Declared excess on 15-Sep-93 and scrapped.
55-0049	464001	B-52D	Retired	2nd BMW	4-Oct-78	CRESTED DOVE. First Wichita built B-52D. Rolled out 7-Dec-55. First flight 14-May-56. One of two B-52D aircraft that set world speed records on 26-Sep-58. Declared excess on Sep-94 and scrapped.
55-0050	464002	B-52D	Destroyed	306th BMW	21-Dec-72	Combat loss over North Vietnam during Operation LINEBACKER II. Shot down by a SA-2 SAM near Bach Mai. Total 0 fatalities/6 survivors (all POWs).
55-0051	464003	B-52D	Retired	97th BMW	14-Nov-78	CRESTED DOVE. Declared excess in Dec-94 and scrapped.

Tail #	CN	Lineage	Status	Last Unit	Date	Comments
55-0052	464004	B-52D	Retired	416th BMW	21-Sep-78	CRESTED DOVE. Landed at U-Tapao RTNAF, Thailand on 22-Nov-72 after being hit by a SA-2 SAM over North Vietnam. To MASDC on 21-Sep-78. Declared excess in May-92 and scrapped.
55-0053	464005	B-52D	Retired	319th BMW	22-Aug-78	CRESTED DOVE. Declared excess in Sep-94 and scrapped.
55-0054	464006	B-52D	Retired	7th BMW	28-Nov-78	CRESTED DOVE. Declared excess in Dec-94 and scrapped.
55-0055	464007	B-52D	Retired	22nd BMW	21-Nov-78	CRESTED DOVE. Declared excess in Dec-94 and scrapped.
55-0056	464008	B-52D	Destroyed	7th BMW	3-Jan-73	Crashed in the South China Sea after being hit by a SA-2 SAM over North Vietnam. Two engines, electrical, and hydraulics lost. Crew flew aircraft out over the sea and ejected. Total 0 fatalities/6 survivors.
55-0057	464009	B-52D	Display	7th BMW	11-Oct-83	Displayed at Maxwell AFB, AL.
55-0058	464010	B-52D	Destroyed	43rd SW	12-Dec-74	Crashed into the ocean seven miles south of Andersen AFB, Guam during a night landing. Attitude gyro failure caused the pilot to enter a steep bank that was unrecoverable when the crew discovered the problem (after coming out of the clouds). Total 4 fatalities/2 survivors (pilot apparently rode it in).
55-0059	464011	B-52D	Retired	7th BMW	3-May-82	Declared excess in Dec-94 and scrapped.
55-0060	464012	B-52D	Destroyed	484th BMW	13-Jan-64	Crashed into Savage Mountain near Barton, MD during a ferry flight enroute from Westover AFB, MA, where it landed after an Operation CHROME DOME patrol over Europe, to Turner AFB, GA. Tail fin and rudder assembly tore off during severe turbulence during a blizzard. Two nuclear weapons found relatively intact. Total 3 fatalities/ 2 survivors.
55-0061	464013	B-52D	Destroyed	96th BMW	21-Dec-72	Combat loss over North Vietnam during Operation LINEBACKER II. Shot down by a SA-2 SAM near Bach Mai. Total 3 fatalities/3 survivors.
55-0062	464014	B-52D	Display	7th BMW	15-Sep-83	Displayed at K.I. Sawyer Heritage Air Museum, MI.
55-0063	464015	B-52D	Display	7th BMW	15-Sep-82	Displayed at Southwest Aerospace Museum, Ft. Worth, TX. Scrapped on site when the museum closed in the early 1990s.
55-0064	464016	B-52D	Retired	92nd BMW	12-Sep-78	CRESTED DOVE. Declared excess in Sep-94 and scrapped.
55-0065	464017	B-52D	Destroyed	42nd MW	16-Sep-58	Crashed during flight 10 miles south of St Paul, MN due to flight control failure and tail separation. Total 7 fatalities/1 survivor.
55-0066	464018	B-52D	Retired	22nd BMW	20-Oct-82	Declared excess in Dec-94 and scrapped.
55-0067	464019	B-52D	Display	7th BMW	5-Nov-82	Arrived at MASDC on 5-Nov-82. Later displayed at Pima Air & Space Museum, Tucson, AZ, Named "The Lone Star Lady".
55-0068	17184	B-52D	Display	43rd SW	30-Jan-84	First Seattle-built B-52D. Displayed at USAF History & Traditions Museum, Lackland AFB, TX.
55-0069	17185	B-52D	Retired	7th BMW	8-Oct-82	Declared excess in Dec-94 and scrapped.
55-0070	17186	B-52D	Retired	7th BMW	10-Nov-82	Declared excess in Dec-94 and scrapped.
55-0071	17187	B-52D	Display	7th BMW	29-Aug-83	Arrived at MASDC on 29-Aug-83. Later displayed at USS Alabama Battleship Memorial, Mobile, AL.

Tail #	CN	Lineage	Status	Last Unit	Date	Comments
55-0072	17188	B-52D	Retired	379th BMW	22-Aug-78	CRESTED DOVE. Declared excess in Sep-94 and scrapped.
55-0073	17189	B-52D	Retired	22nd BMW	18-Oct-82	Declared excess in Dec-94 and scrapped.
55-0074	17190	B-52D	Retired	7th BMW	29-Apr-82	Declared excess in Dec-94 and scrapped.
55-0075	17191	B-52D	Retired	22nd BMW	14-Oct-82	CRESTED DOVE. Declared excess in Dec-94 and scrapped.
55-0076	17192	B-52D	Retired	22nd BMW	23-Jun-77	Declared excess in Sep-94 and scrapped.
55-0077	17193	B-52D	Retired	7th BMW	25-Jul-83	Declared excess and scrapped.
55-0078	17194	B-52D	Destroyed	22nd BMW	30-Oct-81	Crashed during a low-level training mission from March AFB, CA. Aircraft was flying at 400 feet altitude before it hit a 20-foot sand dune nine miles east of La Junta, CO. Total 8 fatalities/ 0 survivors.
55-0079	17195	B-52D	Retired	22nd BMW	12-Oct-82	Declared excess and scrapped.
55-0080	17196	B-52D	Retired	96th BMW	26-Oct-82	Declared excess in Dec-94 and scrapped.
55-0081	17197	B-52D	Retired	93rd BMW	15-Aug-78	CRESTED DOVE. Declared excess in Sep-94 and scrapped.
55-0082	17198	B-52D	Destroyed	42nd BMW	10-Jan-57	Crashed near Andover, New Brunswick (Canada) 10 miles south of Loring AFB, ME during a flight test. Total 8 fatalities/1 survivor.
55-0083	17199	B-52D	Display	7th BMW	6-Oct-83	One of two B-52 aircraft to shoot down an aircraft when A1C Albert Moore shot down a MiG-21 over North Vietnam during Operation LINEBACKER II on 24-Dec-72. The other was also a B-52D (56-0676). Displayed at United States Air Force Academy, Colorado Springs, CO.
55-0084	17200	B-52D	Retired	22nd BMW	20-Sep-83	Declared excess in Mar-95 and scrapped.
55-0085	17201	B-52D	Display	7th BMW	25-Aug-83	Displayed at Museum of Aviation, Robins AFB, GA.
55-0086	17202	B-52D	Retired	7th BMW	4-Oct-82	Declared excess in Mar-95 and scrapped.
55-0087	17203	B-52D	Retired	7th BMW	1-Nov-82	Declared excess in Mar-95 and scrapped.
55-0088	17204	B-52D	Retired	22nd BMW	15-Oct-82	Declared excess in Dec-94 and scrapped.
55-0089	17205	B-52D	Destroyed	28th BMW	3-Apr-70	Crashed during landing at Ellsworth AFB, SD. Caught fire and skidded into a storage building with 150,000 gallons of jet fuel. One engine ran for 40 minutes following the crash. Firefighters and rescue personnel performed valiantly to free the crew resulting in 23 Airman's Medals, 4 Civilian Awards, and Meritorious Services Medals being awarded. The navigator was freed from the broken off nose section after nearly one hour. A fire truck rammed and broke off the tail turret to free the gunner. Total 0 fatalities/ 9 survivors.
55-0090	17206	B-52D	Retired	22nd BMW	21-Oct-82	Declared excess and scrapped.
55-0091	17207	B-52D	Retired	22nd BMW	8-Oct-82	Declared excess in Oct-93 and scrapped.
55-0092	17208	B-52D	Retired	7th BMW	25-Jul-83	Declared excess in Mar-95 and scrapped.
55-0093	17209	B-52D	Destroyed	42nd BMW	29-Jul-58	Crashed in a field at the Harry Moore Farm three miles south of Limestone (Loring) AFB, ME. Elevator trim run-away after touch and go. Total 8 fatalities/1 survivor.
55-0094	17210	B-52D	Display	Boeing	11-Jul-84	CRESTED DOVE. Displayed at the Kansas Aviation Museum, Wichita, KS.

B-52 STRATOFORTRESS: THE IRON FIST OF STRATEGIC AIR COMMAND

Tail #	CN	Lineage	Status	Last Unit	Date	Comments
55-0095	17211	B-52D GB-52D	Display	7th BMW	4-May-82	Retired on 4-May-82 to Chanute AFB, IL and used as a ground trainer. Displayed at the Valiant Air Command Warbird Museum, Titusville, FL (nose and cockpit section only).
55-0096	17212	B-52D	Retired	22nd BMW	7-Nov-78	CRESTED DOVE. Declared excess in Dec-94 and scrapped.
55-0097	17213	B-52D	Destroyed	4258th SW	15-Oct-72	Damaged beyond repair at U-Tapao RTAFB, Thailand. Crashed during emergency landing. Total 0 fatalities/6 survivors. Scrapped at U-Tapao in Feb-73.
55-0098	17214	B-52D	Destroyed	4170th SW	15-Dec-60	Crashed at Larson AFB, WA after colliding with a KC-135 from Fairchild AFB, WA during aerial refueling over Montana. The boom pierced the wing which failed and caught fire during landing. All crew escaped. The KC-135 landed safely at home station. Total 0 fatalities/10 survivors.
55-0099	17215	B-52D	Destroyed	43rd SW	26-Feb-82	Damaged beyond repair at Andersen AFB, Guam. Moved to fire training area. Declared excess in Oct-93 and scrapped.
55-0100	17216	B-52D	Display	43rd SW	12-Feb-74	Displayed as ARC LIGHT Memorial at Andersen AFB, Guam. The aircraft was replaced with 56-0586 on 9-Jun-83 due to corrosion and 56-0586 was marked with the 55-0100 tail number. The aircraft was then moved to the west side of Andersen and dismantled to comply with SALT on 16-Jul-86.
55-0101	17217	B-52D	Retired	7th BMW	6-Oct-82	Declared excess in Dec-94 and scrapped.
55-0102	17218	B-52D	Destroyed	42nd BMW	26-Jun-58	Caught fire during ground servicing at Loring AFB, ME. Crew chief failed to pull to a circuit breaker which caused a spark and ignited the fuel. Total 0 fatalities. Damaged beyond repair.
55-0103	17219	B-52D	Destroyed	306th BMW	19-Nov-68	Crashed during take-off at Kadena AB, Okinawa during Operation ARC LIGHT. The pilot aborted take-off and the aircraft went off the end of the runway, down an embankment, hit a ditch, and came to rest on the perimeter road on top of a security police pick up (driver escaped). Fuel released from the wings that were torn loose. Fire started from fuel spilling over hot brakes, ruptured hydraulic lines, and electrical connections. Fully loaded with bombs and some exploded while the rest were disarmed. Windows shattered in nearby homes from the explosion. Total 2 fatalities/5 survivors (all survived crash, but EWO and Gunner later died from injuries.
55-0104	17220	B-52D	Retired	96th BMW	27-May-82	Declared excess in Mar-95 and scrapped.
55-0105	17221	B-52D	Display	43rd SW	3-Oct-83	Arrived at MASDC on 3-Oct-83. Later displayed at War Service Memorial, Seoul, South Korea.
55-0106	17222	B-52D	Retired	22nd BMW	28-Nov-78	CRESTED DOVE. Declared excess in Sep-94 and scrapped.
55-0107	17223	B-52D	Retired	22nd BMW	13-Oct-82	Declared excess in Oct-93 and scrapped.
55-0108	17224	B-52D	Destroyed	4170th SW	10-Nov-64	Crashed 60 miles south of Glasgow AFB, MT on a night low-level training mission from Larson AFB, WA. The aircraft was flying on a low-level route and

Tail #	CN	Lineage	Status	Last Unit	Date	Comments
						cleared one knoll at 2,550 feet. It then crashed into a second knoll 300 feet later and disintegrated. Total 7 fatalities/0 survivors.
55-0109	17225	B-52D	Retired	2nd BMW	24-Jun-77	CRESTED DOVE. Declared excess in Feb-94 and scrapped.
55-0110	17226	B-52D	Destroyed	96th BMW	22-Nov-72	Combat loss over North Vietnam during Operation LINEBACKER I. Hit by a SA-2 SAM near Vinh. First B-52 loss due to enemy action. Caught fire and lost all four engines on one side. Crashed in the jungle 15 miles south of Nakhon Phanom, Thailand. Total 0 fatalities/6 survivors. Crew ejected/bailed out and was rescued by an HH-53 SAR crew.
55-0111	17227	B-52D	Retired	7th BMW	5-Oct-82	Declared excess in Mar-95 and scrapped.
55-0112	17228	B-52D	Destroyed	Boeing	17-Dec-73	To Boeing Wichita on 25-Jun-73 for destructive testing and scrapped.
55-0113	17229	B-52D	Retired	7th BMW	4-May-83	Declared excess in Oct-93 and scrapped.
55-0114	17230	B-52D	Destroyed	99th BMW	9-Dec-60	Crashed enroute during a night flight from Westover AFB, MA to the Watertown bomb run in New York state. The aircraft rolled over and lost altitude causing the Navigator to believe the aircraft was breaking up. He ejected without pilot direction (although he says the bailout light illuminated). The rapid decompression and loud explosion caused the pilot to order the crew to abandon the aircraft. The instructor pilot was the only remaining crew member and was unable to fly the aircraft since there were no pilot seats remaining. He subsequently bailed out through the open bomb bay. The aircraft crashed 50 miles later in Plainfield, VT. Total 1 fatality/7 survivors (Gunner's body was found seven months later)
55-0115	17231	B-52D	Destroyed	306th BMW	3-Dec-68	Crashed on landing at Kadena AB, Okinawa during Operation ARC LIGHT. Overshot the runway and was destroyed by fire. Total 0 fatalities/7 survivors.
55-0116	17232	B-52D	Destroyed	93rd BMW	13-Jan-73	Damaged beyond repair during an emergency landing at Da Nang AB, South Vietnam during Operation ARC LIGHT. Received battle damage over North Vietnam. Total 0 fatalities/6 survivors. Aircraft scrapped at Da Nang on 29-Mar-73.
55-0117	17233	B-52D	Retired	28th BMW	3-Oct-78	CRESTED DOVE. Declared excess in Apr-94 and scrapped.
55-0673	464020	B-52D	Retired	96th BMW	4-Nov-82	Declared excess and scrapped.
55-0674	464021	B-52D	Retired	7th BMW	4-Oct-83	Declared excess and scrapped.
55-0675	464022	B-52D	Retired	22nd BMW	22-Oct-82	Declared excess and scrapped.
55-0676	464023	B-52D	Destroyed	306th BMW	19-Jul-69	Crashed during take-off in heavy storms at U-Tapao RTNAF during Operation ARC LIGHT. Pilot attempted to abort but was too late and crashed near the north end of the runway. Both forward landing gear collapsed, and the aircraft broke apart at the front wheel well. The aircraft caught fire, and all crew escaped. An HH-43B was attempting to extinguish the fire when they noticed the gunner's turret

Tail #	CN	Lineage	Status	Last Unit	Date	Comments
						in place and believed he was still onboard (he had previously escaped). Suddenly, seven of the bombs onboard the aircraft exploded, destroyed the helicopter, and killed two Thai guards in a nearby bunker. Total 0 fatalities/6 survivors (two HH-43B crew and two Thai guards killed).
55-0677	464024	B-52D	Display	7th BMW	26-Oct-83	Displayed at the Michigan Flight Museum, Willow Run Airport, Ypsilanti, MI.
55-0678	464025	B-52D	Retired	2nd BMW	25-Oct-78	CRESTED DOVE. Declared excess and scrapped.
55-0679	464026	B-52D GB-52D	Display	22nd BMW	1-Nov-82	CRESTED DOVE. Used as a GB-52D ground trainer at March AFB, CA beginning on 1-Nov-82. Removed from service in 1992. Displayed at March Field Air Museum, March Air Reserve Base, CA.
55-0680	464027	B-52D	Retired	2nd BMW	14-Sep-78	CRESTED DOVE. Declared excess and scrapped.
56-0580	17263	B-52D	Retired	22nd BMW	19-Oct-82	CRESTED DOVE. Declared excess and scrapped.
56-0581	17264	B-52D	Retired	93rd BMW	17-Aug-78	CRESTED DOVE. Declared excess and scrapped.
56-0582	17265	B-52D	Retired	416th BMW	19-Sep-78	CRESTED DOVE. Declared excess and scrapped. Nose and cockpit section sent to the Museum of the U.S. Air Force for display. Scrapped in 1996.
56-0583	17266	B-52D	Retired	42nd BMW	29-Aug-78	CRESTED DOVE. Declared excess and scrapped.
56-0584	17267	B-52D	Destroyed	22nd BMW	26-Dec-72	Combat loss during Operation LINEBACKER II. Hit by a SA-2 SAM over Kinh, North Vietnam. Attempted landing and go-around at U-Tapao RTNAF, Thailand with four engines out on one side. The crew decided to attempt landing because the gunner was injured and may not be able to bailout. Total 4 fatalities/2 survivors (copilot and gunner).
56-0585	17268	B-52D	Display	7th BMW	14-Sep-83	Displayed at AFFTC Museum Airpark, Edwards AFB, CA. Removed and scrapped in Sep-16.
56-0586	17269	B-52D GB-52D	Display	43rd SW	9-Jun-83	Removed from service and used as a ground trainer at Andersen AFB, Guam. Displayed at the Arc Light Memorial on Andersen (replaced 55-0100). This aircraft also succumbed to corrosion and was removed in 2014. It was not replaced.
56-0587	17270	B-52D	Retired	7th BMW	3-Nov-82	Declared excess in Dec-94 and scrapped.
56-0588	17271	B-52D	Retired	7th BMW	5-May-82	Declared excess in Sep-94 and scrapped.
56-0589	17272	B-52D GB-52D	Display	7th BMW	30-Nov-78	CRESTED DOVE. Damaged by a SA-2 SAM during Operation LINEBACKER I mission from U-Tapao RTNAF, Thailand on 23-Apr-72. Landed at Da Nang AB, South Vietnam where it was repaired and RTS on 1-Sep-73. Used as a ground trainer at Sheppard AFB, TX. Displayed at Sheppard AFB Air Park until 2012. Replaced by B-52G (58-0200).
56-0590	17273	B-52D	Retired	22nd BMW	9-Nov-78	CRESTED DOVE. Declared excess and scrapped.
56-0591	17274	B-52D	Destroyed	Boeing	23-Jun-59	To Boeing for testing. Supported atomic tests as effects aircraft during Operation HARDTACK I at Eniwetok from 1-Apr-58 to 18-Jul-58. Named "Tommy's Tigator". Crashed 32 miles west of Burns, OR in the Ochoca National Forest during a low-level flight test from Boeing Seattle due to horizontal stabilizer failure. Total 5 fatalities/0 survivors.

Tail #	CN	Lineage	Status	Last Unit	Date	Comments
56-0592	17275	B-52D	Retired	22nd BMW	7-Dec-78	CRESTED DOVE. Declared excess on 9-Mar-94 and scrapped. Nose and cockpit section to Tinker AFB, OK for use as a ground trainer on 21-Apr-01.
56-0593	17276	B-52D	Destroyed	509th BMW	10-May-69	Crashed in the Pacific Ocean shortly after take-off at Andersen AFB, Guam during Operation ARC LIGHT. Crashed at night over water at low altitude. Possible pilot disorientation due to attitude gyro or structural failure. Total 6 fatalities/0 survivors.
56-0594	17277	B-52D	Destroyed	22nd BMW	19-Oct-78	Crashed in a farmer's field shortly after take-off in light fog from March AFB, CA. Multiple engine failure caused the aircraft to dive into the ground. Total 5 fatalities/ 1 survivor.
56-0595	17278	B-52D	Destroyed	22nd BMW	7-Jul-67	Served as test aircraft for the Quail decoy. Later crashed over the South China Sea 20 miles offshore near Binh Dinh Province, South Vietnam during an Operation ARC LIGHT mission from Andersen AFB, Guam. Mid-air collision with another B-52D (56-0627) while changing formation lead. Total 3 fatalities/3 survivors. Fatalities included General Crumm, 3rd AD commander, whose wife and daughter were on Guam waiting for him to finish his final mission before returning stateside.
56-0596	17279	B-52D	Retired	7th BMW	28-May-82	Declared excess in Oct-93 and scrapped.
56-0597	17280	B-52D	Destroyed	92nd BMW	12-Dec-57	Crashed on take-off from Fairchild AFB, WA. Reverse wired trim motors caused the aircraft to climb straight up, stall, and nose over straight down. Total 8 fatalities/1 survivor (gunner).
56-0598	17281	B-52D	Retired	7th BMW	29-Jan-76	CRESTED DOVE. Declared excess in Sep-94 and scrapped.
56-0599	17282	B-52D	Destroyed	7th BMW	27-Dec-72	Combat loss during Operation LINEBACKER II. Hit by a SA-2 SAM and lost all four engines on left wing. Crashed in Thailand. Total 0 fatalities/6 survivors. All crew ejected and rescued by helicopter.
56-0600	17283	B-52D	Retired	7th BMW	8-Nov-82	Declared excess in Mar-95 and scrapped.
56-0601	17284	B-52D	Destroyed	454th BMW	8-Jul-67	Crashed at Da Nang AB, South Vietnam during an Operation ARC LIGHT mission from Andersen AFB, Guam. Diverted to Da Nang due to electrical problems and complete hydraulic failure. The pilot approached the runway for a no-flaps landing. The aircraft touched down and then bounced about 6,000 feet down the runway before it touched down again. It overran the runway and crashed into a minefield. The mines detonated and blew up the aircraft. All crew killed except the gunner who was in the tail section that separated from the aircraft. Total 5 fatalities/1 survivor.
56-0602	17285	B-52D	Retired	22nd BMW	12-Sep-83	Declared excess in Mar-95 and scrapped.
56-0603	17286	B-52D GB-52D	Retired	416th BMW	12-Sep-78	CRESTED DOVE. Arrived at MASDC on 12-Sep-78. To Lowry AFB, CO on 23-Jan-80 and used for ground training. Scrapped at Lowry in Oct-88.
56-0604	17287	B-52D	Retired	416th BMW	26-Sep-78	CRESTED DOVE. Received SA-2 SAM damage

Tail #	CN	Lineage	Status	Last Unit	Date	Comments
						over North Vietnam on 5-Nov-72. RTS using vertical fin from another B-52D (55-0097). Declared excess in Mar-95 and scrapped.
56-0605	17288	B-52D	Destroyed	7th MW	27-Dec-72	Combat loss over North Vietnam during Operation LINEBACKER II. Hit by a SA-2 SAM and exploded. Also, possible hit by a K-13 Anti-Aircraft Missile (AAM) from a MiG-21 (also destroyed by the B-52 explosion). Parts on display at Lenin Park near Hanoi. Total 2 fatalities/4 survivors.
56-0606	17289	B-52D	Retired	22nd BMW	5-May-82	Declared excess in Mar-95 and scrapped.
56-0607	17290	B-52D	Destroyed	92nd BMW	1-Apr-60	Crashed before take-off at Fairchild AFB, WA. Aircraft was number seven in a MITO take-off. When the pilot released the brakes a loud explosion was heard, and the left wing fell off. All crew egressed safely before the aircraft caught fire. Total 0 fatalities/9 survivors (gunner fractured both elbows).
56-0608	17291	B-52D	Destroyed	93rd BMW	18-Dec-72	Combat loss over North Vietnam during an Operation LINEBACKER II mission from U-Tapao RTNAF, Thailand. Hit by a SA-2 SAM and crashed in Huu Tiep Lake (now called B-52 Lake). Wreckage remains as a North Vietnam war memorial. Total 2 fatalities/4 survivors (became POWs).
56-0609	17292	B-52D	Retired	96th BMW	27-Nov-78	CRESTED DOVE. Declared excess in Apr-94 and scrapped.
56-0610	17293	B-52D	Destroyed	28th BMW	11-Feb-58	Crashed short of the runway during approach at Ellsworth AFB, SD. Fuel pump screen iced over and caused total engine failure. Total 2 fatalities/6 survivors.
56-0611	17294	B-52D	Retired	379th BMW	24-Aug-78	CRESTED DOVE. Declared excess in Sep-94 and scrapped.
56-0612	17295	B-52D	Display	22nd BMW	25-Oct-82	To Eglin AFB, FL on 4 April 1977 for ECM testing. Displayed at Castle Air Museum, Atwater, CA.
56-0613	17296	B-52D	Retired	92nd BMW	5-Sep-78	CRESTED DOVE. Declared excess in Sep-94 and scrapped.
56-0614	17297	B-52D	Retired	7th BMW	12-Sep-83	Declared excess and scrapped.
56-0615	17298	B-52D	Retired	2nd BMW	5-Oct-78	CRESTED DOVE. Declared excess and scrapped.
56-0616	17299	B-52D	Destroyed	28th BMW	5-Aug-71	To OC-AMA, Tinker AFB, OK on 21-May-71. Used for destructive test and scrapped.
56-0617	17300	B-52D	Retired	7th BMW	20-Sep-83	Declared excess in Mar-95 and scrapped.
56-0618	17301	B-52D	Retired	22nd BMW	30-Nov-78	CRESTED DOVE. Declared excess in May-95 and scrapped.
56-0619	17302	B-52D	Retired	410th BMW	24-Aug-78	CRESTED DOVE. Declared excess in Sep-94 and scrapped.
56-0620	17303	B-52D JB-52D NB-52D	Retired	AFSWC	29-Nov-71	Never saw operational service with SAC. Dedicated to test duties at the AFSWC, Kirtland AFB, NM on 15-Nov-57. Redesignated JB-52D on 25-Apr-58. Redesignated NB-52D on 4-Jan-62. Dropped nuclear weapons during testing in the Pacific. It was the only B-52D not painted with camo paint scheme. Declared excess on 25-Oct-93 and scrapped.
56-0621	17304	B-52D	Retired	7th BMW	1-Oct-82	Declared excess and scrapped.

Tail #	CN	Lineage	Status	Last Unit	Date	Comments
56-0622	17305	B-52D	Destroyed	7th BMW	20-Dec-72	Combat loss during an Operation LINEBACKER II mission from U-Tapao RTNAF, Thailand. Hit by a SA-2 SAM over North Vietnam. Crashed 35 km east of Khon Kaen, Thailand. Total 4 fatalities/2 survivors (became POWs until Mar-73).
56-0623	17306	B-52D	Retired	96th BMW	9-Nov-78	CRESTED DOVE. Declared excess in Sep-94 and scrapped.
56-0624	17307	B-52D	Retired	28th BMW	31-Aug-78	CRESTED DOVE. Declared excess in Sep-94 and scrapped.
56-0625	17308	B-52D	Destroyed	306th BMW	31-Mar-72	Crashed 1/4-mile short of the runway during an emergency landing at McCoy AFB, FL. Engine #7 caught fire in flight and spread to the wing. Total 7 fatalities/ 0 survivors. Eight people were injured on the ground (10-year-old boy died in the hospital three days later).
56-0626	17309	B-52D	Retired	96th BMW	16-Oct-78	CRESTED DOVE. Declared excess in Dec-94 and scrapped.
56-0627	17310	B-52D	Destroyed	454th BMW	7-Jul-67	Crashed over the South China Sea 20 miles offshore near Binh Dinh Province, South Vietnam during an Operation ARC LIGHT mission from Andersen AFB, Guam. Mid-air collision with another B-52D (56-0595) while changing formation lead. Total 3 fatalities/4 survivors.
56-0628	17311	B-52D	Destroyed	96th BMW	9-May-82	Damaged beyond repair at Dyess AFB, TX.
56-0629	17312	B-52D	Display	22nd BMW	9-Nov-82	Displayed at the Barksdale Global Power Museum, Barksdale AFB, LA.
56-0630	17313	B-52D	Destroyed	70th BMW	27-Jul-69	Crashed during take-off at Andersen AFB, Guam during Operation ARC LIGHT when the wing failed and completely separated from the aircraft. It made a violent bank off the cliff at the end of the runway and crashed into the ocean. Total 8 fatalities/0 survivors. This resulted in an immediate action TCTO to inspect the lower wing spars and cap for cracking between the external stores pylon and inboard engine struts. The entire B-52D fleet received major wing overhauls under PACER SPEED.
56-0631	17314	B-52E JB-52E B-52E	Retired	22nd BMW	13-Jun-69	First Seattle-built B-52E. First flew 3-Oct-57. To Boeing on 7-Oct-57 and used for various tests. To ARDC at Wright-Patterson, OH on 19-Aug-58. Redesignated JB-52E on 25-Aug-58. To Kirtland AFB, NM on 1-Jul-59 for additional testing. To 4043rd SW at Wright-Patterson on 23-Dec-60. Redesignated B-52E on 9-Mar-61. Declared excess 18-Oct-72 and scrapped.
56-0632	17315	B-52E NB-52E	Retired	Boeing	26-Jun-74	Used as test vehicle at Boeing and redesignated NB-52E. Used for prototyping improvements of landing gear, engines, and other major subsystems for subsequent B-52 fleet improvements. Tested a more aerodynamic engine cowling for the J57-P-43WA engines used on the B-52G. Configured for multiple Flight Control tests. Declared excess and

Tail #	CN	Lineage	Status	Last Unit	Date	Comments
						scrapped.
56-0633	17316	B-52E	Destroyed	11th BMW	9-Dec-58	Crashed six miles north of Altus AFB, OK after possible incorrect stabilizer trim. Total 8 fatalities/1 survivor.
56-0634	17317	B-52E	Retired	22nd BMW	8-Dec-69	Declared excess on 13-Oct-93 and scrapped.
56-0635	17318	B-52E	Retired	22nd BMW	4-Feb-70	Declared excess on 8-Feb-94 and scrapped.
56-0636	17319	B-52E NB-52E	Retired	P&W	30-Jul-81	Arrived at MASDC from Bradley Field, CT. Loaned to P&W for Boeing 747 JT9D engine test and redesignated NB-52E on 13-Dec-67. Declared excess and scrapped on 2-Aug-94.
56-0637	17320	B-52E GB-52E	Retired	22nd BMW	27-Jan-70	Retired at Andersen AFB, Guam and used for ground training. Flipped on its back during Typhoon Pamela on 21-May-76 then used for fire training.
56-0638	17321	B-52E	Retired	22nd BMW	13-Mar-70	Declared excess on 8-Feb-94 and scrapped.
56-0639	17322	B-52E	Retired	22nd BMW	5-Mar-70	Declared excess on 8-Feb-94 and scrapped.
56-0640	17323	B-52E	Retired	22nd BMW	5-Feb-70	Declared excess on 8-Feb-94 and scrapped.
56-0641	17324	B-52E	Retired	22nd BMW	26-May-69	Declared excess on 18-Oct-72 and scrapped.
56-0642	17325	B-52E	Retired	22nd BMW	12-May-69	Declared excess on 18-Oct-72 and scrapped.
56-0643	17326	B-52E	Retired	96th BMW	2-Jun-69	Declared excess on 18-Oct-72 and scrapped.
56-0644	17327	B-52E	Retired	22nd BMW	23-Jan-70	Declared excess on 6-Jan-94 and scrapped.
56-0645	17328	B-52E	Retired	96th BMW	25-Nov-69	Declared excess on 15-Sep-93 and scrapped.
56-0646	17329	B-52E	Retired	96th BMW	19-Nov-69	Declared excess on 15-Sep-93 and scrapped.
56-0647	17330	B-52E	Retired	96th BMW	17-Jun-69	Declared excess on 18-Oct-72 and scrapped.
56-0648	17331	B-52E	Retired	22nd BMW	12-Mar-70	Declared excess on 8-Feb-94 and scrapped.
56-0649	17332	B-52E	Retired	96th BMW	17-Nov-69	Declared excess on 13-Jun-73 and scrapped.
56-0650	17333	B-52E	Retired	22nd BMW	21-Jan-70	Declared excess and scrapped in Jan-94.
56-0651	17334	B-52E	Retired	22nd BMW	19-Jan-70	Declared excess and scrapped in Sep-94.
56-0652	17335	B-52E	Retired	96th BMW	12-Jan-70	Declared excess on 20-Sep-94 and scrapped.
56-0653	17336	B-52E	Retired	22nd BMW	12-Dec-69	Declared excess on 6-Jan-94 and scrapped.
56-0654	17337	B-52E	Retired	96th BMW	7-Jun-69	Declared excess and scrapped.
56-0655	17338	B-52E	Destroyed	6th SAW	19-Nov-63	Destroyed by fire during maintenance at Walker AFB, NM. Total 1 fatality (trapped under collapsed wing and burned).
56-0656	17339	B-52E	Retired	22nd BMW	11-Feb-70	Declared excess on 10-Jan-94 and scrapped.
56-0657	464028	B-52D	Display	7th BMW	12-Aug-83	Displayed at the South Dakota Air and Space Museum, Ellsworth AFB, SD.
56-0658	464029	B-52D	Retired	7th BMW	29-Apr-82	Declared excess and scrapped in Mar-95.
56-0659	464030	B-52D	Display	96th BMW	25-May-82	Arrived at MASDC on 25-May-82. To Davis-Monthan AFB, AZ for display in 1989. Displayed at the Air Warrior Park from Jun-91 until removed in 2011. Scrapped in 2012.
56-0660	464031	B-52D	Retired	7th BMW	23-Aug-83	Declared excess and scrapped in Mar-95.
56-0661	464032	B-52D	Destroyed	92nd BMW	8-Sep-58	Exploded and crashed after mid-air collision with another B-52D (56-0681) during approach at Fairchild AFB, WA. Total 7 fatalities/1 survivor.
56-0662	464033	B-52D	Destroyed	7th BMW	28-Jun-82	Damaged beyond repair at Carswell AFB, TX by ground fire caused by leaking LOX mixing with leaking hydraulic fluid ignited by electrical spark. Forward fuselage severely damaged. Total 0 fatalities. Scrapped at Carswell in Apr-84.

Tail #	CN	Lineage	Status	Last Unit	Date	Comments
56-0663	464034	B-52D	Retired	7th BMW	25-May-82	Declared excess and scrapped in Mar-95.
56-0664	464035	B-52D GB-52D	Retired	43rd SW	1-Oct-78	Used for ground training at Andersen AFB, Guam. Declared excess and scrapped at Andersen sometime after 1992.
56-0665	464036	B-52D	Display	97th BMW	1-Sep-78	CRESTED DOVE. Hit by a SA-2 SAM over North Vietnam during Operation LINEBACKER I on 9-Apr-72. Experienced fuel leaks and lost two engines but managed to land at Da Nang. Repaired and RTS. Displayed at the National Museum of the U.S. Air Force, Wright-Patterson AFB, OH.
56-0666	464037	B-52D	Display	43rd SW	24-Aug-83	Arrived at MASDC on 24-Aug-83. Declared excess and scrapped in Dec-94. Vertical fin displayed at the National Museum of the Mighty Eighth Air Force, Savannah, GA. Named "Many Sixes".
56-0667	464038	B-52D	Retired	7th BMW	12-Sep-83	Declared excess and scrapped in Mar-95.
56-0668	464039	B-52D	Retired	96th BMW	2-Nov-82	Declared excess and scrapped in Oct-93.
56-0669	464040	B-52D	Destroyed	306th BMW	21-Dec-72	Combat loss during Operation LINEBACKER II mission from Andersen AFB, Guam. Hit by an SA-2 SAM over North Vietnam. Crew ejected (except radar nav) over Laos and were rescued. Total 1 fatality/5 survivors.
56-0670	464041	B-52D	Retired	96th BMW	3-May-82	Declared excess and scrapped in Mar-95.
56-0671	464042	B-52D	Retired	22nd BMW	1-Oct-82	Declared excess and scrapped in Mar-95.
56-0672	464043	B-52D	Retired	7th BMW	3-May-83	Declared excess and scrapped.
56-0673	464044	B-52D	Retired	22nd BMW	14-Sep-78	CRESTED DOVE. Declared excess and scrapped in Sep-94.
56-0674	464045	B-52D	Destroyed	96th BMW	26-Dec-72	Combat loss over North Vietnam during an Operation LINEBACKER II mission from U-Tapao RTNAF, Thailand. Hit by a SA-2 SAM or possibly K-13 AAM from a MiG-21. Total 2 fatalities/4 survivors (became POWs until Mar-73).
56-0675	464046	B-52D	Retired	22nd BMW	5-Oct-78	CRESTED DOVE. Declared excess and scrapped in Apr-94.
56-0676	464047	B-52D	Display	43rd SW	12-Oct-83	First B-52 credited as MiG Killer when SSgt Samuel O. Turner shot down a MiG-21 on 18-Dec-72. The other was also a B-52D (55-0083). Displayed at Armed Forces & Aerospace Museum, Fairchild AFB, WA.
56-0677	464048	B-52D	Destroyed	96th BMW	30-Jul-72	Crashed shortly after take-off from U-Tapao RTNAF, Thailand during Operation ARC LIGHT. A lightning strike knocked out the instruments and started a fire in left wing. Crashed 65 miles northwest of Ubon RTAFB. Total 5 fatalities/1 survivor.
56-0678	464049	B-52D	Retired	22nd BMW	12-Oct-78	CRESTED DOVE. Damaged by enemy fire over North Vietnam during Operation LINEBACKER II on 18-Dec-72. Landed at U-Tapao RTNAF, Thailand where it was repaired and RTS. Declared excess and scrapped in Sep-94.
56-0679	464050	B-52D	Retired	22nd BMW	12-Oct-82	Declared excess and scrapped in Sep-93.
56-0680	464051	B-52D	Retired	7th BMW	30-Sep-78	CRESTED DOVE. Used as weapons loads trainer at Carswell AFB, TX. Scrapped in Apr-84.

Tail #	CN	Lineage	Status	Last Unit	Date	Comments
56-0681	464052	B-52D	Destroyed	92nd BMW	8-Sep-58	Crashed after mid-air collision with another B-52D (56-0661) and explosion during approach at Fairchild AFB, WA. Total 5 fatalities/ 3 survivors.
56-0682	464053	B-52D	Display	96th BMW	19-Oct-78	CRESTED DOVE. Declared excess and scrapped in May-92. Nose and cockpit section to the National Museum of the U.S. Air Force for display.
56-0683	464054	B-52D	Display	7th BMW	30-Aug-83	Displayed at Whiteman AFB, MO. Previously displayed at Pease AFB, NH.
56-0684	464055	B-52D	Retired	7th BMW	5-May-83	Declared excess and scrapped on 25-Oct-93.
56-0685	464056	B-52D GB-52D	Display	96th BMW	15-Sep-82	Used as ground trainer at Dyess AFB, TX. Later displayed at Dyess Linear Air Park, Dyess AFB, TX.
56-0686	464057	B-52D	Retired	96th BMW	22-Aug-83	Declared excess and scrapped in Mar-95.
56-0687	464058	B-52D	Display	7th BMW	20-Feb-84	Flight to McCoy AFB, FL was the last B-52D flight. Displayed at B-52 Memorial Park, Orlando International Airport (formerly McCoy AFB), Orlando, FL.
56-0688	464059	B-52D	Retired	7th BMW	30-Sep-78	CRESTED DOVE. Used as weapons loads trainer at Carswell AFB, TX. Scrapped in Apr-84.
56-0689	464060	B-52D	Display	7th BMW	8-Oct-83	Displayed at the Imperial War Museums, Duxford, England. Housed in the American Air Museum building with a B-29A (44-61748).
56-0690	464061	B-52D	Retired	7th BMW	28-Oct-82	Declared excess and scrapped in Oct-83.
56-0691	464062	B-52D	Retired	2nd BMW	10-Oct-78	CRESTED DOVE. Declared excess and scrapped in Sep-93.
56-0692	464063	B-52D GB-52D	Display	7th BMW	11-Oct-83	Displayed at the Kelly Field Heritage Airpark, San Antonio, TX. Previously used as ground maintenance trainer and then displayed at Tinker AFB, OK (South side).
56-0693	464064	B-52D	Retired	92nd BMW	7-Sep-78	CRESTED DOVE. Declared excess and scrapped in Sep-94.
56-0694	464065	B-52D	Retired	43rd SW	22-Aug-83	One of two B-52D aircraft that set world speed records on 26-Sep-58. Declared excess and scrapped in Mar-95.
56-0695	464066	B-52D	Display	7th BMW	5-Oct-83	Displayed at the Charles B. Hall Airpark, Tinker AFB, Ok.
56-0696	464067	B-52D	Display	7th BMW	16-Aug-83	Displayed at the Travis AFB Heritage Center, Travis AFB, CA. Named "Twilight D'Lite".
56-0697	464068	B-52D	Retired	7th BMW	20-Sep-83	Declared excess and scrapped in Mar-95.
56-0698	464069	B-52D	Retired	7th BMW	9-Nov-82	Declared excess and scrapped in Dec-94.
56-0699	464070	B-52E	Retired	96th SAW	22-May-69	First Wichita-built B-52E. First flew 17-Oct-57. Declared excess on 18-Oct-72 and scrapped.
56-0700	464071	B-52E	Retired	22nd BMW	5-Jun-69	Declared excess on 18-Oct-72 and scrapped.
56-0701	464072	B-52E	Retired	22nd BMW	27-May-69	Declared excess on 18-Oct-72 and scrapped.
56-0702	464073	B-52E	Retired	96th SAW	10-Jun-69	Declared excess on 18-Oct-72 and scrapped.
56-0703	464074	B-52E	Retired	96th SAW	15-May-69	Declared excess on 18-Oct-72 and scrapped.
56-0704	464075	B-52E	Retired	22nd BMW	12-Feb-70	Declared excess on 1-Feb-94 and scrapped.
56-0705	464076	B-52E	Retired	96th SAW	24-Nov-69	Declared excess on 6-Jan-94 and scrapped.
56-0706	464077	B-52E	Retired	96th SAW	9-Jan-70	Declared excess on 13-Sep-94 and scrapped.
56-0707	464078	B-52E	Retired	22nd BMW	10-Mar-70	Declared excess on 10-Jan-94 and scrapped.
56-0708	464079	B-52E GB-52E	Retired	17th BMW	16-Feb-67	To Chanute AFB, IL and used for ground training. Later scrapped in-place.

Tail #	CN	Lineage	Status	Last Unit	Date	Comments
56-0709	464080	B-52E	Retired	96th SAW	4-Jun-69	Declared excess on 18-Oct-72 and scrapped in Oct-93.
56-0710	464081	B-52E	Retired	22nd BMW	9-Jun-69	Declared excess on 18-Oct-72 and scrapped.
56-0711	464082	B-52E	Retired	22nd BMW	23-Jun-69	Declared excess on 18-Oct-72 and scrapped in Sep-94.
56-0712	464083	B-52E	Retired	22nd BMW	16-May-69	Declared excess on 18-Oct-72 and scrapped.
57-0014	17408	B-52E	Retired	11th SAW	19-Jan-67	Declared excess and scrapped.
57-0015	17409	B-52E	Retired	96th SAW	14-Nov-69	Declared excess and scrapped in Sep-94.
57-0016	17410	B-52E	Retired	96th SAW	12-Aug-69	Declared excess on 18-Oct-72 and scrapped in Feb-94.
57-0017	17411	B-52E	Retired	22nd BMW	4-Mar-70	Declared excess on 20-Sep-94 and scrapped.
57-0018	17412	B-52E	Destroyed	6th SAW	30-Jan-63	Crashed in the Sangre de Christo mountains, ten miles west of Mora, NM during flight from Walker AFB, NM when the tail broke off in turbulence. Total 2 fatalities/4 survivors (all ejected safely except EWO and gunner).
57-0019	17413	B-52E	Destroyed	Boeing	1-Dec-65	Used for teardown inspection at Boeing Wichita, KS to assess fatigue and corrosion condition of the fleet. Scrapped in-place.
57-0020	17414	B-52E	Retired	22nd BMW	28-Jan-70	Declared excess and scrapped in 1994.
57-0021	17415	B-52E	Retired	22nd BMW	2-Feb-70	Declared excess on 8-Feb-94 and scrapped.
57-0022	17416	B-52E	Retired	22nd BMW	21-Jun-69	Declared excess on 18-Oct-72 and scrapped in Feb-94.
57-0023	17417	B-52E	Retired	22nd BMW	25-Jun-69	Declared excess on 18-Oct-72 and scrapped in Jan-94.
57-0024	17418	B-52E	Retired	96th SAW	18-Nov-69	Declared excess and scrapped in Oct-93.
57-0025	17419	B-52E	Retired	96th SAW	19-Jun-69	Declared excess on 18-Oct-72 and scrapped in Feb-94.
57-0026	17420	B-52E	Retired	22nd BMW	26-Jun-69	Declared excess on 18-Oct-72 and scrapped in Sep-93.
57-0027	17421	B-52E	Retired	22nd BMW	26-Jan-70	Declared excess on 13-Sep-94 and scrapped.
57-0028	17422	B-52E	Retired	17th BMW	28-Jun-67	Declared excess on 7-Apr-69 and scrapped.
57-0029	17423	B-52E	Retired	70th BMW	16-Jan-67	Declared excess on 7-Apr-69 and scrapped.
57-0030	17424	B-52F JB-52F B-52F	Retired	93rd BMW	14-Dec-67	First Seattle-built B-52F. First flew 6-May-58. To Boeing on 12-May-58 and used for various tests. Redesignated JB-52F and transferred to North American Aviation on 27-Feb-59 for AGM-28 Hound Dog testing. Committed to operational service on 20-Aug-62 and redesignated B-52F. Declared excess on 7-Apr-69 and scrapped.
57-0031	17425	B-52F	Retired	93rd BMW	12-Jul-71	Declared excess and scrapped in Sep-94.
57-0032	17426	B-52F	Retired	97th BMW	17-Nov-78	CRESTED DOVE. Declared excess and scrapped in Jan-94.
57-0033	17427	B-52F	Retired	92nd BMW	26-Sep-78	CRESTED DOVE. Declared excess and scrapped in Jan-94.
57-0034	17428	B-52F	Retired	93rd BMW	5-Sep-78	CRESTED DOVE. Declared excess and scrapped in Mar-94.
57-0035	17429	B-52F	Retired	93rd BMW	30-Nov-78	CRESTED DOVE. Declared excess and scrapped in Mar-94.

Tail #	CN	Lineage	Status	Last Unit	Date	Comments
57-0036	17430	B-52F	Destroyed	4228th SW	15-Oct-59	Crashed after colliding with a KC-135 (57-1513) during refueling over Hardinsberg, KY during airborne alert (Operation CHROME DOME). Both aircraft were from Columbus AFB, MS. Two nuclear weapons on board. Both recovered intact (one partially burned). Total 4 fatalities/ 4 survivors. Four tanker crew killed.
57-0037	17431	B-52F	Retired	93rd BMW	18-Aug-69	Declared excess and scrapped in Sep-94.
57-0038	17432	B-52F JB-52F GB-52F	Display	OC-ALC	31-Dec-74	Used for test at Eglin AFB, FL and designated JB-52F on 4-Dec-63. Used for ground training at Tinker AFB, OK and designated GB-52F on 26-Jan-70. Displayed at Joe Davies Heritage Airpark, Air Force Plant 42, Palmdale, CA. Previously displayed at Oklahoma City Fairgrounds from 1974 to 2006. Only preserved B-52F.
57-0039	17433	B-52F	Retired	93rd BMW	19-Jul-71	Declared excess and scrapped in Sep-94.
57-0040	17434	B-52F	Retired	320th BMW	10-Jan-68	Declared excess on 7-Apr-69 and scrapped.
57-0041	17435	B-52F	Destroyed	93rd BMW	21-Oct-69	Crashed short of the runway at Castle AFB, CA during approach. The aircraft was on a training flight and the copilot incorrectly transferred fuel to the wings which shifted the center of gravity. The aircraft pitched down causing the aircraft to crash into a cotton field. Total 0 fatalities/9 survivors (several crew suffered broken bones and the gunner suffered a broken back).
57-0042	17436	B-52F GB-52F	Display	93rd BMW	29-Apr-73	To Chanute AFB, IL on 29-Apr-73 and used for ground trainer. Later displayed at Museum of Flying, Santa Monica, CA and then Yanks Air Museum, Chino, CA (nose and cockpit section only).
57-0043	17437	B-52F	Destroyed	4228th SW	23-Dec-63	Crashed after the initial climb out at Columbus AFB, MS. Aircraft entered the clouds and came out inverted and diving to the ground. Possible attitude indicator failure and pilot disorientation. Total 9 fatalities/0 survivors.
57-0044	17438	B-52F	Retired	93rd BMW	27-Jan-67	Declared excess on 7-Apr-69 and scrapped.
57-0045	17439	B-52F	Retired	42nd BMW	28-Sep-78	CRESTED DOVE. Scrapped in Dec-94.
57-0046	17440	B-52F	Retired	93rd BMW	30-Jun-69	Scrapped in Sep-94.
57-0047	17441	B-52F	Destroyed	320th BMW	18-Jun-65	Combat loss over the Pacific Ocean during Operation ARC LIGHT. Aircraft crashed 250 miles off the Demilitarized Zone after mid-air collision with another B-52F (57-0179) while waiting for tanker rendezvous. Total 3 fatalities/3 survivors.
57-0048	17442	B-52F GB-52F	Retired	416th BMW	31-Aug-78	CRESTED DOVE. To Lowry AFB, CO on 6-Mar-80 and used for ground training. Scrapped in-place when base closed.
57-0049	17443	B-52F	Retired	93rd BMW	8-Jul-68	Declared excess on 7-Apr-69. Later scrapped.
57-0050	17444	B-52F GB-52F	Retired	93rd BMW	8-Jul-73	To Andersen AFB, Guam and used as ground trainer. Later scrapped in-place.
57-0051	17445	B-52F	Retired	93rd BMW	5-Dec-78	CRESTED DOVE. Scrapped in Mar-97.
57-0052	17446	B-52F	Retired	93rd BMW	7-Sep-78	CRESTED DOVE. Scrapped in Apr-94.
57-0053	17447	B-52F	Retired	93rd BMW	27-Jun-69	Scrapped in Feb-94.

B-52 Stratofortress: The Iron Fist of Strategic Air Command

Tail #	CN	Lineage	Status	Last Unit	Date	Comments
57-0054	17448	B-52F	Retired	93rd BMW	8-Jul-69	Scrapped in Sep-94.
57-0055	17449	B-52F	Retired	93rd BMW	13-Sep-71	Scrapped in Dec-94.
57-0056	17450	B-52F	Retired	7th BMW	7-Apr-71	Declared excess on 20-Sep-94 and scrapped.
57-0057	17451	B-52F	Retired	93rd BMW	26-Aug-71	Scrapped in Dec-94.
57-0058	17452	B-52F	Retired	42nd BMW	17-Oct-78	CRESTED DOVE. Scrapped in Mar-94.
57-0059	17453	B-52F	Retired	93rd BMW	11-Aug-71	Scrapped in Dec-94.
57-0060	17454	B-52F	Retired	93rd BMW	26-Jul-71	Scrapped in Jan-94.
57-0061	17455	B-52F	Retired	93rd BMW	22-Jul-71	Scrapped in Sep-94.
57-0062	17456	B-52F	Retired	93rd BMW	4-Aug-71	Scrapped in Dec-94.
57-0063	17457	B-52F	Retired	28th BMW	28-Sep-78	CRESTED DOVE. Scrapped in Dec-94.
57-0064	17458	B-52F	Retired	93rd BMW	2-Aug-71	Scrapped in Dec-93.
57-0065	17459	B-52F	Retired	93rd BMW	9-Aug-71	Scrapped in Jan-94.
57-0066	17460	B-52F	Retired	93rd BMW	11-Jun-69	Scrapped.
57-0067	17461	B-52F	Retired	93rd BMW	7-Jul-69	Scrapped in Sep-94.
57-0068	17462	B-52F	Retired	93rd BMW	21-Nov-67	Declared excess on 7-Apr-69. Later scrapped.
57-0069	17463	B-52F	Retired	93rd BMW	2-Oct-78	CRESTED DOVE. Scrapped in Feb-94.
57-0070	17464	B-52F	Retired	93rd BMW	30-Jan-67	Declared excess on 7-Apr-69. Later scrapped.
57-0071	17465	B-52F GB-52F	Retired	28th BMW	6-Jul-77	CRESTED DOVE. Delivered in place to Boeing on 19-Feb-59 and used for various tests. To Sheppard AFB, TX and used for ground trainer on 6-Jul-77. Later scrapped in-place.
57-0072	17466	B-52F	Retired	93rd BMW	21-Nov-78	CRESTED DOVE. Scrapped in Mar-94.
57-0073	17467	B-52F	Destroyed	Boeing	1-Oct-64	Delivered in place at Boeing on 20-Mar-59 and used for static test. Scrapped in-place on 1-Oct 64.
57-0095	464084	B-52E	Retired	96th SAW	19-Nov-69	Scrapped in Feb-94.
57-0096	464085	B-52E	Retired	22nd BMW	29-Jan-70	Declared excess 15-Sep-93. Scrapped in Feb-94.
57-0097	464086	B-52E	Retired	22nd BMW	27-Jan-70	Declared excess 13-Oct-93. Scrapped in Feb-94.
57-0098	464087	B-52E	Retired	22nd BMW	23-Jan-70	Declared excess and scrapped on 13-Sep-94.
57-0099	464088	B-52E	Retired	22nd BMW	30-Jan-70	Declared excess and scrapped on 20-Sep-94.
57-0100	464089	B-52E	Retired	96th SAW	7-Jan-70	Declared excess and scrapped on 8-Feb-94.
57-0101	464090	B-52E	Display	96th SAW	20-May-69	Arrived at MASDC on 20-May-69. Declared excess on 18-Oct-72 and scrapped. Forward fuselage displayed at the Pearl Harbor Aviation Museum, Honolulu, HI.
57-0102	464091	B-52E	Retired	96th SAW	23-May-69	Declared excess on 18-Oct-72 and scrapped.
57-0103	464092	B-52E	Retired	96th SAW	5-Jan-70	Declared excess and scrapped on 13-Sep-94.
57-0104	464093	B-52E	Retired	96th SAW	12-Jun-69	Declared excess on 18-Oct-72 and scrapped.
57-0105	464094	B-52E	Retired	22nd BMW	10-Mar-70	Declared excess and scrapped on 8-Feb-94.
57-0106	464095	B-52E	Retired	22nd BMW	11-Mar-70	Declared excess and scrapped on 10-Jan-94.
57-0107	464096	B-52E	Retired	22nd BMW	16-Jun-69	Declared excess on 18-Oct-72 and scrapped.
57-0108	464097	B-52E	Retired	22nd BMW	11-Feb-70	Declared excess and scrapped on 1-Feb-94.
57-0109	464098	B-52E	Retired	96th SAW	19-May-69	Declared excess on 18-Oct-72 and scrapped.
57-0110	464099	B-52E	Retired	96th SAW	21-May-69	Declared excess 18-Oct-72. Scrapped in 1981.
57-0111	464100	B-52E	Retired	11th SAW	17-Apr-67	Declared excess on 7-Apr-69 and scrapped.
57-0112	464101	B-52E	Retired	22nd BMW	10-Feb-70	Declared excess and scrapped on 10-Jan-94.
57-0113	464102	B-52E	Retired	96th SAW	17-Jan-67	Declared excess on 7-Apr-69 and scrapped.
57-0114	464103	B-52E	Retired	6th SAW	18-Jan-67	Declared excess on 7-Apr-69 and scrapped.
57-0115	464104	B-52E	Retired	96th SAW	12-Jan-70	Declared excess and scrapped on 13-Oct-93.
57-0116	464105	B-52E	Retired	22nd BMW	18-Jun-69	Declared excess on 18-Oct-72 and scrapped.
57-0117	464106	B-52E	Retired	6th SAW	10-Jan-67	Declared excess on 7-Apr-69 and scrapped.

B-52 Stratofortress: The Iron Fist of Strategic Air Command

Tail #	CN	Lineage	Status	Last Unit	Date	Comments
57-0118	464107	B-52E	Retired	96th SAW	8-Jan-70	Initially delivered to Boeing and used for test. Declared excess and scrapped on 8-Feb-94.
57-0119	464108	B-52E JB-52E	Retired	AFFTC	31-Dec-80	Redesignated JB-52E on 5-Oct-66. Used for flight testing at AFFTC, Edwards AFB, CA including the GE TF39 engine for the C-5 Galaxy. Towed to the photo range near Rogers Dry Lake south of Edwards and retired on 31-Dec-80 (estimated date). Broken up in 1991 to satisfy START.
57-0120	464109	B-52E	Retired	96th SAW	3-Sep-69	Declared excess and scrapped.
57-0121	464110	B-52E	Retired	96th SAW	6-Jan-70	Declared excess and scrapped on 15-Sep-93.
57-0122	464111	B-52E	Retired	22nd BMW	28-May-69	Declared excess on 18-Oct-72 and scrapped.
57-0123	464112	B-52E	Retired	22nd BMW	2-Mar-70	Declared excess and scrapped on 6-Jan-94.
57-0124	464113	B-52E	Retired	22nd BMW	11-Jun-69	Declared excess on 18-Oct-72 and scrapped.
57-0125	464114	B-52E	Retired	22nd BMW	29-May-69	Declared excess on 18-Oct-72 and scrapped.
57-0126	464115	B-52E	Retired	22nd BMW	7-Feb-70	Declared excess and scrapped on 1-Feb-94.
57-0127	464116	B-52E	Retired	22nd BMW	24-Jun-69	Declared excess on 18-Oct-72 and scrapped.
57-0128	464117	B-52E	Retired	96th SAW	12-Jan-70	Declared excess and scrapped.
57-0129	464118	B-52E	Retired	22nd BMW	3-Mar-70	Declared excess and scrapped on 1-Feb-94.
57-0130	464119	B-52E	Retired	22nd BMW	4-Feb-70	Declared excess and scrapped on 8-Feb-94.
57-0131	464120	B-52E	Retired	22nd BMW	6-Mar-70	Declared excess and scrapped on 1-Jan-94.
57-0132	464121	B-52E	Retired	22nd BMW	10-Dec-69	Declared excess and scrapped.
57-0133	464122	B-52E	Retired	96th SAW	13-May-69	Declared excess on 18-Oct-72 and scrapped.
57-0134	464123	B-52E	Retired	96th SAW	15-Mar-66	Scrapped at Tinker AFB, OK (structural issues).
57-0135	464124	B-52E	Retired	22nd BMW	14-May-69	Declared excess on 18-Oct-72 and scrapped.
57-0136	464125	B-52E	Retired	22nd BMW	20-Jan-70	Declared excess and scrapped on 13-Oct-93.
57-0137	464126	B-52E	Retired	11th SAW	21-Apr-67	Declared excess on 7-Apr-69 and scrapped.
57-0138	464127	B-52E	Retired	22nd BMW	3-Jun-69	Declared excess on 18-Oct-72 and scrapped.
57-0139	464128	B-52F	Retired	93rd BMW	30-May-73	First Wichita-built B-52F. First flew on 14-May-58. Declared excess and scrapped.
57-0140	464129	B-52F	Retired	42nd BMW	20-Sep-78	CRESTED DOVE. Scrapped in Dec-94.
57-0141	464130	B-52F	Retired	93rd BMW	4-Jan-67	Declared excess on 7-Apr-69 and scrapped.
57-0142	464131	B-52F	Retired	93rd BMW	24-Oct-78	CRESTED DOVE. Declared excess and scrapped. Nose and cockpit section used for ground training at Goodfellow AFB, TX.
57-0143	464132	B-52F	Retired	93rd BMW	14-Jul-71	Declared excess and scrapped in Sep-94.
57-0144	464133	B-52F	Retired	93rd BMW	2-Nov-67	Declared excess on 7-Apr-69 and scrapped.
57-0145	464134	B-52F	Retired	93rd BMW	11-Oct-78	CRESTED DOVE. Declared excess and scrapped in Mar-94.
57-0146	464135	B-52F	Retired	7th BMW	16-Jan-68	Declared excess on 7-Apr-69 and scrapped.
57-0147	464136	B-52F	Retired	379th BMW	29-Aug-78	CRESTED DOVE. Declared excess and scrapped in Feb-94.
57-0148	464137	B-52F	Retired	93rd BMW	15-Aug-78	CRESTED DOVE. Declared excess and scrapped in Apr-94.
57-0149	464138	B-52F	Destroyed	93rd BMW	8-May-69	Brigadier General Jimmy Stewart flew on this aircraft from Guam as an observer for his last combat mission in 1966. Crashed in 1969 due to landing gear failure while practicing touch and goes at Castle AFB, CA. Aircraft skidded 5,000 feet down the runway and exploded. Crew escaped before the explosion. Total 0 fatalities/5 survivors.

B-52 STRATOFORTRESS: THE IRON FIST OF STRATEGIC AIR COMMAND

Tail #	CN	Lineage	Status	Last Unit	Date	Comments
57-0150	464139	B-52F	Retired	93rd BMW	17-Aug-78	CRESTED DOVE. Declared excess and scrapped in Apr-94.
57-0151	464140	B-52F	Retired	93rd BMW	8-Sep-71	Declared excess and scrapped in May-92.
57-0152	464141	B-52F	Retired	93rd BMW	18-Aug-71	Declared excess and scrapped in May-92.
57-0153	464142	B-52F	Retired	93rd BMW	22-Sep-71	Declared excess and scrapped in Jan-94.
57-0154	464143	B-52F	Retired	93rd BMW	7-Nov-78	CRESTED DOVE. Declared excess and scrapped in Jan-94.
57-0155	464144	B-52F	Retired	93rd BMW	11-Jul-69	Declared excess and scrapped in Sep-94.
57-0156	464145	B-52F	Retired	93rd BMW	26-Jan-67	Declared excess on 7-Apr-69 and scrapped.
57-0157	464146	B-52F	Retired	7th BMW	3-Jan-67	Declared excess on 7-Apr-69 and scrapped.
57-0158	464147	B-52F	Retired	93rd BMW	7-Aug-68	Declared excess on 7-Apr-69 and scrapped.
57-0159	464148	B-52F	Retired	93rd BMW	16-Aug-71	Declared excess and scrapped.
57-0160	464149	B-52F	Retired	93rd BMW	1-Jul-69	Declared excess and scrapped in Sep-94.
57-0161	464150	B-52F	Retired	93rd BMW	24-Sep-71	Declared excess and scrapped in Sep-94.
57-0162	464151	B-52F	Retired	93rd BMW	15-Sep-71	Declared excess and scrapped in Dec-94.
57-0163	464152	B-52F	Retired	7th BMW	2-Jul-69	Declared excess 7-Apr-69. Scrapped in 1993.
57-0164	464153	B-52F	Retired	320th BMW	18-Apr-67	Declared excess on 7-Apr-69 and scrapped.
57-0165	464154	B-52F	Retired	92nd BMW	21-Sep-78	CRESTED DOVE. Declared excess and scrapped.
57-0166	464155	B-52F	Destroyed	4134th SW	14-Mar-61	Crashed approximately 15 miles west of Yuba City, CA during an Operation CHROME DOME mission from Mather AFB, CA. Cabin pressurization failed and shattered a window. Crew descended to a lower altitude. Increased fuel consumption caused the aircraft to run out of fuel and crew evacuated. Two nuclear weapons ejected from the aircraft when it impacted the ground. Explosives did not detonate. Total 0 fatalities/ 8 survivors. Fireman killed on the ground after being stuck by a passing car at the crash site.
57-0167	464156	B-52F	Retired	93rd BMW	6-Jan-67	Declared excess on 7-Apr-69 and scrapped.
57-0168	464157	B-52F	Retired	AFSWC	29-Jun-72	Arrived at MASDC from the AFSWC, Kirtland AFB, NM. Used for various nuclear weapons testing. Declared excess and scrapped in Sep-94.
57-0169	464158	B-52F	Retired	97th BMW	19-Oct-78	CRESTED DOVE. Declared excess and scrapped in Feb-94.
57-0170	464159	B-52F	Retired	93rd BMW	14-Nov-78	CRESTED DOVE. Declared excess and scrapped in Mar-94.
57-0171	464160	B-52F	Retired	2nd BMW	7-Dec-78	CRESTED DOVE. Declared excess and scrapped in May-92.
57-0172	464161	B-52F	Destroyed	93rd BMW	9-Oct-69	Crashed during practice touch and goes at Castle AFB, CA. Pitched up during overshoot and pilot lost control. Crashed and exploded on impact 1,000 feet beyond end of runway and skidded into ammunition dump. Total 6 fatalities/ 0 survivors.
57-0173	464162	B-52F	Destroyed	7th BMW	29-Feb-68	Crashed in the Gulf of Mexico near Matagorda Island, TX during a practice bomb run from Carswell AFB, TX. Possible stab trim failure. Total 8 fatalities/0 survivors.
57-0174	464163	B-52F	Retired	96th SAW	30-Aug-71	Declared excess and scrapped in May-92.
57-0175	464164	B-52F	Retired	93rd BMW	14-Jul-69	Declared excess and scrapped in Sep-94.

Tail #	CN	Lineage	Status	Last Unit	Date	Comments
57-0176	464165	B-52F	Retired	93rd BMW	9-Jul-69	Declared excess and scrapped in Sep-94.
57-0177	464166	B-52F	Retired	93rd BMW	1-Sep-71	Declared excess and scrapped in Dec-92.
57-0178	464167	B-52F	Retired	93rd BMW	28-Jul-71	Declared excess and scrapped in Sep-94.
57-0179	464168	B-52F	Destroyed	7th BMW	18-Jun-65	Combat loss over the Pacific Ocean during Operation ARC LIGHT. Aircraft crashed 250 miles off the Demilitarized Zone after mid-air collision with another B-52F (57-0047) while waiting for tanker rendezvous. Total 2 fatalities/4 survivors.
57-0180	464169	B-52F	Retired	93rd BMW	30-Jun-69	Declared excess and scrapped in Mar-94.
57-0181	464170	B-52F	Destroyed	7th BMW	9-Oct-67	Arrived at MASDC on 29-Jun-67. To Boeing on 9-Oct-67 and tested to destruction.
57-0182	464171	B-52F	Retired	93rd BMW	20-Sep-71	Declared excess and scrapped in Apr-94.
57-0183	464172	B-52F	Retired	AFSWC	27-Jun-72	Arrived at MASDC from the AFSWC, Kirtland AFB, NM. Used for various nuclear weapons tests. Declared excess and scrapped in Dec-94.
57-6468	464173	B-52G JB-52G B-52G	Display	320th BMW	10-Jul-89	First B-52G rolled out 23-Jul-58. First flight 31-Aug-58. To ARDC at Wright-Patterson AFB, OH on 1-Nov-58 for testing. Redesignated JB-52G on 13-Nov-58. To Boeing on 28-Feb-59 for testing. To Ramey AFB, PR on 22-Sep-61 and redesignated B-52G. Displayed at Zorinsky Memorial Air Park, Offutt AFB, NE (STRATCOM Gate).
57-6469	464174	B-52G GB-52G	Retired	43rd BMW	30-Nov-89	To Boeing on 7-Nov-58 for testing. To AFFTC at Edwards AFB, CA on 15-Mar-59 for ARDC testing. To Boeing 3-Aug-59 for testing. To Sheppard AFB, TX and used as ground trainer. Scrapped at Sheppard in 1994.
57-6470	464175	B-52G JB-52G B-52G	Retired	93rd BMW	23-Oct-90	To Boeing on 17-Dec-58 for testing. To AFSC at Eglin AFB, FL on 5-Jun-62 for test. Redesignated JB-52G on 19-Apr-63. To Ramey AFB, PR on 1-Nov-63 and redesignated B-52G. Declared excess and scrapped.
57-6471	464176	B-52G	Retired	2nd BMW	29-Jul-92	To Boeing on 4-Dec-58 for TF-33 engine test. Sometimes referred to as YB-52H. Returned to B-52G on 25-Jan-63. AMARG current inventory as of 29-Apr-25.
57-6472	464177	B-52G JB-52G B-52G	Retired	2nd BMW	8-Jul-92	To Boeing (Seattle) on 15-Nov-58 for test. To North American Aviation on 3-Apr-59 for AGM-28 test. Redesignated JB-52G. To Eglin AFB, FL on 21-Feb-62 and redesignated B-52G. AMARG current inventory as of 29-Apr-25.
57-6473	464178	B-52G JB-52G B-52G	Retired	93rd BMW	25-Feb-93	To Boeing on 12-Dec-58 for AGM-48 SKYBOLT test. Redesignated JB-52G and then returned to B-52G on 22-Jul-63. Forward fuselage to McConnell AFB, KS in 2007. Remainder in AMARG current inventory as of 29-Apr-25.
57-6474	464179	B-52G	Retired	379th BMW	15-Oct-91	To Boeing on 7-Jan-59 for ASG-21 DFCS test. Declared excess and scrapped in Apr-96.
57-6475	464180	B-52G	Retired	2nd BMW	20-Aug-91	To Boeing on 9-Jan-59 for various tests. One of seven B-52G aircraft that participated in Operation SENIOR SURPRISE, which was the first combat

Tail #	CN	Lineage	Status	Last Unit	Date	Comments
						launch of AGM-86C CALCMs. Declared excess and scrapped.
57-6476	464181	B-52G	Retired	93rd BMW	2-Nov-93	Declared excess and scrapped on 20-Oct-20.
57-6477	464182	B-52G JB-52G B-52G	Retired	93rd BMW	20-Sep-90	To ARDC Eglin AFB, FL on 15-Apr-59 for AGM-48 SKYBOLT test. Redesignated JB-52G and then returned to B-52G on 26-Feb-63. AMARG current inventory as of 29-Apr-25.
57-6478	464183	B-52G	Retired	320th BMW	27-Jul-89	Declared excess and scrapped on 29-Apr-97.
57-6479	464184	B-52G	Destroyed	92nd BMW	16-Oct-84	Crashed during a low-level training mission from Fairchild AFB, WA. Flew into Hunt's Mesa in Arizona. Hit the ridge and the aircraft broke apart. Total 2 fatalities/ 5 survivors.
57-6480	464185	B-52G	Retired	2nd BMW	5-Aug-92	AMARG current inventory as of 29-Apr-25.
57-6481	464186	B-52G	Destroyed	456th BMW	20-Dec-72	Combat loss during Operation LINEBACKER II from Andersen AFB, Guam. Near hit by two SA-2 SAMs over North Vietnam. Severe shrapnel damage. Lost all electrical power with first SAM and rudder/elevator control with second. Pilot struggled to keep the aircraft in the air for 40 minutes, but it finally crashed 16 km southwest of Nakhon Phanom, Thailand after the crew ejected. Total 0 fatalities/6 survivors.
57-6482	464187	B-52G	Destroyed	93rd BMW	16-Dec-82	Crashed on take-off during MITO training at Mather AFB, CA. Aircraft was operating out of Mather due to runway closure at home station of Castle AFB, CA when B-52G (59-2597) caught fire and burned on the runway. Aircraft was second in the MITO take-off. It took off too soon and was caught in the turbulence of the lead aircraft. The pilot throttled back and flamed out the engines with no ability to restart due to the low altitude. Total 9 fatalities/0 survivors (EWO ejected but landed in the fireball).
57-6483	464188	B-52G	Retired	2nd BMW	19-Sep-91	Declared excess and scrapped in Apr-97. "Rajin Cajun" nose art panel removed and displayed at the National Museum of the U.S. Air Force.
57-6484	464189	B-52G	Retired	42nd BMW	6-Jul-89	Sustained substantial damage on 28-Dec-84 when the front landing gear retracted during take-off roll at Loring AFB, ME. Emergency landing gear switch guards were improperly installed. Aircraft RTS as "Phoenix" due to the excessive work required to bring it back. Declared excess and scrapped on 29-Apr-97.
57-6485	464190	B-52G	Retired	93rd BMW	20-Dec-90	AMARG current inventory as of 29-Apr-25.
57-6486	464191	B-52G	Retired	93rd BMW	15-Aug-91	AMARG current inventory as of 29-Apr-25.
57-6487	464192	B-52G	Retired	2nd BMW	9-May-91	Declared excess and scrapped.
57-6488	464193	B-52G	Retired	42nd BMW	29-Jul-93	AMARG current inventory as of 29-Apr-25.
57-6489	464194	B-52G	Retired	93rd BMW	20-Mar-90	Declared excess and scrapped. "Express Delivery" nose art panel removed and displayed at the National Museum of the U.S. Air Force.
57-6490	464195	B-52G	Retired	2nd BMW	1-Dec-92	AMARG current inventory as of 29-Apr-25.
57-6491	464196	B-52G	Retired	2nd BMW	9-Aug-90	AMARG current inventory as of 29-Apr-25.
57-6492	464197	B-52G	Retired	379th BMW	15-Dec-92	AMARG current inventory as of 29-Apr-25.

Tail #	CN	Lineage	Status	Last Unit	Date	Comments
57-6493	464198	B-52G	Destroyed	68th BMW	3-Sep-75	Crashed due to structural failure near Aiken, SC during a flight from Seymour-Johnson AFB. The crew noticed a fuel leak from the right outboard tank and experienced handling difficulty from fuel imbalance. The right wing broke between engine nacelles after about five minutes, and a separated piece of the wing then sheared off the tail. The aircraft inverted and broke apart over a 10-mile area. Total 3 fatalities/4 survivors.
57-6494	464199	B-52G	Destroyed	72nd BMW	5-Jul-67	Crashed after take-off at Ramey AFB, Puerto Rico when the life raft behind the copilot seat inflated and forced him into the control yoke. The aircraft nosed over and dove into the ocean. Total 4 fatalities/3 survivors.
57-6495	464200	B-52G	Retired	2nd BMW	26-Aug-92	AMARG current inventory as of 29-Apr-25.
57-6496	464201	B-52G	Destroyed	456th BMW	20-Dec-72	Combat loss over North Vietnam during Operation LINEBACKER II from Andersen AFB, Guam after being hit by a SA-2 SAM. Another SAM had just missed the aircraft. Total 2 fatalities/4 survivors (all became POWs until Mar-73).
57-6497	464202	B-52G	Retired	93rd BMW	20-Jan-94	Declared excess and scrapped on 29-Apr-99.
57-6498	464203	B-52G	Retired	2nd BMW	3-Nov-92	Declared excess and scrapped. "Ace in the Hole" nose art panel removed and displayed at the National Museum of the U.S. Air Force.
57-6499	464204	B-52G	Retired	93rd BMW	7-Aug-90	Declared excess and scrapped.
57-6500	464205	B-52G	Retired	93rd BMW	11-May-89	Declared excess and scrapped on 1-Dec-20.
57-6501	464206	B-52G	Retired	416th BMW	3-Oct-91	Declared excess and scrapped on 17-Nov-99.
57-6502	464207	B-52G	Retired	42nd BMW	15-Feb-90	Declared excess and scrapped on 29-Apr-97.
57-6503	464208	B-52G	Retired	2nd BMW	19-Aug-92	AMARG current inventory as of 29-Apr-25.
57-6504	464209	B-52G	Retired	2nd BMW	2-Jul-91	Declared excess and scrapped on 18-Jan-01. "Snake Eyes" nose art panel removed and displayed at National Museum of the U.S. Air Force.
57-6505	464210	B-52G	Retired	93rd BMW	27-Aug-90	Declared excess and scrapped on 18-Jan-01. "Ace in the Hole" nose art panel removed and displayed at National Museum of the U.S. Air Force.
57-6506	464211	B-52G	Retired	43rd BMW	8-Mar-90	Declared excess and scrapped on 29-Apr-97.
57-6507	464212	B-52G	Destroyed	319th BMW	27-Jan-83	Exploded and burned on the ground during maintenance at Grand Forks AFB, ND. The fuel transfer valve was popping circuit breakers. The maintenance crew continually reset the breaker during troubleshooting when the valve motor ignited fuel vapors. The maintenance crew was trapped inside and perished. The aircraft completely burned on the ramp leaving only the skeleton. Total 5 fatalities/0 survivors (the crew had also reset the breaker several times in flight which could have led to inflight explosion).
57-6508	464213	B-52G	Retired	2nd BMW	22-Jul-92	AMARG current inventory as of 29-Apr-25.
57-6509	464214	B-52G	Display	2nd BMW	1-Sep-92	Displayed at Barksdale Global Power Museum, Barksdale AFB, LA, Named "Five O Nine II".
57-6510	464215	B-52G	Retired	93rd BMW	11-Oct-90	AMARG current inventory as of 29-Apr-25.

B-52 STRATOFORTRESS: THE IRON FIST OF STRATEGIC AIR COMMAND

Tail #	CN	Lineage	Status	Last Unit	Date	Comments
57-6511	464216	B-52G	Retired	2nd BMW	6-Oct-92	AMARG current inventory as of 29-Apr-25.
57-6512	464217	B-52G	Retired	93rd BMW	25-Apr-91	AMARG current inventory as of 29-Apr-25.
57-6513	464218	B-52G	Retired	93rd BMW	7-Aug-89	Declared excess and scrapped.
57-6514	464219	B-52G	Retired	93rd BMW	18-Oct-90	AMARG current inventory as of 29-Apr-25.
57-6515	464220	B-52G	Retired	2nd BMW	30-Sep-92	AMARG current inventory as of 29-Apr-25.
57-6516	464221	B-52G	Retired	416th BMW	8-Oct-91	AMARG current inventory as of 29-Apr-25.
57-6517	464222	B-52G	Retired	2nd BMW	13-Jul-90	Declared excess and scrapped.
57-6518	464223	B-52G	Retired	2nd BMW	23-Apr-91	Declared excess and scrapped on 18-Jan-01.
57-6519	464224	B-52G	Retired	6510th TW	11-Dec-90	To AMARC on 11-Dec-90 from 6510th Test Wing, Edwards AFB, CA. AMARG current inventory as of 29-Apr-25.
57-6520	464225	B-52G	Retired	366th Wing	27-Jan-94	AMARG current inventory as of 29-Apr-25.
58-0158	464226	B-52G	Display	2nd BMW	31-Jul-91	Displayed at Fairchild AFB, WA. Removed in 1995 and scrapped.
58-0159	464227	B-52G JB-52G B-52G	Retired	379th BMW	10-Oct-91	To ARDC Eglin AFB, FL on 8-Oct-59 and was first the B-52G with AGM-28. Redesignated JB-52G on 1-Dec-59 and reverted to B-52G on 2-Dec-60. AMARG current inventory as of 29-Apr-25.
58-0160	464228	B-52G	Retired	2nd BMW	19-Nov-92	AMARG current inventory as of 29-Apr-25.
58-0161	464229	B-52G	Destroyed	19th BMW	11-Apr-83	Crashed enroute to RED FLAG mission 20 miles north of St George, UT. Inexperienced navigator was on board replacing the original more experienced navigator. Aircraft deviated from course and had to be redirected by AWACS twice. Crashed into the 7,050-foot Square Top Mountain with nose up impact at 6,800 feet. Total 7 fatalities/0 survivors (Assistant DCM who was hitching a ride to his new duty station was killed).
58-0162	464230	B-52G	Retired	93rd BMW	16-Apr-91	Declared excess and scrapped on 9-Apr-96.
58-0163	464231	B-52G	Retired	93rd BMW	14-Apr-94	Declared excess and scrapped on 3-Sep-20.
58-0164	464232	B-52G	Retired	379th BMW	5-Nov-92	Declared excess and scrapped on 16-Jun-14. Nose and wing section to Boeing Avionics Antenna Laboratory, St Charles County Airport, MO.
58-0165	464233	B-52G	Retired	379th BMW	10-Aug-92	AMARG current inventory as of 29-Apr-25.
58-0166	464234	B-52G	Retired	2nd BMW	21-Dec-92	AMARG current inventory as of 29-Apr-25.
58-0167	464235	B-52G	Retired	93rd BMW	28-May-91	AMARG current inventory as of 29-Apr-25.
58-0168	464236	B-52G	Retired	379th BMW	22-Oct-91	Declared excess and scrapped.
58-0169	464237	B-52G	Destroyed	320th BMW	21-Dec-72	Combat loss over North Vietnam during Operation LINEBACKER II from Andersen AFB, Guam after being hit by a SA-2 SAM. Total 5 fatalities/1 survivor (became POW until Mar-73).
58-0170	464238	B-52G	Retired	2nd BMW	10-Nov-92	Declared excess and scrapped on 29-Apr-99.
58-0171	464239	B-52G	Retired	2nd BMW	16-Aug-90	AMARG current inventory as of 29-Apr-25.
58-0172	464240	B-52G	Retired	43rd BMW	30-Jun-89	Declared excess and scrapped.
58-0173	464241	B-52G	Retired	2nd BMW	5-Aug-92	Declared excess and scrapped in Sep-95. "Let's Make a Deal" nose art panel removed and displayed at National Museum of the U.S. Air Force.
58-0174	464242	B-52G	Destroyed	456th BMW	8-Feb-74	Crashed during a night training mission at Beale AFB, CA. Multiple engine failure on one wing caused the aircraft to veer off the runway on take-

Tail #	CN	Lineage	Status	Last Unit	Date	Comments
						off, skid 1,500 feet and overturn, then explode four times and catch fire. Possible fuel valve failure (had been worked on previously). The aircraft had already flown for approximately two hours. It landed to take on another crew member (Capt. Michael Forster who was to give a proficiency check ride to Capt. Paul Baldy). Total 8 fatalities/ 0 survivors (Capt. Baldy was thrown from the aircraft and initially survived but died in the hospital days later).
58-0175	464243	B-52G	Retired	379th BMW	16-Oct-91	AMARG current inventory as of 29-Apr-25.
58-0176	464244	B-52G	Retired	379th BMW	17-Nov-92	Declared excess and scrapped on 29-Apr-99.
58-0177	464245	B-52G	Retired	2nd BMW	5-Sep-91	One of seven B-52G aircraft that participated in Operation SENIOR SURPRISE, which was the first combat launch of AGM-86C CALCMs. Declared excess and scrapped.
58-0178	464246	B-52G	Retired	93rd BMW	18-Dec-90	Involved in ground accident while assigned to the 320th BMW at Mather AFB, CA on 10-Mar-83. Improperly vented wing tank during refueling caused pressure buildup and blew off the wing. The wing was replaced with one from another B-52G (59-2574) and the aircraft was RTS. AMARG current inventory as of 29-Apr-25.
58-0179	464247	B-52G	Retired	2nd BMW	10-Dec-92	AMARG current inventory as of 29-Apr-25.
58-0180	464248	B-52G	Destroyed	72nd BMW	1-Feb-60	First B-52G to be lost. Crashed during touch and goes at Ramey AFB, Puerto Rico due to incorrect trim setting. Total 7 fatalities/0 survivors.
58-0181	464249	B-52G	Retired	2nd BMW	23-Sep-92	Declared excess and scrapped on 17-Nov-99.
58-0182	464250	B-52G JB-52G B-52G	Retired	379th BMW	10-Jun-92	To Boeing on 1-Aug-59 for test. To ARDC Eglin AFB, FL on 22-Aug-61 for test. Redesignated JB-52G during test and later reverted to B-52G. To Edwards AFB, CA from 2-Jun-72 to 26-Jun-75 for testing. To Boeing from 28-Oct-75 to 12-Jul-77 for AGM-86A ALCM test. Painted gloss white overall during ALCM test period. "What's Up Doc" nose art panel removed and displayed at the National Museum of the U.S. Air Force. AMARG current inventory as of 29-Apr-25.
58-0183	464251	B-52G	Display	2nd BMW	7-Sep-91	One of seven B-52G aircraft that participated in Operation SENIOR SURPRISE, which was the first combat launch of AGM-86C CALCMs. Arrived AMARC on 9-Jul-91 from Barksdale AFB, LA. Departed to Pima Air & Space Museum, Tucson, AZ on 7-Sep-91 for display.
58-0184	464252	B-52G	Retired	2nd BMW	26-Sep-91	AMARG current inventory as of 29-Apr-25.
58-0185	464253	B-52G	Display	2nd BMW	16-Jul-91	One of seven B-52G aircraft that participated in Operation SENIOR SURPRISE, which was the first combat launch of AGM-86C CALCMs. Displayed at the Air Force Armament Museum, Eglin AFB, FL.
58-0186	464254	B-52G	Retired	43rd BMW	29-Mar-90	Declared excess and scrapped on 29-Apr-97.
58-0187	464255	B-52G	Destroyed	4241st SW	24-Jan-61	Crashed during landing at Seymour-Johnson AFB, NC. Aircraft experienced structural failure and fuel

Tail #	CN	Lineage	Status	Last Unit	Date	Comments
						leak of the right wing during airborne alert (Operation CHROME DOME). The wing failed when flaps were engaged during landing (Resulted in implementation of ECP 1050 across the fleet). Two nuclear weapons broke loose. One parachuted safely to the ground and one sank to at least 40 feet in a marsh, and one was never recovered. Total 3 fatalities/5 survivors.
58-0188	464256	B-52G	Destroyed	380th SW	21-Jan-68	Crashed near Thule AB, Greenland during an airborne alert (Operation CHROME DOME) mission. Faulty heater caused electrical fire and knocked out power. The crew ejected and the aircraft crashed on the sea ice. Four nuclear weapons detonated (non-nuclear) and caused widespread contamination. Operation CRESTED ICE launched to clean up contaminated ice and snow. Total 1 fatality/6 survivors.
58-0189	464257	B-52G	Retired	2nd BMW	1-Feb-90	Declared excess and scrapped on 17-Nov-99.
58-0190	464258	B-52G	Destroyed	2nd BMW	24-Jul-89	Destroyed by fire during refueling at Kelly AFB, TX. Vent plug was left installed during previous shift and no red streamer was affixed. Pressure built up during refueling and fuel burst out of the aircraft due to structural failure. The fuel ignited from static charge. Total 1 fatality (trapped in the cockpit and tried to escape through ejection hatch but was burned to death).
58-0191	464259	B-52G	Display	93rd BMW	11-Jul-91	Displayed at Hill Aerospace Museum, Hill AFB, UT. Named "Bearin' Arms".
58-0192	464260	B-52G	Retired	93rd BMW	10-Feb-94	AMARG current inventory as of 29-Apr-25.
58-0193	464261	B-52G	Retired	2nd BMW	8-Dec-92	AMARG current inventory as of 29-Apr-25.
58-0194	464262	B-52G	Retired	2nd BMW	24-Oct-91	AMARG current inventory as of 29-Apr-25.
58-0195	464263	B-52G	Retired	42nd BMW	18-Nov-93	AMARG current inventory as of 29-Apr-25.
58-0196	464264	B-52G	Destroyed	4241st SW	15-Oct-61	Crashed in the Atlantic Ocean off the coast of Newfoundland during a mock attack exercise SKY SHIELD II flight from Seymour-Johnson AFB, SC. Cause undetermined. Total 8 fatalities/0 survivors.
58-0197	464265	B-52G	Retired	42nd BMW	16-Nov-93	AMARG current inventory as of 29-Apr-25.
58-0198	464266	B-52G	Destroyed	456th BMW	20-Dec-72	Combat loss over North Vietnam during Operation LINEBACKER II from Andersen AFB, Guam after being hit by a SA-2 SAM. Total 3 fatalities/3 survivors (all became POWs until Mar-73).
58-0199	464267	B-52G	Retired	93rd BMW	4-Jun-91	"Specter" nose art panel removed and displayed at the National Museum of the U.S. Air Force. AMARG current inventory as of 29-Apr-25.
58-0200	464268	B-52G GB-52G	Display	379th BMW	26-Jul-91	To Sheppard AFB, TX and used as ground trainer. Displayed at Sheppard AFB Airpark. Replaced B-52D (56-0589).
58-0201	464269	B-52G	Destroyed	19th BMW	18-Dec-72	Combat loss over North Vietnam during Operation LINEBACKER II from Andersen AFB, Guam after being hit by a SA-2 SAM. First B-52 loss of LINEBACKER II. Total 3 fatalities/3 survivors (all became POWs until Mar-73).

B-52 Stratofortress: The Iron Fist of Strategic Air Command

Tail #	CN	Lineage	Status	Last Unit	Date	Comments
58-0202	464270	B-52G	Retired	42nd BMW	10-Mar-94	AMARG current inventory as of 29-Apr-25.
58-0203	464271	B-52G	Retired	366th Wing	3-Mar-94	AMARG current inventory as of 29-Apr-25.
58-0204	464272	B-52G	Retired	379th BMW	24-Sep-91	Used as test bed for the fly-off between the Boeing ALCM and the Tomahawk cruise missiles, as well as the Phase VI ECM testing. AMARG current inventory as of 29-Apr-25.
58-0205	464273	B-52G	Retired	2nd BMW	15-Nov-90	Declared excess and scrapped.
58-0206	464274	B-52G	Retired	42nd BMW	1-Feb-94	AMARG current inventory as of 29-Apr-25.
58-0207	464275	B-52G	Retired	93rd BMW	25-Jul-91	Declared excess and scrapped.
58-0208	464276	B-52G	Destroyed	42nd BMW	20-Jul-70	Destroyed by fire on the ground at Loring AFB, ME. Cause unknown. Total 0 fatalities.
58-0209	464277	B-52G	Destroyed	19th BMW	19-Aug-80	Destroyed by fire on the ground at Robins AFB, GA. Cause unknown. Total 0 fatalities.
58-0210	464278	B-52G	Retired	93rd BMW	30-Mar-93	AMARG current inventory as of 29-Apr-25.
58-0211	464279	B-52G	Retired	2nd BMW	2-Sep-92	AMARG current inventory as of 29-Apr-25.
58-0212	464280	B-52G	Retired	366th Wing	8-Feb-94	AMARG current inventory as of 29-Apr-25.
58-0213	464281	B-52G	Retired	93rd BMW	23-Nov-93	AMARG current inventory as of 29-Apr-25.
58-0214	464282	B-52G	Retired	93rd BMW	12-Apr-94	AMARG current inventory as of 29-Apr-25.
58-0215	464283	B-52G	Destroyed	42nd BMW	4-Sep-69	Crashed on take-off at Loring AFB, ME during ORI MITO due to possible power loss or water ejection system failure. Engine start problems were encountered before take-off, but the pilot managed to get it running and received clearance. The aircraft failed to gain altitude and crashed 2-3 miles from the end of the runway. Total 7 fatalities/0 survivors (nav and radar nav ejected but their chutes failed to open).
58-0216	464284	B-52G	Retired	42nd BMW	8-Mar-94	AMARG current inventory as of 29-Apr-25.
58-0217	464285	B-52G	Retired	379th BMW	15-Nov-91	AMARG current inventory as of 29-Apr-25.
58-0218	464286	B-52G	Retired	366th Wing	18-Jan-94	AMARG current inventory as of 29-Apr-25.
58-0219	464287	B-52G	Destroyed	93rd BMW	11-Feb-88	Overran the runway and crashed after aborted take-off at Castle AFB, CA. Total 0 fatalities/6 survivors.
58-0220	464288	B-52G	Retired	93rd BMW	5-Jul-90	Declared excess and scrapped on 29-Apr-97.
58-0221	464289	B-52G	Retired	93rd BMW	21-Oct-93	AMARG current inventory as of 29-Apr-25.
58-0222	464290	B-52G	Retired	2nd BMW	12-Aug-92	AMARG current inventory as of 29-Apr-25.
58-0223	464291	B-52G	Retired	93rd BMW	31-Jul-90	Declared excess and scrapped.
58-0224	464292	B-52G	Retired	43rd BMW	12-Feb-90	To Edwards AFB, CA from 14-Jan to 26-Aug-60 for test. AMARG current inventory 29-Apr-25.
58-0225	464293	B-52G	Display	93rd BMW	23-May-91	Displayed at Mohawk Valley B-52 Memorial, Rome, NY (Mohawk Valley).
58-0226	464294	B-52G	Retired	42nd BMW	19-Oct-93	AMARG current inventory as of 29-Apr-25.
58-0227	464295	B-52G	Retired	2nd BMW	15-Jul-92	AMARG current inventory as of 29-Apr-25.
58-0228	464296	B-52G	Destroyed	2nd BMW	18-Nov-66	Crashed south of Stone Lake, WI during night low-level mission from Barksdale AFB, LA using TA radar. Flew too low while calibrating radar and clipped the trees causing the crash. SAC later stopped nighttime TA flights. Total 9 fatalities/0 survivors.
58-0229	464297	B-52G	Retired	2nd BMW	12-Nov-92	Declared excess and scrapped on 5-Oct-20.
58-0230	464298	B-52G	Retired	42nd BMW	6-May-93	Declared excess and scrapped on 22-Sep-20.
58-0231	464299	B-52G	Retired	379th BMW	3-Dec-92	AMARG current inventory as of 29-Apr-25.
58-0232	464300	B-52G	Display	42nd BMW	3-Apr-90	Declared excess and scrapped. Nose section used

Tail #	CN	Lineage	Status	Last Unit	Date	Comments
						for ejection seat training at Randolph AFB, TX and then displayed at Hangar 25 Museum, Big Spring, TX.
58-0233	464301	B-52G	Retired	93rd BMW	17-Feb-94	AMARG current inventory as of 29-Apr-25.
58-0234	464302	B-52G	Destroyed	42nd BMW	6-Aug-91	Wing crack discovered during deployment to Andersen AFB, Guam. Parts of the tail section were used to repair another B-52G (58-0248) which was inadvertently hit by an AGM-88 HARM during Operation DESERT STORM. Scrapped in place at Andersen on 23-Jan-92.
58-0235	464303	B-52G	Retired	412th TW	29-Mar-94	Last unit 412th Test Wing (TW), Edwards AFB, CA. AMARG current inventory as of 29-Apr-25.
58-0236	464304	B-52G	Retired	2nd BMW	13-Oct-92	Declared excess and scrapped on 29-Apr-99.
58-0237	464305	B-52G	Retired	379th BMW	29-Oct-91	AMARG current inventory as of 29-Apr-25.
58-0238	464306	B-52G	Retired	2nd BMW	22-Aug-91	One of seven B-52G aircraft that participated in Operation SENIOR SURPRISE, which was the first combat launch of AGM-86C CALCMs. AMARG current inventory as of 29-Apr-25.
58-0239	464307	B-52G	Retired	2nd BMW	9-Sep-92	AMARG current inventory as of 29-Apr-25.
58-0240	464308	B-52G	Retired	93rd BMW	3-May-94	AMARG current inventory as of 29-Apr-25.
58-0241	464309	B-52G	Retired	42nd BMW	2-May-91	AMARG current inventory as of 29-Apr-25.
58-0242	464310	B-52G	Retired	366th Wing	22-Feb-94	AMARG current inventory as of 29-Apr-25.
58-0243	464311	B-52G	Retired	93rd BMW	11-Jun-91	Declared excess and scrapped on 18-Jan-01. "Brute Force" nose art panel removed and displayed at National Museum of the U.S. Air Force.
58-0244	464312	B-52G	Retired	2nd BMW	29-Oct-92	AMARG current inventory as of 29-Apr-25.
58-0245	464313	B-52G	Retired	2nd BMW	20-Oct-92	Declared excess and scrapped on 11-Sep-20.
58-0246	464314	B-52G	Destroyed	97th BMW	18-Dec-72	Combat loss during Operation LINEBACKER II mission from Andersen AFB, Guam after being hit by a SA-2 SAM over North Vietnam. Crew ejected after the aircraft was over Thai airspace. Crashed 30 km southwest of Nam Phong AFB, Thailand. Total 0 fatalities/7 survivors.
58-0247	464315	B-52G	Retired	379th BMW	5-Nov-91	AMARG current inventory as of 29-Apr-25.
58-0248	464316	B-52G	Retired	93rd BMW	25-Jan-94	Inadvertently hit by an AGM-88 HARM fired from an F-4G during Operation DESERT STORM. Aircraft flew to Andersen AFB, Guam where it was repaired and RTS with 93rd BMW at Castle AFB, CA as "In HARM's Way". AMARG current inventory as of 29-Apr-25.
58-0249	464317	B-52G	Retired	379th BMW	10-Sep-91	Declared excess and scrapped on 25-Nov-20. "Urban Renewal" nose art panel displayed at the National Museum of the U.S. Air Force.
58-0250	464318	B-52G	Retired	42nd BMW	7-Oct-93	Declared excess and scrapped on 20-Oct-20. "Screamin' Eagle" nose art panel displayed at the National Museum of the U.S. Air Force.
58-0251	464319	B-52G	Retired	43rd BMW	27-Feb-90	Declared excess and scrapped on 29-Apr-97.
58-0252	464320	B-52G	Retired	2nd BMW	18-Apr-91	Declared excess and scrapped.
58-0253	464321	B-52G	Retired	42nd BMW	4-Nov-93	AMARG current inventory as of 29-Apr-25.
58-0254	464322	B-52G	Retired	93rd BMW	4-Dec-90	Declared excess and scrapped on 29-Apr-97.

Tail #	CN	Lineage	Status	Last Unit	Date	Comments
						"Damage Inc" nose art panel removed and displayed at National Museum of the U.S. Air Force.
58-0255	464323	B-52G	Retired	42nd BMW	28-Oct-93	AMARG current inventory as of 29-Apr-25.
58-0256	464324	B-52G	Destroyed	68th BMW	17-Jan-66	Crashed during airborne alert mission (Operation CHROME DOME) from Seymour-Johnson AFB, NC. Collided with a KC-135 (61-0273) during refueling near Palomares, Spain. Four Mk-28 nuclear weapons were ejected on impact with the ground. Three were found on land and two exploded (non-nuclear). One fell into the Mediterranean Sea and was found 2.5 months later. Total 3 fatalities/ 4 survivors. Four tanker crew killed.
58-0257	464325	B-52G	Retired	42nd BMW	14-Oct-93	Declared excess and scrapped on 20-Oct-20.
58-0258	464326	B-52G	Retired	93rd BMW	9-Nov-93	Declared excess and scrapped on 28-Sep-20.
59-2564	464327	B-52G	Retired	2nd BMW	8-Aug-91	One of seven B-52G aircraft that participated in Operation SENIOR SURPRISE, which was the first combat launch of AGM-86C CALCMs. Declared excess and scrapped on 17-Nov-99.
59-2565	464328	B-52G	Retired	93rd BMW	12-Oct-93	AMARG current inventory as of 29-Apr-25.
59-2566	464329	B-52G	Retired	2nd BMW	16-Sep-92	AMARG current inventory as of 29-Apr-25.
59-2567	464330	B-52G	Retired	379th BMW	22-Oct-92	AMARG current inventory as of 29-Apr-25.
59-2568	464331	B-52G	Retired	379th BMW	23-Nov-92	AMARG current inventory as of 29-Apr-25.
59-2569	464332	B-52G	Retired	366th Wing	15-Feb-94	AMARG current inventory as of 29-Apr-25.
59-2570	464333	B-52G	Retired	366th Wing	24-Jan-94	Declared excess and scrapped on 12-Nov-20.
59-2571	464334	B-52G	Retired	2nd BMW	30-Apr-91	Declared excess and scrapped on 29-Apr-97.
59-2572	464335	B-52G	Retired	366th Wing	11-Jan-94	AMARG current inventory as of 29-Apr-25.
59-2573	464336	B-52G	Retired	42nd BMW	26-Oct-93	Declared excess and scrapped on 18-Jun-14.
59-2574	464337	B-52G	Destroyed	416th BMW	8-May-72	Crashed during landing at Griffiss AFB, NY during heavy rain at night. One engine had failed in flight and was shut down. It was restarted for landing and began to run away without the pilots noticing. The brakes were unable to slow the aircraft, and the drag chute did not deploy. The aircraft overshot the runway and broke apart just aft of the crew compartment. It came to rest with the engine still running and the AGM-28 missiles it was carrying sheared off the pylons. Total 0 fatalities/6 survivors. The aircraft was later used for ECM testing at Rome Research Center and then as static display. Its wing was used to replace one on another B-52G (58-0178) which was damaged due to improper fuel tank venting at Mather AFB, CA in Mar-83. It was finally scrapped in Jul-08.
59-2575	464338	B-52G	Retired	93rd BMW	6-Aug-91	Declared excess and scrapped on 17-Nov-99.
59-2576	464339	B-52G	Destroyed	4038th SW	30-Mar-61	Crashed during refueling near Silver Hill, NC on a Radar-Scored Bombing (RSB) competition flight from Dow AFB, ME. Pilot bled off too much airspeed while approaching the tanker causing the aircraft to pitch up and stall. Total 6 fatalities/ 2 survivors.
59-2577	464340	B-52G	Display	2nd BMW	23-Jul-91	Displayed at Heritage Center, Grand Forks AFB, ND.

Tail #	CN	Lineage	Status	Last Unit	Date	Comments
59-2578	464341	B-52G GB-52G	Retired	42nd BMW	13-Mar-90	To Sheppard AFB, TX and used as a ground trainer. Declared excess and scrapped on 22-Aug-10.
59-2579	464342	B-52G	Display	379th BMW	12-Nov-91	Declared excess on 17-Nov-99 and scrapped. Nose section initially displayed at Southern Utah Air Museum, Washington, UT. Displayed at Tillamook Air Museum, Tillamook, OR since 1-Apr-23.
59-2580	464343	B-52G	Retired	379th BMW	5-Jul-92	Declared excess and scrapped on 17-Nov-99.
59-2581	464344	B-52G	Retired	379th BMW	8-Oct-92	AMARG current inventory as of 29-Apr-25.
59-2582	464345	B-52G	Retired	2nd BMW	27-Aug-91	One of seven B-52G aircraft that participated in Operation SENIOR SURPRISE, which was the first combat launch of AGM-86C CALCMs. "Grim Reaper II" nose art panel removed and displayed at National Museum of the U.S. Air Force. Declared excess and scrapped on 17-Nov-99.
59-2583	464346	B-52G	Retired	379th BMW	1-Oct-92	AMARG current inventory as of 29-Apr-25.
59-2584	464347	B-52G	Display	93rd BMW	23-Sep-91	Displayed at Museum of Flight, Boeing Field, WA.
59-2585	464348	B-52G	Retired	320th BMW	15-Apr-93	Declared excess and scrapped.
59-2586	464349	B-52G	Retired	412th TW	22-Apr-94	Arrived from the 412th Test Wing, Edwards AFB, CA. AMARG current inventory as of 29-Apr-25.
59-2587	464350	B-52G	Retired	93rd BMW	4-Jan-90	Declared excess and scrapped Mar-92.
59-2588	464351	B-52G	Retired	93rd BMW	5-Apr-94	AMARG current inventory as of 29-Apr-25.
59-2589	464352	B-52G	Retired	379th BMW	17-Jun-92	Declared excess and scrapped on 29-Apr-99.
59-2590	464353	B-52G	Retired	379th BMW	13-Jul-92	Declared excess and scrapped on Dec-95.
59-2591	464354	B-52G	Retired	2nd BMW	19-Jun-92	Declared excess and scrapped on 17-Nov-99.
59-2592	464355	B-52G	Retired	43rd BMW	7-Feb-90	Declared excess and scrapped on 29-Apr-97.
59-2593	464356	B-52G	Destroyed	42nd BMW	3-Feb-91	Crashed in the Indian Ocean 13 miles north of Diego Garcia during Operation DESERT STORM due to electrical failure. Improper fuel management caused five engines to flame out. Total 3 fatalities/3 survivors.
59-2594	464357	B-52G	Retired	2nd BMW	15-Oct-92	Declared excess and scrapped on 17-Nov-99. "Memphis Belle III" nose art panel displayed at National Museum of the U.S. Air Force. Declared excess and scrapped on 17-Nov-99.
59-2595	464358	B-52G	Retired	93rd BMW	24-Feb-94	AMARG current inventory as of 29-Apr-25.
59-2596	464359	B-52G	Display	43rd BMW	21-Mar-90	Displayed at the Darwin Aviation Museum, Darwin, Australia. Named "Darwin's Pride".
59-2597	464360	B-52G	Destroyed	93rd BMW	29-Nov-82	Destroyed by fire on the runway after landing at Castle AFB, CA due to hydraulic fluid leaking onto hot brakes. The crew escaped before fuel explosion. Total 0 fatalities/10 survivors.
59-2598	464361	B-52G	Retired	93rd BMW	13-Jan-94	Declared excess and scrapped on 18-Nov-20.
59-2599	464362	B-52G	Retired	93rd BMW	15-Mar-94	AMARG current inventory as of 29-Apr-25.
59-2600	464363	B-52G	Destroyed	416th BMW	8-Jul-72	Crashed in the Pacific Ocean shortly after take-off in bad weather from Andersen AFB, Guam during Operation LINEBACKER I. The forward BNS radome was lost and severed the pitot-static airspeed source, and the aircraft became unstable. Total 1 fatality/5 survivors (radar nav ejected but did not survive descent).

B-52 STRATOFORTRESS: THE IRON FIST OF STRATEGIC AIR COMMAND

Tail #	CN	Lineage	Status	Last Unit	Date	Comments
59-2601	464364	B-52G	Display	379th BMW	3-Jun-92	Displayed at Tactical Air Command Memorial Park, Langley AFB, VA.
59-2602	464365	B-52G	Retired	2nd BMW	27-Oct-92	"Yankee Doodle II" nose art panel removed and displayed at National Museum of the U.S. Air Force. AMARG current inventory as of 29-Apr-25.
60-0001	464366	B-52H	Active	2nd BW		Current status/unit as of 2024. Nuclear role.
60-0002	464367	B-52H	Active	2nd BW		Current status/unit as of 2024. Nuclear role. To Boeing from 8-Mar-61 to 9-Jun-63 for autopilot, test.
60-0003	464368	B-52H JB-52H B-52H	Active	307th BW		Current status/unit as of 2024. Conventional role. To Boeing on 3-Mar-61 for test. To AFFTC Edwards AFB, CA on 4-Aug-61 for test. To Boeing on 20-Jul-62 for test. Redesignated JB-52H on 5-Oct-61 and then back to B-52H in Jul-62 after test.
60-0004	464369	B-52H JB-52H B-52H	Active	5th BW		Current status/unit as of 2024. Nuclear role. To Boeing from 3-Apr-61 to 11-Jan-73 for test. Redesignated JB-52H in Sep-67 and returned to B-52H in Sep-68.
60-0005	464370	B-52H JB-52H B-52H	Active	5th BW		Current status/unit as of 2024. Nuclear role. To Boeing on 3-Apr-61 for test. To AFSC Wright-Patterson AFB, OH on 25-Jul-61 for test. To Boeing on 23-Feb-62 for test. Used for ASQ-38 and ECM testing. Redesignated JB-52H on 2-Oct-61 and returned to B-52H in Feb-62.
60-0006	464371	B-52H JB-52H B-52H	Destroyed	17th BMW	30-May-74	To Boeing on 26-Mar-61 for test. To AFFTC Edwards AFB, CA on 1-Jul-61 for test. Redesignated JB-52H on 8-Sep-61. To Boeing on 15-Jun-62 for test and redesignated B-52H. Used as EVS test bed. Crashed on approach at Wright-Patterson AFB, OH due to rudder and elevator failure. Total 0 fatalities/7 survivors.
60-0007	464372	B-52H	Active	5th BW		Current status/unit as of 2024. Nuclear role. Credited with first combat use of the CBU-105 SFW on 2-Apr-03 during Operation IRAQI FREEDOM.
60-0008	464373	B-52H	Active	2nd BW		Current status/unit as of 2024. Nuclear role. 8AF flagship "Mighty Eighth". One of two B-52H aircraft that flew a 20,062-mile, 47.2-hour flight during the GLOBAL POWER 94-7 exercise from 1-2-Aug-94.
60-0009	464374	B-52H	Active	5th BW		Current status/unit as of 2024. Nuclear role. Experienced two engine failures in flight near RAF Mildenhall, UK. The pilot burned off fuel and returned to base safely. Total 0 fatalities.
60-0010	464375	B-52H	Storage	2nd BW	28-Aug-08	AMARG current inventory as of 29-Apr-25. Conventional role.
60-0011	464376	B-52H	Active	2nd BW		Current status/unit as of 2024. Conventional role.
60-0012	464377	B-52H	Active	5th BW		Current status/unit as of 2024. Nuclear role.
60-0013	464378	B-52H	Active	2nd BW		Current status/unit as of 2024. Nuclear role.
60-0014	464379	B-52H	Storage	2nd BW	11-Dec-08	AMARG current inventory as of 29-Apr-25. Conventional role. One of three aircraft launched from Andersen AFB, Guam on 3-Sep-96 loaded with CALCMs to hit targets in Southern Iraq during Op-

Tail #	CN	Lineage	Status	Last Unit	Date	Comments
						eration DESERT STRIKE making the second longest bombing mission at the time resulting award of the 1996 Mackay Trophy.
60-0015	464380	B-52H	Active	307th BW		Current status/unit as of 2024. Conventional role.
60-0016	464381	B-52H GB-52H	Retired	2nd BW	Unknown	Damaged beyond economical repair. Ground maintenance trainer at Barksdale AFB, LA. Tail used for 60-0001 which was damaged in a hangar accident.
60-0017	464382	B-52H	Active	5th BW		Current status/unit as of 2024. Nuclear role.
60-0018	464383	B-52H	Active	5th BW		Current status/unit as of 2024. Nuclear role.
60-0019	464384	B-52H	Storage	2nd BW	7-Aug-08	AMARG current inventory as of 29-Apr-25. Conventional role.
60-0020	464385	B-52H	Storage	2nd BW	4-Sep-08	AMARG current inventory as of 29-Apr-25. Conventional role. Assigned to 6510th TS at Edwards AFB, CA on 13-Feb-84 to support AGM-129 ACM testing and other black programs. To 5th BMW, Minot AFB, ND on 24-Aug-91.
60-0021	464386	B-52H	Active	2nd BW		Current status/unit as of 2024. Nuclear role.
60-0022	464387	B-52H	Active	2nd BW		Current status/unit as of 2024. Nuclear role.
60-0023	464388	B-52H	Active	5th BW		Current status/unit as of 2024. Conventional role.
60-0024	464389	B-52H	Active	2nd BW		Current status/unit as of 2024. Conventional role.
60-0025	464390	B-52H	Active	2nd BW		Current status/unit as of 2024. Conventional role. Spare aircraft launched from Andersen AFB, Guam on 3-Sep-96 loaded with CALCMs to hit targets in Southern Iraq during Operation DESERT STRIKE.
60-0026	464391	B-52H	Active	5th BW		Current status/unit as of 2024. Nuclear role.
60-0027	464392	B-52H	Destroyed	5th BMW	4-Oct-68	Crashed eight miles short of the runway during approach at Minot AFB, ND due to fuel mismanagement which caused flame-out of No. 1-4 engines. Total 4 fatalities/2 survivors.
60-0028	464393	B-52H	Active	2nd BW		Current status/unit as of 2024. Nuclear role.
60-0029	464394	B-52H	Active	5th BW		Current status/unit as of 2024. Nuclear role.
60-0030	464395	B-52H	Storage	2nd BW	21-Aug-08	AMARG current inventory as of 29-Apr-25. Conventional role.
60-0031	464396	B-52H	Active	307th BW		Current status/unit as of 2024. Nuclear role. Supports 49th Test and Evaluation Squadron to provide B-52 operational testing and nuclear weapons evaluation as part of the ACC 53rd Wing. Aircraft located at Barksdale AFB, LA. Sorties generated by the 307th BW to support 49th missions.
60-0032	464397	B-52H	Active	2nd BW		Current status/unit as of 2024. Nuclear role.
60-0033	464398	B-52H	Active	5th BW		Current status/unit as of 2024. Nuclear role.
60-0034	464399	B-52H	Active	5th BW		Current status/unit as of 2024. Conventional role. Arrived AMARG on 14-Aug-08 from the 5th BW. RTS at the 2nd BW 14-May-19.
60-0035	464400	B-52H	Active	307th BW		Current status/unit as of 2024. Conventional role.
60-0036	464401	B-52H	Active	412th TW		Current status/unit as of 2024. Nuclear role. Modified beginning in Aug-67 to carry D-21 reconnaissance drone under Project SENIOR BOWL. Cur-

Tail #	CN	Lineage	Status	Last Unit	Date	Comments
						rently assigned to AFTC, 412th TW, 419th TS at Edwards AFB, CA supporting developmental test and evaluation.
60-0037	464402	B-52H	Active	5th BW		Current status/unit as of 2024. Nuclear role.
60-0038	464403	B-52H	Active	307th BW		Current status/unit as of 2024. Conventional role.
60-0039	464404	B-52H	Destroyed	410th BMW	1-Apr-77	Crashed during approach during a storm at K.I. Sawyer AFB, MI due to pilot disorientation in the clouds. The aircraft was in steep descent when it exited the clouds although the pilot thought it was straight and level. Total 8 fatalities/0 survivors.
60-0040	464405	B-52H	Destroyed	410th BMW	6-Dec-88	Set world record for speed over a recognized course on 10-11-Jan-62 under Operation PERSIAN RUG. Crashed during take-off while doing touch and goes at K.I. Sawyer AFB, MI. Overheated fuel pump caused explosion and separated the tail section. The aircraft broke into three pieces after crashing (crew compartment, mid-body/wings, tail section) and crew section slid more than 3,000 feet down the runway. Total 0 fatalities/ 8 survivors (3 medically retired).
60-0041	464406	B-52H	Active	307th BW		Current status/unit as of 2024. Conventional role.
60-0042	464407	B-52H	Active	307th BW		Current status/unit as of 2024. Conventional role.
60-0043	464408	B-52H	Storage	2nd BW	2-Oct-08	AMARG current inventory as of 29-Apr-25. Conventional role.
60-0044	464409	B-52H	Active	5th BW		Current status/unit as of 2024. Nuclear role.
60-0045	464410	B-52H	Active	307th BW		Current status/unit as of 2024. Conventional role.
60-0046	464411	B-52H	Storage	2nd BW	23-Oct-08	AMARG current inventory as of 29-Apr-25. Conventional role.
60-0047	464412	B-52H	Destroyed	5th BW	19-May-16	Overran the runway after aborted take-off at Andersen AFB, Guam during routine training mission. Pilot observed bird activity and believed cockpit indications showed insufficient thrust for take-off. He tried to abort but the drag chute failed to deploy and the brakes failed to stop the aircraft. Total 0 fatalities/7 survivors.
60-0048	464413	B-52H	Active	2nd BW		Current status/unit as of 2024. Nuclear role.
60-0049	464414	B-52H	Storage	2nd BW	28-Jul-17	Aircraft received severe fire damage during oxygen servicing on 28-Jan-14 at Barksdale AFB, LA. It was determined beyond economical repair and replaced on 27-Sep-16 with B-52H (61-0007) "Ghost Rider", which was the first B-52 resurrected from AMARG and RTS. AMARG current inventory as of 29-Apr-25. Storage type 4000 "Instructional Fuselage Only".
60-0050	464415	B-52H	Active	412th TW		Current status/unit as of 2024. Nuclear role. To Boeing for OAS and ALCM modifications on 16-Jan-85. To and AFFTC Edwards AFB, CA on 1-Aug-85 for CSRL and other test. First AGM-136A Tacit Rainbow test from a B-52H on 10-Jan-89. First JDAM test on 30-Apr-97. Launch vehicle for first flight of the WAVERIDER unmanned scramjet on 26-May-

Tail #	CN	Lineage	Status	Last Unit	Date	Comments
						10. Currently assigned to AFTC, 412th TW, 419th TS at Edwards AFB, CA supporting developmental test and evaluation.
60-0051	464416	B-52H	Active	307th BW		Current status/unit as of 2024. Conventional role.
60-0052	464417	B-52H	Active	2nd BW		Current status/unit as of 2024. Conventional role.
60-0053	464418	B-52H	Destroyed	2nd BW	21-Jul-08	The aircraft assigned to Barksdale AFB, LA crashed into the Pacific Ocean about 25 miles northwest of Apra Harbor, Guam. The aircraft crashed 15 minutes before it was scheduled to appear in the Guam Liberation Day celebration. The aircraft was only partially recovered, and the accident investigation board concluded that parts designed to limit the horizontal stabilizer movement may have malfunctioned as the aircraft descended from 14,000 feet to 1,000 feet. Total 6 fatalities/ 0 survivors.
60-0054	464419	B-52H	Active	2nd BW		Current status/unit as of 2024. Conventional role. One of three aircraft launched from Andersen AFB, Guam on 3-Sep-96 loaded with CALCMs to hit targets in Southern Iraq during Operation DESERT STRIKE making the second longest bombing mission at the time resulting award of the 1996 Mackay Trophy.
60-0055	464420	B-52H	Active	5th BW		Current status/unit as of 2024. Nuclear role.
60-0056	464421	B-52H	Active	5th BW		Current status/unit as of 2024. Nuclear role.
60-0057	464422	B-52H	Active	307th BW		Current status/unit as of 2024. Conventional role. Flagship for 340th Weapons Squadron assigned to the 57th Wing, part of the Air Force Warfare center. Aircraft located at Barksdale AFB, LA and sorties are generated by the 307th BW to support 340th mission requirements.
60-0058	464423	B-52H	Active	2nd BW		Current status/unit as of 2024. Nuclear role.
60-0059	464424	B-52H	Active	2nd BW		Current status/unit as of 2024. Nuclear role. One of two B-52H aircraft that flew a 20,062-mile, 47.2-hour flight during the GLOBAL POWER 94-7 exercise from 1-2-Aug-94.
60-0060	464425	B-52H	Active	5th BW		Current status/unit as of 2024. Conventional role.
60-0061	464426	B-52H	Active	307th BW		Current status/unit as of 2024. Conventional role.
60-0062	464427	B-52H	Active	2nd BW		Current status/unit as of 2024. Nuclear role.
61-0001	464428	B-52H	Active	5th BW		Current status/unit as of 2024. Nuclear role. An engine fell off the aircraft 25 miles northwest of Minot AFB, ND on 4-Jan-17 due to first stage fan disk failure. Aircraft returned to base safely. Total 0 fatalities/5 survivors and the aircraft was RTS.
61-0002	464429	B-52H	Active	2nd BW		Current status/unit as of 2024. Nuclear role.
61-0003	464430	B-52H	Active	5th BW		Current status/unit as of 2024. Conventional role.
61-0004	464431	B-52H	Active	2nd BW		Current status/unit as of 2024. Conventional role.
61-0005	464432	B-52H	Active	5th BW		Current status/unit as of 2024. Nuclear role.
61-0006	464433	B-52H	Active	2nd BW		Current status/unit as of 2024. Nuclear role.
61-0007	464434	B-52H	Active	5th BW		Current status/unit as of 2024. Nuclear role. Arrived at AMARG on 13-Nov-08 from 5th BW. First B-52 ever resurrected from AMARG and returned to the

Tail #	CN	Lineage	Status	Last Unit	Date	Comments
						active strategic bomber fleet. It was flown from AMARG to 2nd BW, Barksdale AFB, LA on 13-Feb-15. It then entered depot maintenance at Tinker AFB, OK on 14-Dec-15 where it was overhauled and RTS with 5th BW at Minot AFB, ND as "Ghost Rider" on 27-Sep-16.
61-0008	464435	B-52H	Active	307th BW		Current status/unit as of 2024. Conventional role.
61-0009	464436	B-52H	Retired	Boeing	22-Jan-22	Arrived at AMARG on 25-Sep-08 from 2nd BW. To Boeing, Tinker AFB, OK on 22-Jan-22 to support CERP and radar upgrade integration testing.
61-0010	464437	B-52H	Active	307th BW		Current status/unit as of 2024. Nuclear role.
61-0011	464438	B-52H	Active	307th BW		Current status/unit as of 2024. Conventional role. Received minor damage during landing at Castle AFB, CA on 4-Feb-77. Total 0 fatalities/6 survivors and aircraft RTS.
61-0012	464439	B-52H	Active	2nd BW		Current status/unit as of 2024. Nuclear role.
61-0013	464440	B-52H	Active	2nd BW		Current status/unit as of 2024. Nuclear role.
61-0014	464441	B-52H	Active	307th BW		Current status/unit as of 2024. Conventional role. Supports 49th Test and Evaluation Squadron to provide B-52 operational testing and nuclear weapons evaluation as part of the ACC 53rd Wing. Aircraft located at Barksdale AFB, LA. Sorties generated by the 307th BW to support 49th missions.
61-0015	464442	B-52H	Active	2nd BW		Current status/unit as of 2024. Nuclear role.
61-0016	464443	B-52H	Active	2nd BW		Current status/unit as of 2024. Nuclear role.
61-0017	464444	B-52H	Active	307th BW		Current status/unit as of 2024. Conventional role.
61-0018	464445	B-52H	Active	5th BW		Current status/unit as of 2024. Conventional role.
61-0019	464446	B-52H	Active	412th TW		Current status/unit as of 2024. Nuclear role. Currently assigned to AFTC, 412th TW, 419th TS at Edwards AFB, CA supporting developmental test and evaluation.
61-0020	464447	B-52H	Active	2nd BW		Current status/unit as of 2024. Nuclear role.
61-0021	464448	B-52H	Active	307th BW		Current status/unit as of 2024. Conventional role. Modified to carry D-21 reconnaissance drone under Project SENIOR BOWL beginning 12-Dec-66. First B-52H to fly with an operational AN/AAQ-28 Litening II Targeting Pod and made the first B-52 combat drop of an LGB on 11-Apr-03 during Operation IRAQI FREEDOM.
61-0022	464449	B-52H GB-52H	Retired	2nd BW	9-Sep-09	Damaged beyond economical repair. To ground maintenance trainer at Sheppard AFB, TX. Vertical fin removed 17-Nov-09 to comply with SALT.
61-0023	464450	B-52H JB-52H B-52H	Storage	2nd BW	24-Jul-08	To Boeing from 18-May-62 to 20-Feb-67 for test. Redesignated JB-52H during test and later returned to B-52H. Boeing used the aircraft for turbulence tests. On 10-Jan-64, during flight test the entire vertical fin and rudder were ripped off and the aircraft was able to land safely at Blytheville AFB, AR. Total 0 fatalities. Aircraft RTS and remained active until stored at Tinker AFB, OK on 24-Jul-08. First B-52H retired. AMARG current inventory as of 29-Apr-25.

B-52 Stratofortress: The Iron Fist of Strategic Air Command

Tail #	CN	Lineage	Status	Last Unit	Date	Comments
						Conventional role.
61-0024	464451	B-52H	Storage	2nd BW	6-Jan-09	AMARG current inventory as of 29-Apr-25. Conventional role.
61-0025	464452	B-52H NB-52H GB-52H	Retired	NASA	9-May-08	Transferred to NASA at Edwards AFB, CA on 30-Jul-01 as replacement for NB-52B (52-0008). Re-designated NB-52H. Damaged beyond economical repair due to modifications. Authorized one-time flight on 9-May-08 to become ground maintenance trainer at Sheppard AFB, TX. Vertical fin removed to comply with SALT. Given modern white and blue NASA paint scheme.
61-0026	464453	B-52H	Destroyed	92nd BW	24-Jun-94	Crashed during airshow practice at Fairchild AFB, WA. Banked past 90 degrees during a go-around, stalled, clipped a power line and crashed. The pilot flew beyond operational limits and lost control. The aircraft had already completed the practice maneuvers but initiated the go-around due to a KC-135 still on the runway. Total 4 fatalities/ 0 survivors.
61-0027	464454	B-52H	Storage	5th BW	21-Jan-09	AMARG current inventory as of 29-Apr-25. Conventional role.
61-0028	464455	B-52H	Active	307th BW		Current status/unit as of 2024. Nuclear role. Supports 49th Test and Evaluation Squadron to provide B-52 operational testing and nuclear weapons evaluation as part of the ACC 53rd Wing. Aircraft located at Barksdale AFB, LA. Sorties generated by the 307th BW to support 49th missions.
61-0029	464456	B-52H	Active	307th BW		Current status/unit as of 2024. Conventional role.
61-0030	464457	B-52H	Destroyed	319th BMW	2-Nov-67	Crashed at Griffiss AFB, NY after missed approach during bad weather. Departed Westover AFB, MA and had indications of fire and overheat on engines No. 5-6. The pilot attempted to land at Griffiss. The aircraft stalled and rolled over from about 2,000 feet. First B-52H lost. Total 6 fatalities/2 survivors.
61-0031	464458	B-52H	Active	307th BW		Current status/unit as of 2024. Conventional role.
61-0032	464459	B-52H GB-52H	Active	5th BW		Damaged beyond economical repair. Ground maintenance trainer at Minot AFB, ND.
61-0033	464460	B-52H	Destroyed	5th BMW	14-Nov-75	Destroyed by fire and explosion during ground refueling at Minot AFB, ND due to internal boost pump failure and overheat in a wing tank. Total 2 fatalities.
61-0034	464461	B-52H	Active	5th BW		Current status/unit as of 2024. Nuclear role. The first B-52 to be completely powered by synthetic fuel when it made a 7-hour test flight from Edwards AFB, CA on 15-Dec-06.
61-0035	464462	B-52H	Active	5th BW		Current status/unit as of 2024. Nuclear role.
61-0036	464463	B-52H	Active	2nd BW		Current status/unit as of 2024. Conventional role.
61-0037	464464	B-52H	Destroyed	5th BMW	21-Jan-69	Crashed during take-off at Minot AFB, ND due to incorrect trim setting. Aircraft stalled after climbing about 200-300 feet. Total 6 fatalities/0 survivors.
61-0038	464465	B-52H	Active	307th BW		Current status/unit as of 2024. Conventional role.
61-0039	464466	B-52H	Active	5th BW		Current status/unit as of 2024. Nuclear role.
61-0040	464467	B-52H	Active	5th BW		Current status/unit as of 2024. Nuclear role.

Bibliography

Boyne, W. J. (1981). *Boeing B-52: A Documentary History.* London: Jane's Publishing Company Limited.

Callaway, L. (2009). *SAC Bomb Comp - History Chronology and Factoids.* Barksdale AFB, LA: 8AF History Office.

Correll, J. T. (2016, April). 1946: The Year After the War. *Air Force Magazine.*

Davis, L. (1992). *B-52 Stratofortress In Action.* Carrollton, TX: Squadron/Signal Publications, Inc.

Eickhoff, B. (1982, February 1). SAC Trains the Way It Would Fight. *Air Force Magazine.*

Elson, B. M. (1978, December). Bomb/Nav Changes Key to B-52 Update. *Combat Crew*, pp. 16-20.

Fehner, T. R. (2006). Atmospheric Nuclear Weapons Testing, 1951-1963. *Battlefield of the Cold War: The Navada Test Site (DOE/MA-0003).*

Ford, D. (1996, April). B-36: Bomber at the Crossroads. *Air and Space Magazine.*

Goebel, G. (2003). *B-52 At War.*

Head, W. P. (2002). *War From Above The Clouds.* Maxwell AFB, AL: Air University Press.

Hoage, C. (n.d.). *NB-52/B-52A Stratofortress.* Tucson, AZ: Pima Air & Space Museum.

Jenkins, D. R. (2001). *Magnesium Overcast: The Story of the Convair B-36.* North Branch, MN: Speciality Press.

Kaminski, T. (2012). Stratofortress Upgrades. *B-52 Stratofortress: 60 Remarkable Years*, 82-88.

Katz, K. P. (2007). *B-52G/H Stratoforttress In Action.* Carrollton, TX: Squadron/Signal Publications.

Kitfield, J. (2010, February). The Cruise Missile Question. *Air Force Magazine* .

Knaack, M. S. (1988). *Post-World War II Bombers 1945-1973.* Washington DC: Office of Air Force History.

Kohn, R. H. (1988). *Strategic Air Warfare.* Washington, D.C.: Office of Air Force History.

Kopp, D. C. (n.d.). Cruise Missiles Post World War II. *Defence Today.*

Leone, D. (2019, December 20). *Remembering the AGM-69 SRAM.* Retrieved from The Aviation Geek Club: https://theaviationgeekclub.com/remembering-the-agm-69-sram-sac-bombers-short-range-attack-missile/

Leone, D. (2022, February 22). *The Story of the B-52 Stratofortress that Carried the Flashback Test Vehicle.* Retrieved from The Aviation Geek Club: https://theaviationgeekclub.com/the-story-of-the-b-52-stratofortress-that-carried-the-flashback-test-vehicle-the-nuclear-bomb-bigger-than-the-soviet-tsar-bomba/

Mailes, Y. (2016). *IQ Builder: Operation Desert Storm B-52 Specific.* Offutt AFB, NE: Air Force Global Strike Command History Office.

Mandeles, D. M. (1998). *The Development of the B-52 and Jet Propulsion.* Maxwell AFB, AL: Air University Press.

McClary, D. C. (2012, April 2). *U.S. Air Force B-52 On A Low-Level Test Flight Crashes.* Retrieved from History Link: https://www.historylink.org/File/10063

Metz, G. C. (1978, May). R&D Corner. *Combat Crew*, pp. 26-27.

Mortimer, R. (2022). *Boeing B-52H Stratofortress.* Stamford, UK: Key Aero.

Parks, W. H. (1983). Linebacker and the Law of War. *Air University Review*, 2-30.

Peacock, L. (1987). *B-52 Stratofortress.* London: Osprey Publishing Ltd.

Ripley, T., & Shuler, L. G. (2012). Operation Desert Storm. *B-52 Stratofortress: Celebrating 60 Remarkable Years*, 66-71.

Strategic Arms Limitation Treaty. (n.d.). Retrieved from The AMARC Experience.

Swopes, B. R. (2019). *This Day In Aviation, 19 June 1955.* Retrieved from Pratt&Whitney J57-P-1W Archives: https://www.thisdayinaviation.com/tag/pratt-whitney-j57-p-1w/

Tagg, L. S. (2004). *Development of the B-52: The Wright Field Story.* Wright-Patterson AFB, OH: History Office Aeronautical Systems Center, Air Force Materiel Command.

Tirpak, J. A. (1994, April 1). The Secret Squirrels. *Air Force Magazine*.

Notes and Citations

Design and Development

[1] Project A became the parent to the B-17, B-24, and B-29 which were employed heavily in World War II.
[2] (Tagg, 2004, p. 15)
[3] (Tagg, 2004, pp. 6-10, 16)
[4] The light bomber (and ground support) mission was ultimately taken up by Tactical Air Command (TAC) using tactical bombers such as the B-45, B-57, B-66, F-105, F-111, and F-15.
[5] (Tagg, 2004, pp. 16-17)
[6] (Tagg, 2004, p. 17)
[7] (Tagg, 2004, p. 21)
[8] (Tagg, 2004, pp. 26, 48)
[9] (Tagg, 2004, p. 31)
[10] (Tagg, 2004, p. 53)
[11] (Tagg, 2004, pp. 19-20)
[12] Interestingly, both aircraft designs presented far-exceeded the 300,000-pound gross weight that Boeing believed LeMay's requirements would allow.
[13] (Tagg, 2004, p. 88)
[14] (Tagg, 2004, p. 23)
[15] (Tagg, 2004, p. 27)
[16] (Tagg, 2004, p. 29)
[17] (Tagg, 2004, p. 30)
[18] (Tagg, 2004, pp. 27-28)
[19] (Tagg, 2004, pp. 32-33)
[20] This meeting was held after the AAF issued the definitive contract (W33-038 ac-15065) for XB-52 Phase I superseding the letter contract issued 28 June 1946.
[21] (Tagg, 2004, pp. 35-36)
[22] (Tagg, 2004, pp. 38-39)
[23] (Tagg, 2004, p. 43)
[24] (Tagg, 2004, p. 44)
[25] (Tagg, 2004, p. 44)
[26] (Mandeles, 1998, p. 85)
[27] GE had also been developing jets since 1943 and had also developed the J47 with 4,900 pounds of thrust.
[28] (Tagg, 2004, pp. 45-46)
[29] (Tagg, 2004, p. 47)
[30] (Mandeles, 1998, p. 39)
[31] (Tagg, 2004, p. 53)
[32] (Tagg, 2004, pp. 53-54)
[33] Change of the XB-52 to jets meant no further turboprop development in the Air Force until the C-130 (Allison T56) in the early 1950's.
[34] (Tagg, 2004, p. 59)
[35] (Tagg, 2004, p. 62)
[36] (Tagg, 2004, p. 64)
[37] (Mandeles, 1998, p. 91)
[38] (Mandeles, 1998, p. 92)
[39] Murray was still concerned that the YB-60 might be selected for production over the XB-52. He recommended immediate action to ensure the YB-60 was required to meet the same design criteria as the XB-52 – same ground rules for both.
[40] (Knaack, 1988, pp. 218-219)

Models and Variants

[1] (Knaack, 1988, p. 224)
[2] (Katz, 2007, p. 4)
[3] (Knaack, 1988, pp. 221-222)
[4] (Knaack, 1988, p. 222)
[5] (Knaack, 1988, p. 223)
[6] (Tagg, 2004, p. 5)
[7] Engineers realized the B-52 would need some type of augmented take-off thrust system and considered rocket assistance like that used in the B-47. However, P&W developed the water injection system to solve the take-off thrust problem. Initial problems with the system were not fully rectified until the -29WA configuration and previous -29W engines were retrofitted to the new standard.
[8] (Knaack, 1988, p. 246)
[9] (Knaack, 1988, p. 246)
[10] (Knaack, 1988, p. 238)
[11] (Knaack, 1988, p. 241)
[12] Presumably these were aircraft with -19W, -29W, or -29WA engines (53-0380 to 53-0398)
[13] (Knaack, 1988, p. 236)
[14] Currently called the Strategic Air Command and Aerospace Museum and located in Ashland, NE.
[15] This paint was subsequently applied to all models including the B models.
[16] The B-52 was often called Big Ugly Fat Fucker (BUFF) by maintainers and crews. This was cleaned up to Big Ugly Fat Fellow in mixed company.
[17] (Knaack, 1988, p. 254)
[18] (Knaack, 1988, p. 257)

[19] (Knaack, 1988, p. 257)
[20] (Boyne, 1981, p. 114)
[21] (Knaack, 1988, p. 259); (Davis, 1992, p. 17)
[22] (Knaack, 1988, p. 262)
[23] (Boyne, 1981, p. 68)
[24] (Boyne, 1981, p. 76)
[25] Although fitted with ailerons, previous models also had primarily relied on the spoilers for roll control due to the risk of wing twist when the ailerons were deflected. Ailerons were relatively small and included primarily to give the pilot a normal feel on the controls. Elimination of the ailerons initially produced a tendency for Dutch roll along with a slight buffeting or pitch up during air refueling but this was later solved with a modification.
[26] (Boyne, 1981, p. 77)
[27] (Knaack, 1988, p. 273)
[28] (Boyne, 1981, p. 77)
[29] (Knaack, 1988, p. 272)
[30] The J75 was an improved version of the J57 engine used on previous B-52 models and was intended to provide at least 17,000 pounds of thrust each. However, it gained little traction in the aviation industry and the B-52G retained the J57-P-43WB used on the B-52F.
[31] (Knaack, 1988, pp. 272-273)
[32] (Knaack, 1988, pp. 274-275)
[33] (Knaack, 1988, p. 275)
[34] (Boyne, 1981, p. 77)
[35] (Knaack, 1988, p. 276)
[36] (Boyne, 1981, p. 111)
[37] (Knaack, 1988, p. 278)
[38] Thermal curtains were installed in all aircraft windows during nuclear bomb runs to block nuclear heat and flash. However, they also meant the crew was flying completely blind and relying on instruments. EVS gave them an outside view.
[39] (Elson, 1978, p. 16)
[40] (Davis, 1992, p. 48)
[41] AMARC was redesignated as AMARG in May 2007 aligning under the Ogden Air Logistics Complex at Hill AFB, UT. This gave reach-back capability for support activities as well as a group and squadron structure more traditional to the Air Force.
[42] (Knaack, 1988, pp. 283-284)
[43] (Mortimer, 2022, p. 12)
[44] (Boyne, 1981, p. 79)
[45] (Davis, 1992, p. 48)
[46] (Knaack, 1988, p. 285)
[47] (Boyne, 1981, p. 80)
[48] (Knaack, 1988, pp. 287-288)
[49] By this time the OC-AMA had been renamed the OC-ALC.
[50] (Katz, 2007, p. 16)
[51] (Goebel, 2003)
[52] (Katz, 2007, p. 18)
[53] (Kaminski, 2012, p. 85)
[54] (Knaack, 1988, pp. 219-221)
[55] Ultimately, 27 aircraft were designated RB-52B and 23 were delivered as B-52B. The RB-52C models were all redesignated and delivered as B-52C.
[56] The NACA was formed in 1915 to perform research benefitting commercial and military aviation. It was absorbed by the newly formed National Aeronautics and Space Administration (NASA) in 1959.
[57] (Knaack, 1988, p. 232)
[58] (Knaack, 1988, p. 232)
[59] (Hoage)
[60] (Knaack, 1988, p. 243); (Davis, 1992)
[61] (Knaack, 1988, p. 263)
[62] (Knaack, 1988, p. 263)

Armament & Weapons

[1] (Knaack, 1988, p. 246)
[2] (Knaack, 1988, p. 283)
[3] (Knaack, 1988, p. 272)
[4] (Knaack, 1988, pp. 271-272)
[5] (Knaack, 1988, p. 275)
[6] (Davis, 1992, p. 34)
[7] (Kopp, p. 56)
[8] (Davis, 1992, p. 34)
[9] (Mortimer, 2022, p. 11)
[10] (Mortimer, 2022, p. 69)
[11] (Mortimer, 2022, p. 13)
[12] (Mortimer, 2022, p. 13)
[13] (Leone, Remembering the AGM-69 SRAM, 2019)
[14] (Kitfield, 2010)
[15] (Davis, 1992, p. 39)
[16] Some B-52H aircraft could also carry external conventional weapons using the AGM-28 pylons and I-Beam Adapters to support contingencies, but they remained dedicated to the nuclear mission until the B-52G was retired.
[17] (Mortimer, 2022, p. 62)
[18] (Katz, 2007, p. 23)

[19] (Katz, 2007, pp. 34-35)
[20] (Davis, 1992, p. 33)
[21] (Davis, 1992, p. 35)

Organization and Basing

[1] (Correll, 2016, p. 65)
[2] (Correll, 2016, p. 67)
[3] SAC also established the 4th AD at Barksdale AFB, 6th AD at MacDill AFB, 12th AD at March AFB, 14th AD at Travis AFB, and 47th AD at Walker AFB on 10 February 1951; 5th AD in French Morocco on 14 June 1951; and 7th AF in England on 20 March 1951. It inherited 3rd AD on Guam when FEAF was discontinued on 18 June 1954.
[4] (Jenkins, 2001, p. 48)
[5] In some cases, as many as 90 B-47 aircraft were assigned to one base.
[6] (Knaack, 1988, pp. 236-237)
[7] (Knaack, 1988, pp. 50-51)
[8] (Knaack, 1988, p. 251)
[9] (Knaack, 1988, p. 256)
[10] SAC also previously realigned its NAFs on 1 July 1963 to balance the Missile Wings ignoring the previous geographic assignments. The 8th AF received three wings from 15th AF and two from 2nd AF.
[11] U-Tapao had converted from a Forward Operating Base (FOB) to a Main Operating Base (MOB) for B-52D operations in January 1969.
[12] SAC also previously realigned its NAFs effective 1 July 1973 to give ADs a variety of aircraft and missiles rather than just one, and to give missiles to the 2nd AF instead of concentrating in 15th AF.

Firsts and Records

[1] (Davis, 1992, p. 10)
[2] (Knaack, 1988, p. 289)
[3] (Knaack, 1988, p. 290)

Operations

[1] (Jenkins, 2001, p. 52)
[2] LeMay's title changed twice during his tenure in SAC. It was changed from Commanding General to Commander in 1953 and then to Commander-In-Chief Strategic Air Command (CINCSAC) in 1955. He was promoted to General on 29 October 1951 becoming the youngest 4-star since General Grant.
[3] (Ford, 1996, p. 5)
[4] (Ford, 1996, p. 6)
[5] (McClary, 2012)
[6] (Eickhoff, 1982)
[7] (Eickhoff, 1982)
[8] (Goebel, 2003)
[9] (Goebel, 2003)
[10] (Goebel, 2003)
[11] (Davis, 1992, p. 26)
[12] (Goebel, 2003)
[13] (Goebel, 2003)
[14] (Goebel, 2003)
[15] It was not unusual to find crews with 300-400 missions after multiple TDYs in SEA.
[16] B-52D missions flown out of Kadena were kept secret to avoid inflaming Japanese public opinion. A total of 15 aircraft were initially deployed to Kadena (as well as 11 to Andersen) for possible action against North Korea in response to the USS Pueblo incident as part of Operation PORT BOW. JCS subsequently authorized using the aircraft for ARC LIGHT sorties despite American Embassy in Japan objections.
[17] (Head, 2002, p. 47)
[18] (Head, 2002, pp. 47-48)
[19] (Head, 2002, p. 48)
[20] (Head, 2002, pp. 50-51)
[21] (Head, 2002, p. 51)
[22] (Parks, 1983, p. 2)
[23] BULLET SHOT officially began in February 1972 when 8th AF began deploying B-52 bombers to Guam for operations designed to bring the North Vietnamese back to the negotiating table for more fruitful discussion. The operations would finally employ the bombers over strategic targets as originally proposed by LeMay and SAC in 1964.
[24] (Parks, 1983, p. 5)
[25] (Head, 2002, p. 66)
[26] (Head, 2002, p. 65)
[27] (Head, 2002, p. 63)
[28] (Head, 2002, pp. 68-69)
[29] (Kohn, 1988, p. 125)
[30] Other sources indicate as many as 1,242 missiles were shot.
[31] Another nine aircraft were hit but were able to land safely back at U-Tapao.
[32] (Peacock, 1987, p. 42); (Head, 2002, p. 95)
[33] Crews from the 340th BS/97th BMW and 668th BS/416th BMW were also deployed without aircraft.

[34] (Mailes, 2016) Most references indicate the aircraft were based at King Abdul Aziz airport. However, they were actually housed at Prince Abdullah AB (now called King Abdullah AB) which is adjacent to and shares its airfield with the airport.

[35] (Ripley & Shuler, 2012, pp. 68-69)

[36] These aircraft were back filled by moving the six aircraft on Guam to Diego Garcia.

[37] Dubbed Operation SECRET SQUIRREL by aircrews.

[38] (Tirpak, 1994) Although these aircraft were the first to take off for DESERT STORM, they were not the first to hit their targets despite claims from some sources. The first attack aircraft hit their targets at 0302 Iraqi time (H-hour) on 17 January. The B-52s took off from Barksdale at 0636 local time on 16 January but given about 16 hours flight time plus an 8-hour time change they did not arrive on station until about 0630 Iraqi time on 17 January.

[39] The bombers carried a total of 39 missiles but four could not be launched due to software problems. Of the 35 launched one fell short in the desert and was later recovered, and one was lost and never recovered.

[40] The landings were staggered with the first aircraft landing at 33.9 hours and the last touching down at 35.4 hours.

[41] A total of 12 aircraft were launched including three spares. Two spares returned to Wurtsmith about five hours into the mission. The remaining spare continued on as the fourth aircraft in the third cell.

[42] (Davis, 1992, p. 37)
[43] (Mailes, 2016, p. 2)
[44] (Mailes, 2016, p. 2)
[45] (Mailes, 2016, p. 5)
[46] It was about 40 percent of U.S. tonnage.
[47] (Mailes, 2016, p. 4)
[48] (Ripley & Shuler, 2012, p. 70)
[49] (Mortimer, 2022, pp. 38-39)
[50] (Mortimer, 2022, p. 40)
[51] (Mortimer, 2022, p. 40)
[52] (Mortimer, 2022, p. 41)
[53] (Mortimer, 2022, pp. 44-45)
[54] (Mortimer, 2022, p. 43)
[55] (Mortimer, 2022, p. 47)
[56] (Mortimer, 2022, pp. 47-48)
[57] (Mortimer, 2022, pp. 48-49)
[58] (Mortimer, 2022, pp. 59-60)
[59] (Mortimer, 2022, pp. 62-63)
[60] (Mortimer, 2022, pp. 52-53)
[61] (Mortimer, 2022, p. 53)
[62] (Mortimer, 2022, p. 54)
[63] (Mortimer, 2022, p. 57)
[64] (Callaway, 2009, p. 18)
[65] (Eickhoff, 1982)
[66] (Eickhoff, 1982)
[67] (Eickhoff, 1982)
[68] (Eickhoff, 1982)
[69] (Eickhoff, 1982)
[70] This deployment to RAF Marham was preceded by three B-52D (55-0071, 55-0080, and 56-0694) aircraft from the 22nd BMW at March AFB, CA on 23 April 1980.
[71] (McClary, 2012)
[72] (Fehner, 2006, p. 197)
[73] (Fehner, 2006, p. 198)
[74] (Leone, The Story of the B-52 Stratofortress that Carried the Flashback Test Vehicle, 2022)

Displays and Disposition

[1] (Strategic Arms Limitation Treaty, n.d.)

Colors and Markings

[1] (Davis, 1992, p. 48) Although the reference says dark green (34086) was used, a comparison of aircraft photos and FS color charts indicates that olive drab (34088) was more likely used on some or all aircraft.

About the Author

H.J. Campbell is a long-time veteran of the U.S. Air Force and the defense aerospace industry. He completed 20 years in the Air Force including 15 years in SAC. His SAC assignments included the 42nd BMW at Loring AFB, ME, 320th BMW at Mather AFB, CA, 43rd SW at Andersen AFB, Guam, and HQ SAC Maintenance Standardization and Evaluation Team (MSET). After SAC was deactivated in June 1992, he was transferred to HQ Air Mobility Command (AMC) where he supported much needed KC-135 avionics upgrades. After his retirement from the Air Force, he served in the aerospace industry as a program manager and director for modification, maintenance, and logistics programs on KC-135, KC-10, C-130, C-27, and other Air Force aircraft. With this book, another in a planned series about SAC aircraft, he is continuing his "third career" as a military historian and author.

Also available from *Electrikbooks* at Amazon.com, BarnesandNoble.com, and BooksaMillion.com

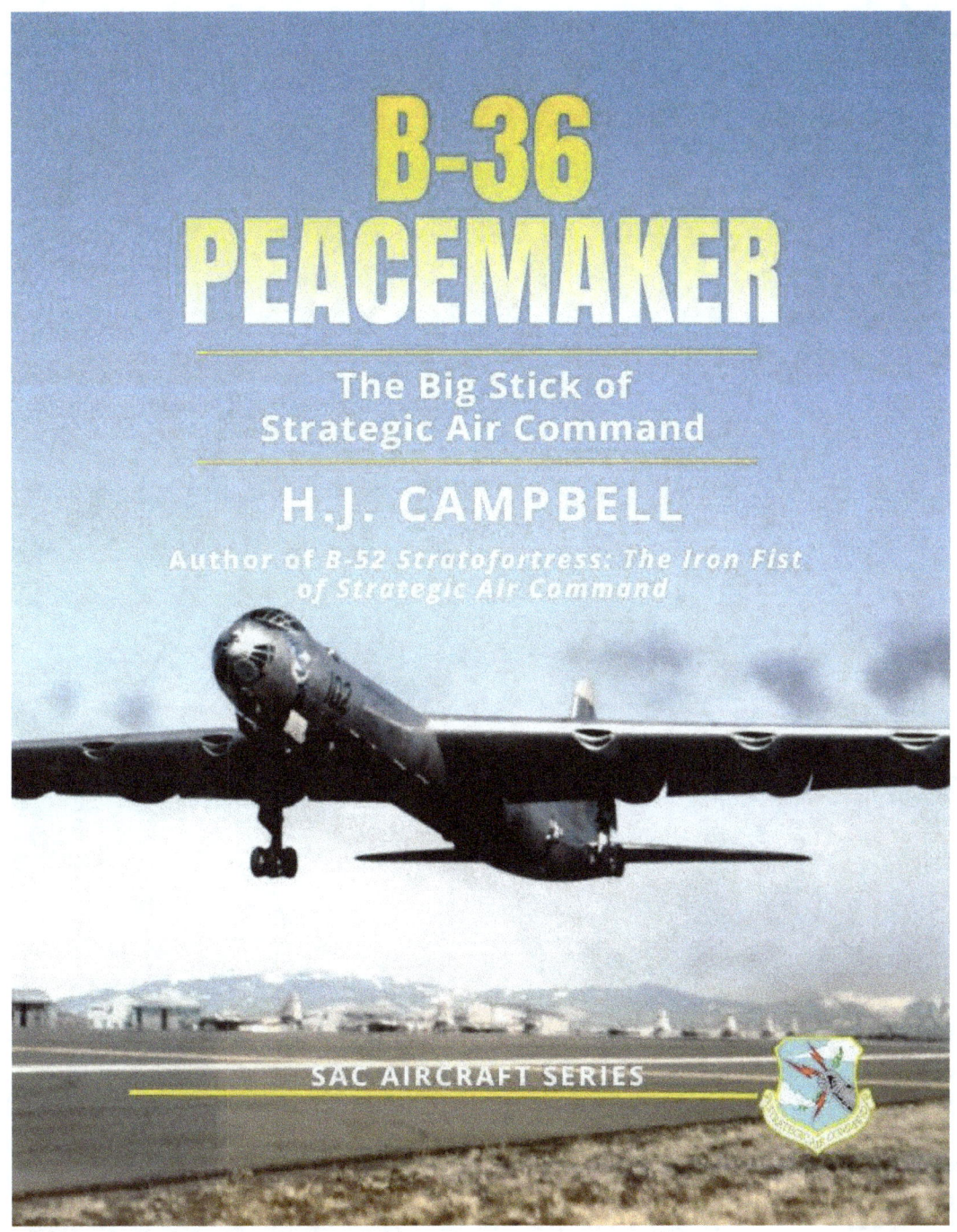